Agriculture Cannot Wait

This work is published by
Academic Foundation, New Delhi
in association with

NAAS

National Academy of
Agricultural Sciences
New Delhi

Agriculture Cannot Wait

NEW HORIZONS IN INDIAN AGRICULTURE

EDITED BY

M S Swaminathan

ACADEMIC FOUNDATION

NEW DELHI

First published in 2007
by

ACADEMIC FOUNDATION
4772-73 / 23 Bharat Ram Road, (23 Ansari Road),
Darya Ganj, New Delhi - 110 002 (India).
Phones : 23245001 / 02 / 03 / 04.
Fax : +91-11-23245005.
E-mail : academic@vsnl.com
www. academicfoundation.com

in association with :

National Academy of Agricultural Sciences, New Delhi
www.naas-india.org

Agriculture Cannot Wait:
New Horizons in Indian Agriculture
editor: M.S. Swaminathan

ISBN 13: 978-81-7188-625-8
ISBN 10: 81-7188-625-6

Typeset by Italics India, New Delhi.
Printed and bound in India.

Dedicated to

the vision of **Jawaharlal Nehru**

who emphasised

in India's tryst with destiny

everything else can wait

but not agriculture.

CONTENTS

Section II

Soil Health Enhancement and Fertiliser Use

Section III

Water Resource Management

Section IV

Agrobiodiversity and Biosafety

Section V

Agricultural Research and Education

Section VI

Globalisation and Agri-Exports

Foreword

The National Academy of Agricultural Sciences (NAAS) was established in 1990 by leading agricultural scientists of our country with the following mission:

> To gain recognition as a credible think tank to provide views of the scientific community on all agriculture related policy issues, to encourage talent and promote excellence in science, making it a powerful instrument for the growth of national economy with a vibrant farm sector.

In accordance with the above mission, the Academy has been sponsoring from time to time consultations on important policy issues in the areas of agricultural research, education, extension and development. Based on such consultations among leading scientists, policy papers have been prepared and widely circulated among all those concerned with generating synergy between technology and public policy. So far 39 policy papers have been issued.

The policy areas covered relate to sustainable livelihood and nutrition security, water resources management, soil health enhancement and fertiliser use, agrobiodiversity and biosafety, agricultural research and education and globalisation and agricultural exports. The texts of all these policy papers have been brought together in this publication for the use of both the academic and development community. This book will also be useful for scholars in agricultural universities and research institutions. I hope it will be widely read and used for policy and investment decisions in the field of agriculture, since many of the suggestions and recommendations contained in the book present a road map for rescuing the fate of farmers and farming from the present agrarian crisis prevailing in several parts of the country.

My sincere gratitude goes to Dr. Uma Kapila and Mr. Raj Kapila of the Academic Foundation for the excellent quality of the publication. It is their untiring efforts that have led to this important contribution to contemporary literature in the field of agriculture. I am also grateful to Dr. J.S.P. Yadav, Editor, NAAS for his important inputs into the design of this book. Finally, I wish to record the sincere appreciation to all the contributors and conveners of the policy papers for the time they so generously spared for this work.

— M.S. Swaminathan

Acronyms

AICRPs	All India Coordinated Research Programmes
AICTE	All India Council of Technical Education
APL	Above Poverty Line
APEDA	Agricultural and Processed Food Product Export Development Authority
APMC	Agricultural Produce Marketing Committee
ASSOCHAM	Associated Chambers of Commerce and Industry of India
ATMA	Agricultural Technology Management Association
BIS	Bureau of Indian Standards
BPL	Below Poverty Line
CADAs	Command Area Development Authorities
CAU	Central Agricultural University
CBD	Convention on Biological Diversity
CD	Compact Disc
CFB	Community Food Bank
CFTRI	Central Food Technological Research Institute
CGIAR	Consultative Group on International Agricultural Research
CII	Confederation of Indian Industry
CIFE	Central Institute of Fisheries Education
CHM	Clearing House Mechanism
CITA	Confederation of Industrial Trade in Agriculture

CSIR	Council of Scientific and Industrial Research
CUs	Central Universities
DARE	Department of Agricultural Research & Education
DAH	Department of Animal Husbandry
DSP	Dispute Settlement Panel
DU	Deemed University
EGS	Employment Guarantee Scheme
EIC	Export-Import Council
EICI	Export-Import Council of India
EXIM Bank	Export-Import Bank
FAO	Food and Agricultural Organization
FAQ	Food Analysis and Quality
FAQs	Frequently Asked Questions
FCI	Food Corporation of India
FICCI	Federation of Indian Chambers of Commerce and Industry
FYM	Farmyard Manure
GATT	General Agreement on Tariff and Trade
GATS	General Agreement on Trade in Services
GDP	Gross Domestic Product
GHP	Good Hygienic Practices
GI	Geographical Indication
GMP	Good Manufacturing Practices
GOI	Government of India
GPH	General Principles of food Hygiene
HACCP	Hazard Analysis and Critical Control Points
HRD	Human Resource Development
IARI	Indian Agricultural Research Institute
ICAR	Indian Council of Agricultural Research

ICMR	Indian Council of Medical Research
ICT	Information and Communication Technology
IDA	Indian Dairy Association
IDBI	Industrial Development Bank of India
IFFCO	Indian Farmers Fertilizer Cooperative Ltd.
IFPRI	International Food Policy Research Institute
IGIDR	Indira Gandhi Institute of Development Research
IIM	Indian Institute of Management
IMR	Infant Mortality Rate
IPNM	Integrated Plant Nutrient Management
IPR	Intellectual Property Rights
IRRI	International Rice Research Institute
ITK	Indigenous & Traditional Knowledge
ITK	Insight Toolkit
IT	Information Technology
ITPGRFA	International Treaty on Plant Genetic Resources for Food and Agriculture
IVRI	Indian Veterinary Research Institute
JNU	Jawaharlal Nehru University
KRICHCO	Krishak Bharti Cooperative Ltd.
LBW	Low Birth Weight
LL.M.	Master's degree in Law
MAT	Mutually Agreed Terms
MH	Maleic Hydrozide
MMR	Maternal Mortality Rate
MSP	Minimum Support Price
MSSRF	M.S. Swaminathan Research Foundation
NAAS	National Academy of Agricultural Sciences
NABARD	National Bank for Agriculture & Rural Development

NAP	National Agricultural Policy
NAARM	National Academy of Agricultural Research Management
NARES	National Agricultural Research and Education System
NARS	National Agricultural Research System
NATP	National Agricultural Technology Project
NCR	National Capital Region
NDDB	National Dairy Development Board
NDRI	National Dairy Research Institute
NGC	New Generation Cooperatives
NGOs	Non-Governmental Organizations
NHDB	National Horticultural Development Board
NRM	Natural Resource Management
NRI	Non-Resident Indian
NSC	National Seeds Corporation
O&M	Organization & Management
ODL	Open and Distance Learning
PDS	Public Distribution System
PFA	Prevention of Food Adulteration Act
PG	Post Graduate
PGR	Plant Germplasm Registration
PIB	Press Information Bureau
PIC	Prior Informed Consent
PPV & FR	Protection of Plant Varieties & Farmers' Rights
PVP	Plant Variety Protection
QMS	Quality Management Systems
QRS	Quick Service Restaurants
R&D	Research & Development
RAWE	Rural Agricultural Work Experience

S&T	Science & Technology
SAARC	South Asian Association for Regional Cooperation
SAUs	State Agricultural Universities
SPS	Sanitary and Phyto-sanitary Standards
SPC	Statistical Process Control
SNF	Solid Non Fat
SQC	Statistical Quality Control
SWOT	Strengths, Weaknesses, Opportunities and Threats
TAC	Technical Advisory Committee
TBT	Technical Barriers of Trade
TFP	Total Factor Productivity
TISCO	Tata Iron and Steel Company Ltd.
TQM	Total Quality Management
TRIPS	Trade Related aspects of Intellectual Property Rights
TVEs	Town and Village Enterprises
UG	Under Graduate
UGC	University Grant Commission
UPOV	International Union for the Protection of New Varieties of Plants
VEZ	Vegetable Export Zone
WHO	World Health Organization
WIPO	World Intellectual Property Organization
WTA	World Trade Agreement
WTO	World Trade Organization

Introduction

M.S. SWAMINATHAN

Since its establishment in 1990, the National Academy of Agricultural Sciences (NAAS) has given priority to the promotion of public and political understanding of science related to farmers and farming. Significant events in our agriculture like the green revolution in wheat and rice and milk revolution owe their origin to mutually reinforcing packages of technology, services and public policies. While scientists can develop yield enhancing technologies, these will not make an impact on production and productivity without appropriate support from public policy and investment. This is why NAAS has from time to time assembled groups of eminent scientists to analyse public policy issues in important areas related to agriculture and prepare papers relevant to policy formulation. The present book includes 39 policy papers issued by the Academy during the last 15 years, a wide range of issues like Sustainable Livelihood and Nutrition Security, Water Resources Management, Soil Health Enhancement and Fertiliser Use, Agro-biodiversity and Biosafety, Agricultural Research and Education, and Globalisation and Agri-Exports. The Academy is indebted to Drs. Uma and Raj Kapila for organising the policy papers into compact groups. It has been a pleasure working with them in organising the policy papers for publication in a book form so that they can reach a wider audience.

The wheat revolution commemorated in 1968 through the issue of a special stamp by Smt. Indira Gandhi is a classic example of the power of a symphony approach in designing and implementing programmes which can help in leapfrogging in the achievement of the desired goals. For example, the farmers in the Indian Fertile Crescent Area comprising Punjab, Haryana and Western UP helped to increase wheat production between 1964 and 1968 by an order comparable to that achieved from the time of Mohenjodaro, i.e., our farmers increased wheat production in four years by a quantity similar to that accomplished during the preceding four thousand years. This is a good illustration of the power of the symphony approach in the organisation of multidimensional research and development programmes.

It was however, clear even during 1966-1968 that unless farmers adopt environmentally sound technologies like integrated nutrient supply, integrated pest management and scientific water management there will be problems in the long-term sustainability of production. I, hence made the following statement in my lecture at the Indian Science Congress held in Varanasi in January 1968:

> Exploitative agriculture offers great dangers if carried out with only an immediate profit or production motive. The emerging exploitative farming community in India should become aware of this. Intensive cultivation of land without conservation of soil fertility and soil structure would lead, ultimately, to the springing up of deserts. Irrigation without arrangement for drainage would result in soils getting alkaline or saline. Indiscriminate use of pesticides, fungicides and herbicides could cause adverse changes in biological balance as well as lead to an increase in the incidence of cancer and other diseases, through the toxic residues present in the grains or other edible parts. Unscientific tapping of underground water will lead to the rapid exhaustion of this wonderful capital resource left to us through ages of natural farming. The rapid replacement of numerous locally adapted varieties with one or two high-yielding strains in large contiguous areas would result in the spread of serious diseases capable of wiping out entire crops, as happened prior to the Irish potato famine of 1854 and the Bengal rice famine in 1942. Therefore, the initiation of exploitative agriculture without a proper understanding of the various consequences of every one of the changes introduced into traditional agriculture, and without first building up a proper scientific and training base to sustain it, may only lead us, in the long run, into an era of agricultural disaster rather than one of agricultural prosperity.

Based on such an analysis, I coined the term "ever-green revolution" to stress the need for the enhancement of productivity in perpetuity without associated ecological harm. I have also been advising farm families to avoid taking to the path of a greed revolution leading to land and water mining. Our farmers hereafter will have to produce more under conditions of diminishing per capita arable land and irrigation water resources as well as expanding biotic and abiotic stresses. This is why the Academy chose issues relating to irrigation water and soil health management very early in its analysis of important policy issues. This wisdom was conveyed by Confucius when he remarked over 2000 years ago, "Despite many accomplishments of mankind, we owe our existence to a six-inch layer of top soil and the fact that it rains."

The need for a rational and transparent land use policy is becoming apparent from the increasing number of conflicts in relation to the acquisition of farm land for industrial and other non-farm uses. Such issues should be dealt with scientifically and not just politically or emotionally. Short-term profit should not undermine long term food security. In a predominantly rural and agricultural country like India, farming is the principal source of work and income security. Food security with home grown food is not only important for strengthening rural livelihood security, but it is also essential for both internal security and external sovereignty.

The policy papers contained in this book will help to give the reader an idea of the enormous progress made in our country in the areas of agricultural science and technology as well as of the gap between scientific know-how and field level do-how. Nevertheless, such accomplishments arising from public good research are not captured in the calculation of technology achievement. I would therefore like to discuss this matter at some length.

Measuring Technology Achievement

The UN Millennium Development Goals adopted unanimously at the UN Millennium Summit held in 2000, represent a Global Common Minimum Programme for sustainable human wellbeing and security. The first among these goals relates to reduction in hunger and poverty by half by 2015. Achievement of these goals will depend heavily on the technological and knowledge empowerment of the socially and economically underprivileged sections of the society. Starting with the Industrial Revolution in Europe triggered by the invention of the steam engine in 1780 by James Watt, technology has been an important factor in the rich-poor divide. If technology has enlarged social and gender divides in the past, the challenge now is to enlist technology as an important instrument for achieving social and gender equity. If technology is to serve this purpose, access to it must be based on the principle of social inclusion. The *antyodaya* or unto the last principle of Mahatma Gandhi is the pathway for inclusive economic growth.

The revolutions in agriculture, health and education witnessed during the last century were mostly the products of public good research supported by governments and philanthropic institutions. The contemporary world is however witnessing a shift from public good research to commercially profitable research supported by the private sector and protected by intellectual property rights (IPR). Even drugs which are important for the control of HIV/AIDS are available only at a cost which the poor cannot afford. This is why in the Doha round of WTO negotiations, there was an agreement that in the case of drugs of importance for controlling diseases like HIV/AIDS, there must be a provision for the compulsory licensing of rights. It is now widely accepted that gross social and gender inequity will be threats to peace and human security. It is hence appropriate that Mr. Kofi Annan, the then Secretary General of the United Nations, made the following observations in a guest editorial in *Science* (Vol. 299, 7 March 2003):

> The unbalanced distribution of scientific activity generates serious problems not only for the scientific community in the developing countries, but for development itself. It accelerates the disparity between advanced and developing countries, creating social and economic difficulties at both national and international levels. The idea of two worlds of science is anathema to the scientific spirit. It will require the commitment of scientists and scientific institutions throughout the world to change that portrait to bring the benefits of science to all.

UNDP has been rendering valuable service through the publication of an Annual Human Development Report (HDR), the brainchild of the late Dr. Mahbul Huq. Nations which undervalue the human resource and overvalue material resources tend to remain poor. The HDR has helped to arouse political, public, professional and media awareness of the factors which enable nations to leapfrog in terms of human development, as for example, China among developing countries, and the State of Kerala within India. Unless political will expressed in speeches is converted into political action for achieving the goals of food, health, literacy and a clean environment for all, Kofi Annan's warning that the forces of violence and war will destroy prospects for sustained economic development will come true.

The UNDP HDR for the year 2001 introduced an index for measuring technology achievement. India ranks very low in this index, in spite of its substantial advances in science and technology. The indicators used by UNDP for measuring technology achievement are largely those covered by intellectual property rights (IPR) and private sector innovations. Unfortunately, the Technology Achievement Index (TAI) of UNDP fails to take into account the contributions of public good research in improving human wellbeing in basic areas like food, water, health and energy security.

The contributions of public good research in India will be clear from the following data:

- Foodgrain production increased from about 45 million tonnes in 1951-52 to over 200 million tonnes at the beginning of this century.
- Productivity of major cereals increased from 700 kg per hectare in 1961-62 to over 1700 kgs. per hectare now.
- The net area under irrigation increased from about 21 million hectares in 1951-52 to over 60 million hectares now; gross irrigated area has also increased by over 300 per cent. Groundwater irrigation has played the lead role in bringing more area under irrigation, thanks to technological advances
- Annual milk production has gone up from about 20 million tonnes in 1950-51 to nearly 100 million tonnes this year (2007), thereby taking India to the first position in the world in milk production. This was achieved through management innovations spearheaded by the National Dairy Development Board, leading to mutually reinforcing linkages among milk production, processing and marketing.
- Both marine and inland fisheries have registered impressive progress: a major contribution to this progress has been made by scientific advances in the production of seed, feed, and induced breeding as well as crafts and gear.
- Science and technology coupled with social engineering have helped to promote conservation, restoration and commercial forestry and the regeneration of coastal mangrove wetlands.
- Significant progress has been made in the development of affordable drugs for the control of malaria, tuberculosis, leprosy, cholera and the other diseases; small pox has been eradicated and leprosy is likely to be eradicated soon.
- Many nutritional disorders like those arising from micronutrient deficiencies have now affordable remedies through a food-cum-fortification approach.
- Rural drinking water supply has been made nearly universal through the design of simple water pumps and the application of remote sensing and hard rock drilling techniques.

- Rural energy systems have gained enormously from scientific work related to the harnessing of biogas, biomass, solar and wind and other forms of renewable energy. Nuclear power generation based on indigenously developed technologies is making an important contribution to energy security.
- Great progress has been made both in space technologies and in the peaceful applications of atomic energy.

Thus, the list of meaningful achievements brought about by public good research is large and impressive. The impact of such research is not generally captured when indicators based upon technologies covered by IPR are used. The task of developing technology achievement indicators based on public good research is not easy, because of the fact that for technology to strike roots and bear fruit, appropriate technology delivery systems and public policies and investment are needed. Also, the benefits cannot be measured only in metric or monetary terms. Non-monetary benefits like a general improvement in human quality of life are equally important.

I hope the present publication will help to stimulate the birth of a new science—*Science for inclusive human development*. It is appropriate that the XI Five-Year Plan aims to promote faster but more inclusive economic growth. It is important that the impressive progress recorded in this publication in such vital areas like agriculture, animal husbandry, fisheries, forestry, irrigation, agro-processing and renewable energy is tapped for alleviating agrarian distress in over 30 districts of India where farmers' suicides are unfortunately prevalent. Can the technologies on the shelf now assist in improving the productivity, profitability, stability and sustainability of the farming systems prevalent in the agrarian hotspot districts? Can they help farm families with small holdings in unirrigated areas to improve their income from every drop of water and through crop-livestock-fish integrated farming systems? We cannot be silent onlookers to the continued coexistence of mountains of affordable technology and acute agrarian distress. If the recommendations contained in this publication are implemented, we can make the era of farmers' suicides history.

Wake-up Call

2007 represents the 60th anniversary of what Jawaharlal Nehru termed as "India's tryst with destiny."

A few days prior to his assassination in January 1948, Mahatma Gandhi said, "Forget the past. Remember every day dawns for us from the moment we wake up. Let us all, every one, wake up now."

Farm families in India constitute over two-thirds of the population. Since farmers are also consumers, the sharp distinction often made in industrialised countries between the interests of farmers and consumers, is not valid in the Indian context. Detailed analyses of the causes of food insecurity in rural and urban areas have revealed that the major cause of under- and malnutrition among children, women and men is the lack of adequate purchasing power to permit access to balanced diets and clean drinking water. Therefore, a three-pronged strategy needs to be introduced to ensure the economic wellbeing and nutrition security of rural families. First, families possessing assets like land, livestock or fish ponds will have to be assisted to enhance the productivity of their resource endowments in an environmentally and economically sustainable manner. The smaller the holding, the greater is the need for marketable surplus. Hence, the maximum emphasis has to be placed on increasing output per units of land, water, nutrients and labour based on technologies which are ecologically and economically sound. For this, we need more research on the development of eco-technologies based on blending traditional ecological prudence with frontier technologies like information and biotechnology and space, nuclear and renewable energy technologies.

Second, nearly a third of the rural population and a large proportion of women earn their livelihood through wage employment. They have no assets like land or livestock or fishpond and are also often illiterate. The challenge in the case of landless agricultural labour is enhancing the economic value of their time and labour by bringing about a paradigm shift from unskilled to skilled work. A massive effort in the area of knowledge and skill empowerment of the women and men constituting the landless labour workforce is essential if economic value is to be added to their time and labour. They will have to be enabled to take to skilled non-farm employment through market-driven micro-enterprises supported by microcredit. Self-help Groups (SHGs) of assetless women and men will have to be made sustainable through backward linkages to credit and technology and forward linkages with management and markets. Common property resources will have to be developed and managed in a manner that they can provide essential support systems in areas such as fodder and feed for stall-fed animal husbandry as well as fuel wood.

At the same time, the unfinished segments of land reform including the distribution of ceiling surplus land to assetless families should be attended to with speed and commitment. The interests of unregistered cultivators, tenants and tribal cultivators will have to be safeguarded. Wherever possible, atleast one acre of land should be provided to each landless labour family.

The third group are rural artisans working in the secondary and tertiary sectors of the economy. Their skills will have to be mobilised to enhance the competitiveness of agriculture through value addition to primary products and diversification of livelihood opportunities. The strategy for the technological upgradation of rural professions should be based on the principle of social inclusion.

Thus, the three, pronged strategy consists of improving the productivity of land, water, livestock and labour in the case of asset owning farm families, converting unskilled agricultural labour into skilled entrepreneurs engaged in organising market-driven non-farm enterprises, and enhancing the skills of families involved in the secondary and tertiary sectors of the rural economy, so that they are able to assist in improving agricultural efficiency and competitiveness and in ending the prevailing mismatch between production and post-harvest technologies.

Sustainable Advances in Small Farm Productivity

Sustainability has to be measured in ecological, economic and equity terms.

First, the ecological foundations essential for sustained advances in productivity, such as soil, water, biodiversity, forests and climate are under severe anthropogenic pressures. The human and farm animal population supporting capacity of the ecosystem has been exceeded in many parts of the country. The quantity and quality of ground water, which is now the dominant source of irrigation water, is fast deteriorating. Although India has over 20 per cent of the world's farm animal population, good grazing lands are practically nonexistent. Compounding current problems, the possibility of adverse changes in precipitation, temperature and sea level due to global warming and climate change is no longer just a theoretical conjecture.

Second, in the area of farm economics, resource flow to the agriculture sector is declining and indebtedness of small and marginal farm families is rising. Input costs are increasing, while factor productivity is declining. Contrary to the general impression of agrarian prosperity in Punjab, the total debt in the farm sector has been estimated to have increased from Rs.5,700 crore in 1996-97 to Rs.11,133 crore in 2002-03. The average farm debt in Punjab now exceeds a lakh of rupees (Rs.1,01,210), out of which more than 40 per cent is provided by non-institutional sources at an interest rate of nearly 24 per cent per annum. The cost-risk-return structure of farming is becoming adverse to over 80 million farming families operating small holdings, since the resource poor families cultivating 1 to 2 ha or less are unable to benefit from the power of scale at both the production and post-harvest phases of farming. There is no level playing field for them in the market and they are being exposed to competition from the capital, technology and subsidy driven agriculture of industrialised nations.

Third, a technology fatigue has further aggravated farmers' problems, since smaller the farm the greater is the need for sustained marketable surplus, in order to have cash income. Linkages between the laboratory and the field have weakened and extension services have often little to extend by way of location, time and farming system specific information and advice. Good quality seeds at affordable prices are in short supply and spurious pesticides and biofertilisers are being sold in the absence of effective quality control systems. Input supply is in a disarray particularly in dry farming areas. Micronutrient deficiencies in the soil as well as problems relating to soil physics are crying for attention. Farmers have no way of getting proactive advice on land use based on meteorological and marketing factors. Though it is now over 12 years since the WTO regime started operating in agriculture, serious attempts are yet to be made to launch in rural areas movements for quality literacy (sanitary and phytosanitary measures and *codex alimentarius* standards of food safety), trade literacy (likely demand-supply and price situation), legal literacy (IPR, Farmers' Rights) and genetic literacy (genetically modified crops). No wonder the prevailing gap between potential and actual yields even with technologies currently on the shelf is wide (Table).

Table

Comparative Crop Productivity

(Kg/Hectare)

Crop	*USA*	*China*	*India*
Maize	8900	4900	2100
Paddy	7500	6000	3000
Soybeans	2250	1740	1050
Seed Cotton	2060	3500	750
Tomato	6250	2400	1430

Source: FAO Yearbook (various issues).

In the area of technology, there is also a need to bridge the growing digital and genetic divides. Post-harvest technology is poor and there is little value addition particularly in the case of fruits, vegetables and spices including a wide range of tubers and medicinal and aromatic plants. Sustainable intensification, ecologically, economically and nutritionally desirable diversification and value addition to the entire biomass are important for raising small and marginal farm families above subsistence level. All this will call for initiating an era of knowledge intensive agriculture. Modern information communication technologies (ICT) afford an opportunity for launching a knowledge revolution in rural India. The torch bearers of this revolution should be rural women and men. Participatory research and knowledge management involving farm women and men should be the principal pathways of research, education and extension. Farmers should be regarded as partners and innovators in bringing about agricultural transformation and not as 'beneficiaries' of government programmes. Such a change in mindset among government functionaries as well as scientists is essential for progress.

Technologies should help in promoting labour diversification and not displacement. Women farmers and labour particularly need to be assisted with implements and equipment which will help to reduce drudgery and the number of hours of work, while adding economic value to each hour of work. Women suffer from a multiple burden on their time due to homemaking and keeping, child rearing and income earning responsibilities. They need appropriate support services like crèches and child care centres and adequate nutrition. The feminisation of agriculture, due to male out-migration, needs specific attention with reference to gender sensitive farm and credit policies.

All educational research and development programmes in agriculture must be engendered.

Revitalisation of small farmer-friendly technologies should be based on sound principles of economics and participatory research and knowledge management. If for example, rice-wheat rotation is not desirable in Punjab from the standpoint of ecological sustainability, the alternative farming systems proposed should be capable of yielding similar income. Agronomic data should not only indicate yield per hectare of land, but also the likely return from every rupee invested and every litre of water used by the farmer. Lateral learning among farm women and men should be fostered, since farmer to farmer learning is based on the principle—"one ounce of practice is worth tons of theory". India is also the home of some of the best farm families in the world. In the midst of many hunger and agro-ecological 'hot spots', there are also numerous farming 'bright spots'. In order to multiply the benefits from the experience and skills of outstanding farm men and women, Farm Schools should be established in their fields. Farmer to farmer learning is very effective because of the faith small farmers have in the economic data provided by fellow farmers.

Priorities in Farm Research and Development

1. Conservation Farming and Green Agriculture

Conservation farming lays stress on the sustainable management of soil and water. It involves soil health enhancement through concurrent attention to the physics, chemistry and microbiology of soils. Scientific water management aims to use the available irrigation water in a conjuctive manner (for example, surface and groundwater) and to maximise income per drop of water. Government support for conservation farming does not attract the trade distorting provisions of WTO.

Green agriculture involves the use of environmentally desirable technologies like IPM, Integrated Nutrient Supply, etc. While organic farming precludes the use of mineral fertilisers, chemical pesticides and genetically modified crop varieties, green agriculture permits their use in such a manner that productivity can be improved in perpetuity without associated ecological harm. To a majority of small farm families who have no livestock, green agriculture is the pathway for an ever-green revolution. Therefore, an optimum blend of green agricultural practices and organic farming methods should be promoted in all major farming systems. There should be internationally accepted certification procedures for the products of both organic farming and green agriculture.

2. *Productivity and Livelihood Enhancement in Rainfed Areas*

The major pathway has to be productivity enhancement and the cultivation of high-value but low-water requiring crops. A 'catalytic intervention' which will help to increase productivity immediately is attention to soil healthcare. The provision of micronutrients like sulfur, zinc and boron can help to increase yield by over 50 per cent in dryland farming areas. Facilities for identifying and remedying micronutrient deficiencies in soils are urgently needed. Soil Health Clinics can be operated by SHGs comprising rural women and men who can issue each farm family with a Soil Health Card. Also, livestock and livelihoods are closely interrelated in semi-arid and arid areas. Hence, the establishment of fodder and feed banks should receive urgent attention. The gap between potential and actual yields is high in dryland farming areas. Bridging the productivity gap in an economically viable manner is the best safety net against farmers' distress in such areas. For accelerating progress in finding lasting solutions to the economic woes of farmers and agricultural labour, it would be useful to form District-level consortia of private and public sector institutions willing to help in improving small farm income and off-farm employment opportunities through appropriate technologies and producer-oriented marketing.

3. *Small Farmers' Horticulture Estates*

The cultivation of fruits, vegetables, flowers, spices, medicinal and aromatic plants is now happening in a big way in several parts of the country. A National Horticulture Mission has been launched to double the production of vegetables and fruits in seven years (from 150 million tonnes per year to 300 million tonnes). Being perishable commodities, horticultural crops need effective infrastructure support in the areas of production, processing, storage, transportation and marketing. In villages adjoining large consumption centres (both for home and export markets), small farmers can be helped to organise Small Farmers' Horticulture Estates in the form of SHGs or cooperatives covering an area of 200 to 500 hectares. In such estates, specialised activities like seed production, tissue culture propagation, production of compost, vermiculture, biofertilisers, biopesticides and e-commerce can be promoted through technological and credit empowerment. Such Estates will confer on farmers cultivating one to two hectares the power of scale both at the production and post-harvest phases of the horticultural enterprise. Low cost greenhouses coupled with fertigation techniques can be promoted, in addition to high-tech horticulture which can be undertaken by farm and home science graduates. The production of good quality, disease-free planting material is important in all clonally propagated species. Also, seeds and planting materials of varieties suitable for processing will have to be provided to farmers in areas where production and processing are linked. Such symbiotic linkages between producers and processors will facilitate sourcing of good quality raw material for the processing industry. The basic principle underlying such Estates is decentralised production supported by key centralised services.

4. *Enhancing the Productivity of Cotton and the Global Competitiveness of the Cotton Textile Industry*

The textile sector is a major employment and income providing sector of the national economy. With the coming to an end of the multi-fibre arrangement on 1 January, 2005, our cotton producers, weavers and the textile industry are encountering both new opportunities and threats. Without enhanced efficiency, it will be impossible to take advantage of the emerging market opportunities. Technological upgrading of all the components of the cotton production–processing–marketing cycle, will be needed to protecting this vital sector of our economy from setbacks. A productivity, quality and value addition revolution is urgently needed in cotton production and processing. The pollution problems associated with small-scale textile units can be solved by farming Environmental Management Cooperatives comprising 50 to 100 small-scale textile units.

5. *Women Farmers and Farm Labour*

There is increasing feminisation of agriculture in families with small and marginal holdings, due to the out-migration of men. The problem is particularly severe in hill areas like Uttaranchal, Jharkhand and the North eastern states. Yet, out of nearly 50 million Kisan Credit Cards issued in the country, less than 5 per cent seem to have been issued to women (precise data are not available). The proposed conferment of land rights to women will help to redress this distressing situation. Meanwhile, joint pattas will have to be issued to make women eligible for institutional credit. Also, women working the whole day in the field require support services like crèches and day care centres. A unique opportunity now exists for engendering the ongoing National Rural Employment Guarantee programme by enlarging the concept of work in the case of women by including such essential activities like running crèches and child care centres, preparing school noon meals, undertaking immunisation of children, providing family planning services, etc. Work will then become a powerful tool for human and social development.

6. Every Village a Knowledge Centre

India's strength in ICT provides uncommon opportunities for taking digital and knowledge connectivity to every village in the country by August 15, 2007. Reaching the unreached and voicing the voiceless will be possible through an integrated ICT strategy involving the internet, community radio, cable TV, cell phone and vernacular press. The fibre-optic network of Bharat Sanchar Nigam Ltd. (BSNL), comprising 30,000 exchanges, covers all the 6,000 blocks of the country. If each exchange is extended to 20 nearby villages, all the 600,000 villages can be covered at minimal expenditure.

The Rural Knowledge Centres or Gyan Chaupals should be located in public spaces like Panchayat Buildings or Village Schools, so that there is social inclusion in access. The National Alliance for Mission 2007: Every Village a Knowledge Centre and the Jamsetji Tata National Virtual Academy for Rural Prosperity are effective instruments for launching knowledge based agriculture and rural non-farm enterprises. Panchayat Raj Institutions will have to play a lead role in the rural knowledge revolution. NABARD, SBI and other financial institutions can support Rural ICT-SHGs, who can maintain and operate the Gyan Chaupals. Connectivity, content, capacity building, care and management of the Gyan Chaupal will all need integrated attention to make the movement socially meaningful and economically sustainable.

The Government of India (Department of Information Technology) has launched a Common Service Centre (CSC) programme designed to take the digital technology to 100,000 villages. The Indian Space Research Organisation (ISRO) is planning to establish one Village Resource Centre (VRC) in every block of the country. The private sector, particularly the Indian Tabacco Company (ITC) plans to cover nearly 50,000 villages will e-chaupals. Thus, the infrastructure needed for the Rural Knowledge Revolution is becoming available. The last mile and the last person connectivity can be achieved through the integrated use of the internet and FM radio or cell phone.

7. Building a Sustainable Nutrition Security System

Nutrition security is best defined as 'physical, economic, social and environment access to balanced diet and clean drinking water for all and for ever'. The Central and State Governments have initiated from time to time a large number of nutrition safety net programmes. An accelerated advance in achieving the goal of nutrition security at the level of every child, woman and man will be possible through the following interactive steps:

- Deliver all nutrition support programmes on a whole life cycle basis and fill up gaps in ongoing programmes, such as the nutritional needs of adolescent girls, pregnant women and 0-2 infants.
- Ensure access to clean drinking water, environmental hygiene, primary healthcare and primary education.
- Take special steps to overcome hidden hunger caused by micronutrient deficiencies like iron, iodine, zinc and Vitamin A.
- Promote the cultivation and consumption of fruits and vegetables as well as of dairy products. Popularise horticultural remedies for the nutritional maladies of the village.
- Facilitate the setting up of local level community Food Banks, comprising locally grown grains like millets and legumes, thereby enlarging the composition of the food basket. Also, promote the setting up of Fodder and Feed Banks, since livestock and livelihoods are intimately related in most parts of the country, particularly in semi-arid and arid areas.
- Generate awareness about the human and farm animal population supporting capacity of the ecosystem, in order to generate interest in population stabilisation.
- Launch a massive movement in rural areas for additional employment and income generations. The ongoing SHG revolution should be made sustainable through backward

linkages with technology and credit and forward linkages with management and markets.

- Provide nutrition support to rural women and men suffering from HIV/AIDS and tuberculosis.

Such a local level Nutrition Security Compact is best designed and managed by SHGs operating under the oversight of the Gram Panchayat and Gram Sabha. We should also plan to launch a Food Guarantee Act by integrating the principles of Employment Guarantee and Food for Work programmes. Such a Food Guarantee Act will help to fulfil Mahatma Gandhi's vision for a hunger free India, when he said in 1946 at Naokhali:

> To a people famishing and idle, the only acceptable form in which God can dare appear is work and promise of food as wages.

8. Employment Generation in Agriculture

In the ultimate analysis, a sustainable end to hunger can be achieved only by providing opportunities for every woman and man to earn their daily bread. A detailed strategy for creating additional skilled jobs through horticulture, cotton, energy plantations, animal husbandry, biofuels and biomass utilisation has to be developed. India is rich in livestock resources. Nearly 20 per cent of the farm animal population are in India. The ownership pattern is more egalitarian in the case of livestock, as compared to land. Crop-livestock integrated production systems can help to enhance both household nutrition security and cash income. Overriding priority should go to fighting the famine of jobs/sustainable livelihood opportunities through the creation of economically rewarding and intellectually stimulating work opportunities in villages. This is the only way to attract and retain educated youth in villages. We need to launch a major non-farm livelihood initiative on the lines of the Township-Village Enterprises of China.

9. Uncommon Opportunities

There are numerous institutional structures already available to Government, like SFAC, NHB, NDDB, Agri-clinics, Agri-business Centres, Food Parks, Agri-export Zones, several Commodity Centred Technology Missions, Watershed and Wasteland Development Programmes, National Rainfed Area Authority, National Fisheries Development Board and National Rural Health Mission. Instead of starting many new schemes, what is needed is the revitalisation and restructuring of existing schemes and institutional structures and the retooling and retaining of existing staff. Also, convergence and synergy among the numerous on-going vertically structured programmes can help to reduce transaction costs and improve the efficiency of delivery.

The initiation of a National Rural Employment Guarantee Scheme together with an expanded Food for Work Programme provides a unique opportunity for launching a multipronged attack on poverty and rural unemployment. Engendered work under this programme can become a catalyst of a long-term sustainable livelihood security movement in the different ecological, hydrological and farmers' distress hotspot regions of the country.

To succeed, the above programmes will have to be implemented in a decentralised manner with authority and accountability being linked at the field level. The 11th Schedule of the Constitution 73rd Amendment Act 1992 on the *panchayats* lists agriculture, including agricultural extension, as the very first responsibility of *panchayats* (Article 243 G). Other items including Animal husbandry, dairying and poultry, fisheries, minor irrigation, water management, watershed development, land improvement, implementation of land reforms and land consolidation and soil conservation are also the responsibilities of *panchayats*. Therefore they should be actively involved in the detailed planning and implementation of programmes in agriculture and allied sectors.

10. Credit and Public Investment

The crisis in our agriculture has arisen because of lack of appropriate public policies as well as adequate public investment in rural infrastructure, particularly in the areas of storage, processing and value addition. Therefore, the cures for the prevailing maladies can be found only in public policies and investment. Spending by Central and State Governments in strengthening the ecological, technological and social foundations for sustained advances in farm productivity has been going down. Most of the money spent by Central and State Governments goes to the salaries of government officers and employees. Consequently, rural infrastructure including power, irrigation, market yards, rural godowns and communication, as well as vital sectors like health and education remain grossly underfunded.

Declining credit-deposit ratios in rural areas, volatility of prices, weak research and extension and neglect of women farmers and farm labour have all added to the woes of the farming community. Indebtedness of farmers is rising not only because of farming related expenditure, but also because of the need for money for healthcare and domestic needs. The public healthcare system in villages is in a state of collapse. Pandemics like HIV/AIDS and tuberculosis are spreading in villages. Because of protein-energy undernutrition as well as micronutrient deficiencies, a purely drug based approach to the control of diseases is not adequate. A nutrition support programme is equally important. Thus, there is need for a holistic approach in addressing the technological, financial, health, educational and marketing problems faced by farm women and men.

The crucial role of women farmers and labour in agriculture is yet to receive recognition. Women play a key role in all the four major components of farming, namely, conservation, cultivation, consumption and commerce. Until the commitment to provide land rights to women is fulfilled, it is important to ensure that joint pattas making women eligible for Kisan Credit Cards and institutional credit, are issued without further delay. It is time that a New Deal for Women in Agriculture is launched. The National Commission on Farmers has offered several suggestions including a Gram Mahila Fund for meeting gender specific needs in villages. If the NCF recommendations are implemented, the New Deal will become a reality.

Another area which needs urgent attention is the economy of scale in farm holdings. The average size of farm holding has declined from 2.69 ha in 1960-61 to 1.41 ha in 1995-96. The process of decline in the size of an operational holding is continuing. There are now about 115.6 million farms. The use of purchased inputs by farmers has multiplied 283 times during 1950-51 to 2000-01. To meet input costs, the rural poor borrow 84 per cent of their credit needs from non-formal sources. The quantities of marketable surplus have multiplied to the tune of 10 times of cereals, 4.6 times of oilseeds, 5.3 times of milk, 15.4 times of poultry products and 7.4 times of fish during the last 50 years. Yet, post-harvest infrastructure is weak. Even now farmers in several parts of the country dry the freshly harvested paddy on paved roads, where there is often heavy vehicular traffic. Drying yards are a luxury in a majority of villages.

Value addition in the case of horticulture products is less than 10 per cent. The gross marketing margin in farm commodities is estimated at Rs.1009 billion, out of which nearly 70 per cent is accounted by marketing cost. About 77 per cent of marketing costs are estimated to be avoidable losses during handling, storage and transport. Quality, labelling, brands, taxes, subsidies, sanitary and phytosanitary (SPS) issues, price volatility, removal of quantitative restrictions on imports and the absence of a level playing field in international trade due to the very high support (nearly 1 billion dollars per day for 10 million farming families) extended to farmers in OECD countries, are all becoming significant factors in agricultural marketing.

11. Enhancing Agricultural Competitiveness: Basket of Choices

Raising the agricultural competitiveness of farmers with small holdings is a major challenge. Methods of conferring the power of scale to small farm families both at the production and post-harvest phases of farming is an urgent necessity. A basket of choices should be available to farm women and men cultivating 1 or 2 ha or below to enhance their income earning capacity. Productivity improvement to increase the marketable surplus available to small farm families, assured and remunerative marketing opportunities, and creating opportunities for landless agricultural labour families for skilled non-farm employment should be the bottom line of public policy for rural India. The basket of choices for enhancing productivity, profitability and sustainability could include the following:

a) Formation of SHGs of farm families to undertake group operations in areas like water harvesting and management, pest management and post-harvest technology. Some examples are: Small Farmers' Horticulture, Cotton and Poultry Estates.

b) Promotion of contract farming to ensure assured marketing outlets on the basis of a well-defined Code of Conduct; the Code of Conduct for Contract Farming should include provision of support to small producers in the areas of technology and input supply and fair and assured price for the produce. The contract should represent a win-win situation both for the producer and the purchaser.

c) Revitalising and restructuring various government schemes like Agri-clinics, Agri-business centres, Rural Godowns, Small

Farmers Agri-Business Consortium (SFAC), National Horticulture Board, etc. so as to make them farmer-centric.

d) Helping small farmers to farm either cooperatives or producer companies

e) Promoting the active involvement of PRIs and local bodies to foster Watershed/Command Area Communities and making the Watershed or the irrigation command area the point of convergence and integration of all relevant Technology Missions like those relating to oilseeds, pulses, maize, cotton, horticulture, milk, etc. Convergence and synergy among the numerous Technology Missions now in progress will improve their utility and impact and also help to reduce overall transaction costs. All the Missions could be integrated under an umbrella set up, which could be termed 'National Federation of Farm Technology Missions'. The ongoing Missions operating on parallel lines appear to have very little accountability, since in spite of their existence, imports of pulses and oilseeds are going up year after year. Pulses and oilseeds are important nutrition and income providing crops in rainfed areas and farmers in dry farming areas will continue to suffer in poverty and deprivation unless the proposed National Federation of Farm Technology Missions extends to them the needed help at the right time and place. Such a National Federation is best chaired by a practicing farm woman or man, who has a proven record of unleashing the power of creativity in small farm management. Its major aim should be to enhance farm productivity and agrarian and rural prosperity.

12. Critical Role of Education and Healthcare

Value addition to primary products, turning unskilled into skilled labour and using digital technology in villages, require for effective implementation an educated population, with minimum level of Class X education. Unless all children in our villages, especially girls, Dalits and rural poor get educated, the goals of social equity and agrarian and rural prosperity cannot be achieved. Until formal literacy becomes universal, the 'techniracy' movement (i.e., learning the latest technical skills through work experience) should be spread. For this Krishi Vigyan Kendras should be developed into 'Krishi aur Udyog Vigyan Kendras', with equal emphasis on production and post-harvest technologies. Farm Schools can also be established in the fields of innovative farmers. A desirable structure of KVKs for the 21st century is indicated in the figure.

Figure

Krishi Vigyan Kendras in the 21st Century

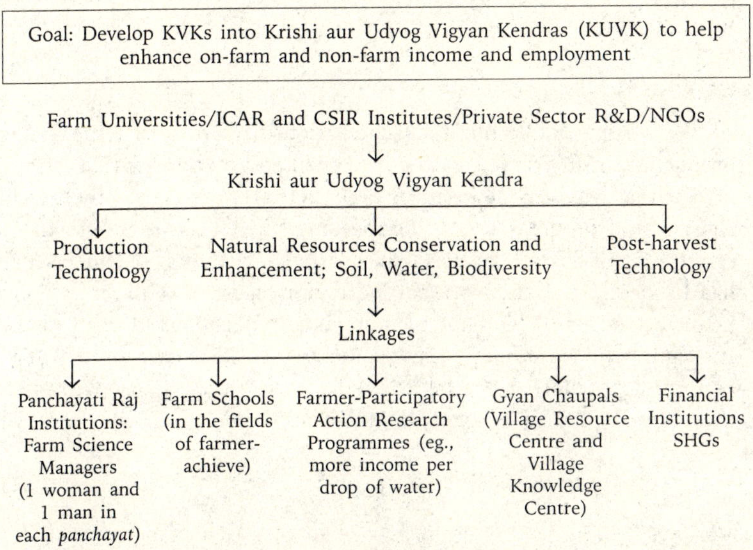

The motto of our farm universities and home science colleges should be 'every scholar an entrepreneur'. This will call for appropriate changes in curricula as well as pedagogic methodologies.

Along with education, healthcare facilities for the rural poor need to be strengthened. There is need to increase, substantially public investment both in elementary education and healthcare in rural India.

Anticipatory Research: Preparing for Climate Change

Climate change leading to changes in temperature, precipitation, Co_2 levels and sea level are likely to affect different parts of the world differently. It is however clear that poor nations and the poor in all nations will suffer most, since they lack the coping capacity to withstand the adverse consequences of climate change. Recent reports indicate that as a result of the melting of ice caused by higher temperatures, a new chain of islands which were under the ice have now become visible. Greenland and the Artic Glaciers are melting and floods are likely to be more common in North India. It will be prudent to prepare computer stimulation models of different weather probabilities and prepare contingency plans to meet the challenges. Anticipatory research is also needed to protect coastal communities through bio-shields and genetic shields. This area of anticipatory research should receive high priority.

Conclusion

We need to shift our focus in agriculture from just measuring production in metric terms to the women and men for whom farming is both a way of life and the principal means to livelihood. Agricultural progress should be measured in terms of growth in the real income of farm families. The National Commission on Farmers has proposed a draft, National Policy for Farmers. This draft policy takes into account the many recommendations contained in this book. If such a policy is adopted, the income and work security of two-thirds of India's population can be assured. *Gram Swaraj* is the pathway to the *Poorna Swaraj* envisaged by Mahatma Gandhi.

Section I

Sustainable Livelihood and Nutrition Security

1

Employment Opportunities in Farm and Non-farm Sectors through Technological Interventions with Emphasis on Primary Value Addition

Introduction

Generation of adequate work opportunities for the growing labour force has been a central objective of development planning since independence. Productive employment, apart from being a means to poverty reduction and economic well-being, is an end in itself, because it is a basic source of human dignity and self-respect. Although India was able to keep, on the whole, open unemployment quite low, the trends in recent years, call for renewed efforts for creating additional employment opportunities, particularly in the rural areas.

(i) Despite acceleration in the growth of the economy during the nineties, the pace of creation of jobs has remained low. This has happened mainly due to deceleration in the employment elasticity in the agricultural sector. Employment to GDP elasticity for agricultural sector during 1993-1994 to 1999-2000 was as low as 0.01. In the base year (2001-02) for Tenth Five Year Plan projections, there were 35 million unemployed, besides several underemployed.

(ii) Nearly 56.7 per cent of the total workforce is engaged in agriculture, 12.1 per cent in manufacturing, and 31.2 per cent in the service sector. The change in the sector-wise employment during the nineties, shows that the number of people employed in manufacturing and service sectors increased, whereas there was no change in those employed in

---

Policy Paper 37, September 2006.

agriculture. Out of 20.91 million additional jobs created between 1993-1994 and 1999-2000, 51 per cent was in THR (trade, hotels and restaurants) sector, 28 per cent in manufacturing sector, 19 per cent in construction sector and 18 per cent in TSC (transport, storage and communication) sector. In the CSPS (community, social and personal services) sector, there was a decline of 20 per cent (4.14 million jobs).

(iii) There are 115.6 million farm holdings, out of which 62 per cent (71.2 million) operate on less than one hectare of land, with an average of 0.40 hectare each. The income of tiny landholders is lower than that of landless families. Most of these are poor and food insecure. Technology for primary value addition can be a boon for these families.

(iv) According to NSSO survey (1999-2000), there are 25 million non-agricultural informal sector enterprises in the rural areas (apart from 19 million in urban areas). Nearly 94 per cent (23.6 million) of rural enterprises, are 'own account' enterprises, which operate with family labour. Nearly 40 million persons, are engaged in these enterprises. Out of 25 million rural non-farm enterprises, 9.6 million are doing some kind of manufacturing, 8.6 million are doing trade and repairs work, and 2.0 million are engaged in transport, storage and communication. Technological support can substantially help rural families dependent on these enterprises. While looking at the technological options for these families, one other fact which needs to be kept in view is that 45 per cent (11.2 million) of these enterprises are located within household premises, 32 per cent (8 million) outside households but with fixed location and 23 per cent (5.8 million) operate as street vendors, mobile shops or work at changing construction sites.

(v) On the other hand, the agricultural production sector has shown remarkable growth and the linkages of the farm production sector with the rest of the economy have increased manifold, both in terms of increase in the use of purchased farm inputs and volume of farm surpluses available for marketing and processing. During the past 50 years, the use of purchased inputs by the farmers has multiplied 283 times from Rs. 245 crores in 1950-51 to Rs. 69390 crores in

2000-01. Purchased inputs mostly supplied by the non-farm sector, now account for 72 per cent of total inputs (other than land and labour) as against 20 per cent in 1950-51.

(vi) Apart from the increase in demand for inputs, there has been considerable increase in marketed surplus of agricultural commodities, generating demand for handling, transportation, storage, grading, packaging, processing and retailing services. This happened on account of both increase in output as well as increase in the marketed surplus output ratios of all the farm products. The latter is estimated to have gone up from 33.4 per cent in 1950-1951 to 64.1 per cent in 1999-2000 and further to around 70 per cent, currently. The quantities marketed have multiplied to the tune of 10 times that of cereals, 4.6 times of oilseeds, 5.3 times of milk, 15.4 times of poultry products and 7.4 times of fish. In terms of value, the marketed surplus of farm products in 2000-01 was 103.4 times of that in 1950-51, valued at Rs. 4037 billion (Rs. 403740 crores). Handling of such huge surpluses emerging from 115.6 million farm holdings provides tremendous opportunities for employment in the rural non-farm as well as the farm sector. The value addition in agricultural commodities in India, is considerably lower than that in several developing countries.

The gross marketing margin in agricultural commodities is estimated as Rs. 1009 billion, consisting of Rs. 151 billion as statutory charges, Rs. 207 billion as net margins and Rs. 651 billion as marketing cost. About 77 per cent of the marketing costs, amounting to Rs. 500 billion are estimated as avoidable losses during handling, transport and storage. Investment in primary value addition activities can save these losses and generate employment in the rural non-farm sector.

A Special Group, constituted by the Planning Commission, has, in its report submitted in 2002, identified several areas for creation of additional 50 million jobs by the end of the year 2006-07. These potential areas are: (i) Agriculture including diversification from cereals to oilseeds and pulses; Watershed development for rainfed areas; Horticulture; Own-farm water management; Agro clinics and seed production; Farm machinery; Greening India Programme (wasteland development, joint forest management and agroforestry);

Development of medicinal plants and energy plantations (9.5 million). (ii) Special Employment Generation Programmes like Sampoorna Gramin Rozgar Yojana (SGRY); Swarn Jayanti Gram Swarozgar Yojana (SGSY); Pradhan Mantri Gram Sadak Yojana (PMGSY); and Prime Minister's Rozgar Yojana (PMRY) in SSI and KVIC related areas (7.1 million). (iii) Manufacturing sector including small-scale industries like coir; handlooms; power looms; handicrafts; and sericulture and wool (7.1 million); and large-scale manufacturing (6.3 million). (iv) Services including construction activities (6.3 million); trade, hotels and restaurants (11.2 million); transport, storage and communications (5.5 million); financial sector (1.9 million); and community services (0.5 million) with a total of 25.4 million.

In addition to the ongoing special employment generation programmes, some other new initiatives in the recent past for creating employment in rural areas are: (a) National Employment Guarantee Act, now in operation in 200 districts; (b) scheme for promotion of rural godowns with facilities for cleaning, grading and packaging of products; (c) scheme for creation of rural infrastructure for handling and processing of high value horticulture and livestock products; (d) promotional schemes of National Horticulture Mission; (e) promotional schemes of Ministry of Food Processing Industries; (f) amendment in State Agricultural Produce Market Acts, for promotion of private and cooperative markets; (g) promotion of contract farming; and (h) promotion of private sector participation in use of wastelands. Availability of adoptable and sustainable technologies and dissemination/transfer of these technologies to the entrepreneurs is a critical factor in realising the potential of projections of employment made by the Planning Commission. The network of Krishi Vigyan Kendras (KVKs) is expanding and by the end of Tenth Five Year Plan, all the districts in the country will have a KVK each, which is an important link between research/science institutions and farmers/extension organisations. The science and technology research organisations (ICAR and CSIR institutes) have evolved several technologies, which are adoptable and sustainable. The institutional framework for technology transfer, credit support and training is also available. What is needed is the delineation of these technologies for the rural areas and dovetailing the institutional support system for their widespread adoption.

With a view to identifying the technological interventions with high potential of employment creation in the rural areas, both in the farm and non-farm sectors, the National Academy of Agricultural Sciences organised a Symposium.[1] The following recommendations are the outcome of the deliberations in the Symposium and subsequent write-ups received from the participants.

Recommendations

The recommendations have been grouped into six sections *viz.* (1) foodgrains, (2) fruits and vegetables, (3) sugarcane, (4) livestock, (5) bio-fuel production, and (6) biotechnology applications.

1. Foodgrains

1.1 Maize

Maize is grown in almost all parts of the country. During the last 10 years, the annual production of maize has varied between 11 and 15 million tonnes. Nearly 55 per cent of the maize grains are used for food purposes, 14 per cent for livestock feed, 18 per cent for poultry feed, 12 per cent for starch and 1 per cent for seed. The mature maize kernel, contains 70 to 75 per cent starch, 8 to 10 per cent protein and 4 to 5 per cent oil. The endosperm is largely composed of starch (about 90 per cent) and the germ contains high level of protein and oil. Currently, very little of maize is dry milled (about 0.5 million tonnes) where separation of germ is also carried out. Potential exists to process about 4 to 5 million tonnes of maize by dry milling to obtain about 0.5 million tonnes of maize germ, from both white and yellow coloured maize.

(a) Processing of Maize for Use as Food

Major portion of maize is converted to flour and semolina in plate grinders. The grain contains 4 to 5 per cent oil, most of which is

1. Symposium held on January 14-15, 2006, at the Central Food Technological Research Institute (CFTRI), Mysore. The Symposium was held under the Convenership of Dr. V. Prakash, Director, CFTRI and Prof. Shabd S. Acharya, Honorary Professor, Institute of Development Studies, Jaipur and Fellow of NAAS. The Symposium brought together 35 experts including the scientists of CFTRI and representatives of government and non-government organisations, engaged in technological interventions in the rural areas. After the introductory remarks by the Convener and presentation of background paper by the Co-Convener, each participant made presentation on the potential of technological interventions for creating employment opportunities in farm and non-farm sectors. The participants suggested several primary processing and value addition activities through available technologies.

present in its germ. Germ accounts for about 10 per cent of the maize kernel and has oil content of 25 to 30 per cent. It is like an oilseed having a good potential to produce, nearly 1.5 to 2 lakh tonnes of oil to supplement the oil requirements of the country. The deoiled germ cake has a protein content of about 25 per cent, which can be used to produce supplementary and health foods. The degermed grits can be further processed to produce high value consumer products, such as maize flakes, extruded snacks and fermented beverages using low fat, low fibre grits. These grits should be the principal focus of dry milling process. Other maize products (germ, flour and meal) have a variety of food and feed uses. Large-scale processing of maize is mostly confined to wet milling for which about 5 per cent of the maize is utilised. Wet milling aims at producing starch and other products for industrial applications. However, a good part of starch, dextrins and sweeteners find their way into a variety of convenience and snack foods. A product of maize with good potential is the high fructose syrup. There are about 19 units in the organised sector producing about 3.8 lakhs tonnes of starch (worth a market value of Rs. 440 crores in 2000-01), and liquid glucose and dextrose, using about 4 to 5 lakh tonnes of maize. Demand for starch is poised to rise substantially. Annual growth rate for starch and other value-added food products from maize is expected to be around five per cent. There is scope for production of food grade starch at the village level in maize producing areas.

(b) Processing of Maize for Use as Feed

It is estimated that about 42 million tonnes of animal feed and around 14 million tonnes of poultry feed are produced and used in the country. Out of this, around 3 million tonnes of animal feed and 4 million tonnes of poultry feed are produced by organised feed manufacturers. Maize being an important component of feed formulations, the manufacturers need good quality maize grits for making compound feeds.

Mostly, dry milling process is used to obtain grits. In dry milling system, maize grains (with less than 17 per cent moisture) are subjected to milling to produce grits, meal, flour and germ. Grits are subsequently reduced to the desired size. There is another tempering-degerming process (also called TD process or partially wet degerming process), in which water is added to maize to increase the moisture content to 20 to 23 per cent. It is then tempered and milled to produce grits, meal, flour, bran and germ.

(c) New Maize Processing Methods

In India, currently, substantial quantity (about 4 to 5 million tonnes) of maize is ground flour in rural sector and the rest for feed purpose. Virtually, no germ is recovered in this process. Due to high content of germ and high oil in it, the resultant products have poor shelf-life. However, some quantity (30000-40000 tonnes annually) is also processed in India by large-scale dry milling methods using specially designed degerming equipments like Beal degermer and Entoleter which are imported units. There are no indigenous dry milling plants operating in the country. However, Central Food Technological Research Institute (CFTRI) has developed following three types of improved maize milling systems:

(i) Mini Grain Mill: The system of mini grain mill consists of an improved plate grinder in which moisture-conditioned maize is ground, aspirated and then sifted through a deck of sieves to obtain different fractions of the milled grain like bran, coarse semolina, fine semolina and flour. It has a capacity of 50 to 60 kg per hour.

(ii) Maize Mill: This is a small maize mill, with a capacity of 300 to 400kg/hr. In this system, the maize grain is first moistened, conditioned, passed through modified hullers, aspirated and then, ground. This is then sifted to obtain various fractions, *viz.*, bran, meal, grits of different sizes, germ and semolina.

(iii) 1Q-D Type Dry Milling Plant: It is one tonne per hour capacity maize mill. It is an indigenously developed dry maize milling system incorporating all the concepts that exist in a large capacity modern dry milling system. The operations involved are cleaning, destoning, water conditioning and degerming in stages, size separation and gravity separation of the milled products. Various products and by-products obtained include bran, meal, germ, large grits, medium grits and small grits. The grits can be further ground to semolina and flour, and germ is taken for oil extraction. Value-added products for downstream processing of products obtained from the dry milling technology include products like maize flour and semolina, snack foods, noodles, flaked products and *roti* flour. Apart from this, the germ recovered can be used for oil extraction that can contribute to the edible oil pool of the country.

(d) Employment Potential

It is estimated that 750 to 1000 maize mills at one TPH capacity can be established in the country. The products from these mills *viz.*, big and medium grits, small grits, bran, meal and germ could be used by subsidiary units for further value addition. The low fat, low fibre big and medium grits can be used in the manufacture of maize flakes with better shelf life. The small grits can be used in the brewery industry, and can also be extruded with different formulations to give a variety of products. Maize germ is an excellent source of oil for cooking purposes. The deoiled germ and meal from the CFTRI maize mill could be used in the manufacture of cattle feed. Maize bran could also be used as a source of dietary fibre in bakery industries. Each CFTRI maize mill can create employment for atleast 10 persons.

For fully exploiting the potential, there is a need for setting up of maize parks, at least one each in the major maize producing states of the country. The maize parks can have an integrated approach wherein good quality maize procured from the farmers could be processed for value addition into products like maize flakes, RTE flakes, extruded maize products, maize *roti* flour, maize germ oil extraction, feed formulations, and bakery products, all under one roof.

Not only is the technology for manufacture of these products available in the country (mainly at CFTRI), but the machinery is also available. The milling machinery developed at CFTRI Mysore, could be gainfully utilised for processing of maize to prepare grits, flour, germ separation, and these products would be suitable for various food and added uses. With the diversification of products from maize, it would boost the production of good quality maize in the country, directly benefiting the grower. Also, a wide variety of products at competitive prices would be available to the consumer. The technology and machinery used for processing of maize could, with minor modifications, be used for value addition to other grains also. This would ensure working of the mills, all round the year.

1.2 Sorghum and Millets

India ranks second in the world for sorghum production and first with respect to many regionally important crops like millets and pseudocereals. Much of the production of these crops is by the small and marginal landholders, living in marginal environments. Sorghum and millets (finger millet (*ragi*), foxtail millet (*navane*), kodo millet (*varagu*), little millet (*kutki*), barnyard millet (*sawan*), and proso millet (*cheena*)) are crucial to the food security of many poor people in the semi-arid tropics. There are other grains, which are not exactly cereals, but are consumed the way cereals are consumed and are classified as 'pseudocereals'. Buckwheat, amaranthus and job's tears, fall in this category.

It is reported that more than 400 million people in the world depend on millets for sustenance. It may be noted, that the dietary patterns, largely dependent on rice and wheat, have led to widespread nutritional deficiency. The prevalence of micronutrients deficiency even in the affluent population is a matter of serious concern. In order to alleviate this problem and improve nutritional security, small millets and pseudocereals can play a major role. Sorghum and millets have chemical composition, similar to or better than rice and wheat in some respects. In fact, they contain high fibre, non-starchy polysaccharides and starch with some unique characteristics. Protein quality and essential amino acid profile of some of the millets is better than many of the cereals. In recognition of this, these grains are now considered as 'Nutritious or Nutri grains'.

Sorghum and millets are grown in diverse soils, varying rainfall regimes and in areas widely differing in thermo and photoperiods. The resilience exhibited by these crops is helpful in adjusting themselves to different kinds of ecological niches. Another important property of most of the millets is that they can be stored without getting infested for years together in the rural areas. These grains hold important place in the grain economy of our country, but utilisation of these cereals is limited on account of absence of appropriate processing technologies to produce shelf-stable primary products. Lack of recognition of the potentialities and limited efforts in expanding the market beyond the traditional consumers, have also restricted their widespread consumption.

Primary processing converts the grains into products for their direct utilisation for food purposes or make them available in the ready-to-eat form. Value addition during primary processing, as it stands now, is not very high, but the potential is large. With increasing consumer awareness, the potentialities of value addition could be realised, which provides considerable employment possibilities. Opportunities in the processing of sorghum and millets include, cleaned *jowar* and millets; pearled grains and products;

refined flours suitable for special foods/bakery; shelf stable flours for conventional and composite flour based foods; rollable *roti* flours; ready to flake materials; *jowar* and millet *rotis*, flakes, RTE flakes; malt for food and brewing; and popped RTE foods. Some of the new small-scale processing technologies pertaining to sorghum and millets are as below:

(i) Mini Grain Mill

It is a grain processing unit that comprises mainly the plate mill, commonly used to pulverise the cereals for preparation of flour for conventional foods. It has about 100 kg per hour capacity and enables to produce semolina of desired size and flour, nearly free from the coarse seed coat matter. It is a versatile mill, suitable for preparation of milling fraction from wheat, maize, sorghum and millets.

(ii) Finger Millet Malt

Finger millet or *ragi* is unique among tropical cereals for its suitability to prepare malt for food as well as for brewing. Malting methodology standardised at CFTRI, Mysore, could be practised conveniently to prepare the malt flour suitable for weaning food, milk based beverage and some other special foods. The malt flour also finds extensive usage as amylase rich food.

(iii) Popped RTE Products

Popping of cereal grains to prepare ready-to-eat products would be very useful to diversify their utilisation in the value-added foods, especially in supplementary feeding programmes. Popped grains can be used as snack food also. Popped products are stable for longer period compared to many other products from these germs. Popping, being a simple dry treat processing technique, can be adopted at the household to industrial level conveniently.

1.3 Wheat

Wheat, which is widely grown in five continents, is the most important cereal grain produced, consumed and traded in the world today. It is a highly versatile food product, as it can be stored safely for long periods of time and transported in bulk, over long distances. Utilisation of wheat includes food, feed, seed and other industrial uses. Use of wheat as food accounts for 67 per cent of total consumption. Wheat is consumed as food in numerous forms, all of which involve some degree of processing.

Wheat produced in India, belongs to three species *viz.*, *Triticum aestivum* (95 per cent), *Triticum durum* (4 per cent) and *Triticum dicoccum* (1 per cent). Varieties of *Triticum aestivum* are used for bakery products, *Triticum durum* for pasta products and *Triticum dicoccum* for traditional products. Wheats are also classified as hard, medium-hard and soft, based on their protein content and suitability for different purposes. Soft wheats are suited for cake, biscuit and pastry, medium-hard wheats for unleavened breads and noodles, and hard wheats for bread. Extra-hard durum wheats are suitable for pasta products. Most Indian wheat is soft or medium-hard, best suited for making *chapati, nan, parontha* and *roti*. Consumers usually take their wheat to small flourmills where it is milled into whole meal flour called *atta,* for making traditional products. In India, more than 50 million tonnes of wheat is being processed every year into whole-wheat flour (*atta*) in the unorganised sector. The *chakkies* or disc mills are spread all over the country in the urban, semi-urban and rural regions.

The wheat milling process aims to break open the wheat grain, remove the outer bran layer and the wheat germ, and then grind the endosperm to the fineness of semolina (*suji*) and flour. The different types of flours that can be produced are patent flour, baker's flour, first clear flour, and second clear flour. Flour numbers indicate their suitability for the production of bread, biscuits, crackers and cakes. In India, the milling industry is producing wheat flour (*maida*), semolina, resultant *atta* and bran.

The roller flour milling industry in India, comprises more than 820 mills and most of them are having a grinding capacity in the range of 80 to 200 tonnes per day. The installed capacity of the total milling industry in the country is more than 15 million tonnes, but the actual grinding is about 8 million tonnes which shows a capacity utilisation of only 50 per cent. Some flour mills produce whole wheat flour and market it in both bulk as well as unit packs.

There is a vast potential for the production of whole wheat *atta* at the village level, as the demand for the *atta* exists in all the households of semi-urban and urban regions. This will help in generation of employment in the rural areas. Since the *atta* is shelf stable for three months, production, storage and transportation will not cause any problem. The total cost of machinery and equipments necessary for primary processing of wheat into whole-*atta* is approximately, Rs. 2.7 lakhs and the estimated electricity consumption for processing of wheat into whole wheat flour is about 100 units per tonne. The number of personnel required is seven. The production capacity of the unit is about 2.5 tonnes per eight hour shift.

1.4 Pulses

India is the largest producer of pulses (around 14.5 million tonnes annually) in the world. Nearly 11 million tonnes of pulses are converted to *dal* (dehusked and split pulse); of the remaining 3.5 million tonnes, a large proportion is utilised as whole grain for cooking and rest is used for seed purposes. Conversion of pulse grains into *dal* is one of the important food processing industries, as pulses in India are consumed mostly as *dal*.

Although a large quantity of pulses are processed by medium- and small-scale industries, about 1.5 million tonnes of pulses are still processed in the rural sector without proper machinery, which not only affects the availability of *dal* in rural sector due to loss of *dal* (*dal* yield–65-68 per cent), but also results in the yield of inferior quality product (removal of husk–93-95 per cent). This inferior quality *dal* fetches 20 per cent less value in the market than the average quality *dal* and hence, it is generally sold in the rural market only, denying good quality *dal* to rural consumers.

CFTRI has developed a mini *dal* mill which can process major pulses (*Tur, Chana*) into *dal* at a lower processing cost (Rs. 20 to 24 per quintal as against Rs. 70 to 75 in large- scale mill). The capacity of the mill is 100 kg per hour and it requires only 1.5 HP power. The mill gives a *dal* yield of 75 per cent with almost 99 per cent removal of husk and minimum breakage (2 to 3 per cent). It can also be adopted at farm level in the rural sector and can give self-employment to the rural men or women. A team of two/three workers can operate the mini *dal* mill (self-employed) and there is scope for engaging another two persons for allied works like pre-milling treatment of pulses. The mill can provide better facilities for milling of pulses at the rural level and also value addition to the by-products (husk, brokens and powder) through separation and utilisation of edible material from the by-products. Woman entrepreneurs and organisations, self-help groups, and pulse growing farmers can be the beneficiaries of installation of mini *dal* mill.

There are 11 pulse growing states in the country, but most of the pulse processing centres are away from the main pulse productive areas. There is, therefore, a scope for setting up around 8000 mini *dal* mills in the rural areas of the country, providing direct job opportunities for 30,000 rural men/women, while another 10,000 persons can be indirectly benefited because of upliftment of pulse processing activity in the rural sector. At least, 5000 small *dal* mills could be set up in major pulse growing states during the Eleventh Five Year plan. Indirect benefit of better milling facility will be available to 15,000 villages. The availability of *dal* will increase by 50,000 tonnes in the rural sector alone because of less breakage and improved dehusking.

Another area of employment potential is, value addition to about 2.5 million tonnes by-products obtained annually from *dal* mills in India. At present, these by-products are sold as cattle feed. Analysis of commercial samples of by-products has revealed that around 35 per cent of cotyledon material can be recovered using appropriate machinery for destoning, size separation and air classification. The recovered cotyledon material can be used in traditional products like *papad, vada, chakli, sev,* and *halwa*. This can enhance the availability of pulses in the country. By-product recovery units can be set up near a major pulse processing centre where a cluster of *dal* mills exist [like Indore (MP), Gulberga (Karnataka), Akola (Maharashtra), Virudunagar (Tamil Nadu) etc.].

There is a scope for setting up 5000 such by-product recovery units in semi-urban sector which can give an income of about Rs. 2 lakhs per year per unit and an employment generation to about 25,000 persons. The net value addition to the total by-products of *dal* milling sector in the country can be of the order of Rs. 100 crores.

2. *Fruits and Vegetables*

India ranks second in production of fruits and vegetables in the world. It produces a wide range of fruits, of which mango, banana, citrus, guava and apple, account for 75 per cent of the total fruit production in the country. In recent years, papaya, *amla, ber,* fig, date palm, watermelon, pomegranate and custard apple have also shown marked improvement in output, availability and market demand. India produces about 70 different varieties of leafy, fruity and starchy tuber varieties of vegetables. Potatoes, onions, tomatoes and brinjals, account for bulk of the vegetables produced in the country.

Obviously, production of such a wide spectrum of fruits and vegetables provides immense scope for their proper harvesting, storage, transportation and marketing, every stage involving deployment of manpower. Establishing primary processing centre in the producing areas of selected fruit and vegetable can serve to protect the crop from perishability, reduction in bulk, easy transportation, storage and trade, including value addition.

The potential of employment generation through primary processing of fruits and vegetables is presented under four broad groups *viz.*, (i) basic processing; (ii) minimal processing; (iii) processing of under-utilised fruits; and (iv) other products.

2.1 Basic Processing

(a) Dehydration

Dehydration represents the simplest technology of value addition and extension of storage life of fruits or vegetables, by removal of water content to a residual moisture of around five per cent. This requires the installation of mechanical driers at the rural centres. Owing to the problem of availability of electricity, solar driers can be used to remove moisture levels of fresh commodities. But solar driers cannot fine-tune the residual moisture to the range of 5 to 8 per cent. Another dimension of dehydration is osmo-dehydration, which encounters the use of sugar/salt solution depending on the desired fruit or vegetable. Use of mechanical driers after osmotic removal of water, ensures fine quality product and microbiological safety. Osmo-dried fruit products maintain excellent texture and flavour with added sugar. Dehydration has many advantages of reducing the bulk, reducing the freight charges and enhancing the storage life. Dehydrated products could be reconstituted in water at the room temperature or with mild heat, to fresh-like fruit or vegetable, assuring easy consumption. This is applicable in the case of a number of fruits (like mango, pineapple and jackfruit), vegetables and mushrooms.

The fresh commodities (either fruits or vegetables) require washing, grading, and sizing followed by some pre-treatment before subjecting them to the choice of dehydration. This can contribute to the development of quality product for domestic as well as export market. Further, the processed produce requires hygienic handling, packaging, sealing and storage till consumption. All of these post-harvest operations are labour intensive and help in linking the producers with processing centres, and delivering the items of preferred consumption to the urban masses. Dehydrated lime and mango pickles, are classic examples. The commodities available in the harvest season are brine soaked and dried, followed by spicing to form packs, which can be reconstituted overnight to fresh pickles. Such dry pickles offer the advantages of lightness, reduction in volume and freight charges, extended storage life and quality after reconstitution, akin to the fresh pickle. Dehydrated mushroom is another illustrated example. Here, 90 per cent of water content of the fresh mushroom is reduced to a residue of about 5 per cent, using mechanical hot air driers with pre-treatment to prevent discoloration. The dry mushroom can also find useful application as a flavouring agent in a number of dishes like soups, drinks, *chutneys, pulavs,* and fried products. In case of mushrooms too, both production and processing require manual assistance, particularly in cleaning, washing, grading, sizing, drying, packing, sealing and storing under defined conditions. In fact, in mushroom industry both in production and processing, manpower cost is more than 20 per cent of the net project cost, and hence, is a lucrative business. It is well known that the international trade of specific major varieties of mushrooms is in dehydrated form.

(b) Canning and Aseptic Packaging

Canning unit serves the purpose of holding the primary processed product under sterile conditions till it is transported to and used in the secondary processing centres. As canning ensures sterility of the product, it is very convenient for exporters. Classic examples of canning are, canned slices of mango, pineapple and mushrooms, mango pulp and green peas. The canning units are labour intensive and have great potential of employment creation.

In some varieties of citrus, the juice, when heated, turns bitter due to condensation of lactones into limonin. The aseptic packaging is very useful for such juices. The juice can be aseptically filled in packages of 50 kg for storage at optimum low temperature and for long distance transport. In aseptic packaging, as well, manpower requirement is quite high.

2.2 Minimal Processing of Fruits and Vegetables

CFTRI has done considerable work on minimal processing of vegetables. Conditions have been established to minimally process as many as 27 vegetables and store under optimum low temperature conditions. These vegetables include, ash gourd, beet root, beans, bitter gourd, carrot, cabbage, cauliflower, cluster beans, coccinia drumsticks, cucumber, field beans, green peas, green chillies, knol khol, okra, onion, plantain, ridge gourd, snake gourd, tomato, turnip and leafy vegetables such as coriander, curry mint, fenugreek and spinach. The technology has been successfully transferred to five industrial entrepreneurs, which in turn serve to link the rural production centres with processing centres.

The basic steps involved in minimal processing include washing of the vegetables harvested at optimum maturity, their grading, peeling, trimming, slicing/dicing/shredding, pre-treatment, surface drying, modified atmosphere packaging, storage at optimum low temperature, and marketing. The method confers several advantages, namely, fresh-like vegetables, free of wastage, in ready-to-cook form, reduction in bulk, extension of storage life, easy transportation and value addition up to 60 per cent. Eventually, these steps generate lot of employment.

2.3 Processing of Under-Utilised Fruits

Several fruits like *ber*, wood-apple, *jamoon*, *amla*, and custard apple are grown under natural conditions in the forest areas or waste lands without much care by the human beings. These fruits contain useful compounds of biofunctionality such as prominent antioxidant activity, antiplatelet aggregation, and peroxidase inhibition. Careful collection of these fruits and utilisation in many *ayurvedic* preparations can fetch considerable income to the poor families. CFTRI is continuing to do more research on the extension of the storage life in fresh and processed form of such fruits.

Wood-apple is a classic example of under-utilisation, due to lack of information about standards of maturity, harvesting and ripening indices. It is mostly grown in forest areas, but requires very careful harvesting, since the fruits that fall from tall trees develop hairline cracks that lead to microbiological infection of the pulp inside. This is a fruit, just like coconut, with a hard shell outside. Several products such as juice, jam and jelly have been developed from this fruit. Similarly, *jamoon* can find useful applications in the preparation of several products like juice, concentrate, jam and jelly. The fruit and its products are known for many *ayurvedic* benefits to the human body. *Amla* is well known for its tannin contents with related antioxidant potential and its use in pickle preparation. The fruit pulp in fresh form finds immense application in preparation of *ayurvedic* products.

Custard apple is another fruit, which is little exploited for processing. The fruit peel and seeds of custard apple are reportedly used for insecticide preparation, and the pulp is thrown as waste for making compost. The pulp is rich in sugar, minerals and antioxidant compounds, and can find useful food applications. CFTRI has developed eight products, for which national and international patents have been filed. The products, through alternative methods of processing, are free from bitterness, discolouration and off flavour. Similarly, methods have been developed at CFTRI for preparing products like nectar, ready-to-serve beverages, jam and jelly from palmirah palm.

There is an immense potential of fruits and vegetables from the forests and the uncultivated lands, that can serve to generate employment and income of the forest-dependent poor and tribal populations. The products in fresh/processed form can contribute to the biomedicinal properties gearing towards improvement of human health. This requires concerted measures to increase their production, quality through varietal multiplication, careful harvest, storing, processing and development of market linkages both in the domestic and foreign markets. Since such fruits are hardier varieties and come out of natural growth with little agronomic care, they deserve special attention as they can generate considerable employment in the difficult areas.

2.4 Other Products

(a) Preparation of Grape Raisin

Fully ripe seedless grapes with high sugar content (about 22° brix) are used for making raisin. The grapes are exposed to sulphur dioxide fumes and dried. The dry raisins are then freed from stalks and pedicles, and kept for moisture equilibrium. These can be packed in polyethylene bags, which can be further placed in cardboard boxes.

(b) Osmo-Air Dried Fruits

It is a novel process for fruit drying wherein the dry fruit is comparable to fresh fruit in colour, flavour and texture. Fruits like apricots, *ber*, pineapple, jackfruit and mango, can be used for drying by this method. The process involves operations like selection of fruits, cleaning, washing, peeling, curing and slicing/dicing. The prepared fruit slices are steeped in sugar solution to remove about half of the water by osmosis. The slices are then drained, dried in a hot air drier and packed in flexible pouches.

(c) Jam, Jelly and Marmalade

Jam is prepared by boiling fruit pulp with sugar to a moderate consistency. Commercially prepared product has 45 parts of fruit pulp for every 55 parts of sugar and contains about 68 per cent soluble solids. Jelly is prepared by boiling clear fruit extract with sugar and additives to a stage at which a clear gel forms. Marmalade is a fruit jelly wherein the fruit slices or peels are suspended. It is generally prepared from oranges and lemons.

(d) Fruit Toffee

Fruit toffee is more nutritious than sugar boiled confectionery. It is made from fruit pulp and other ingredients. The process involves cooking the fruit pulp to one third volume. Other ingredients are added and the whole mass is spread as a thin sheet of 0.75 to 1 cm. thickness. The cooled sheet is cut with a toffee cutter and dried. The toffee is wrapped individually with decoratively printed cellophane.

(e) Bleached Dry Ginger

Fresh ginger is cleaned thoroughly to remove the adhering soil and dirt. The outer skin is scrapped off using SS knives or sharp-edged bamboo device. It is washed, partially sun-dried and then soaked in limewater overnight and dried in the sun. The process of dipping and drying is repeated two or three times and finally dried to a moisture level of 10 to 12 per cent and packed in polythene-lined gunny bags.

(f) Spice Powders

It is a process for making ready spice/curry powders for *sambhar, rasam, pulao* and other preparations. The dried clean spices are powdered to 40 to 50 mesh size. The powder is cooled to room temperature and sifted. The spice powder is placed in airtight container and fumigated. The powder is packed in flexible pouches for marketing. For curry powder preparation, the cleaned dry spices are given a mild roasting (optional step), mixed as per recipe and ground to 40 to 50 mesh, cooled to room temperature, sifted, fumigated and packed.

3. Sugarcane-Jaggery

Jaggery or unrefined sugar (*Gur*), which is the concentrated form of sugarcane juice, is one of the important traditional food components of Indian culinary. Jaggery is nutritionally superior to refined sugar in terms of minerals and neutraceuticals. In India, nearly 35 per cent of about 250 million tonnes of sugarcane produce is crushed for jaggery preparation. Current annual production of jaggery in the country is around 15 million tonnes. In many of the Indian food preparations, jaggery is preferred to sugar because of its specific sensory and textural characteristics.

Although there is considerable technological advancements with respect to production of sugar from sugarcane, preparation of jaggery has still remained at the cottage industrial level as a traditional practice. Normally, preparation of jaggery from sugarcane involves crushing of canes in iron or steel rolls and simmering the expelled juice in open pan for evaporation of water. During boiling of sugar-cane juice, the floating impurities are separated using perforated ladle. Some of the dissolved impurities are made to coagulate and float over the surface of the juice by addition of external source of coagulants. Traditionally, natural coagulant such as lady finger root extract was being used but nowadays, most of the jaggery manufacturers have shifted to chemical additives. Although, addition of chemicals improves the texture and appearance, it invariably affects the taste and keeping quality of the product. There are reports of jaggery samples containing 1000 ppm of SO_2. High proportion of SO_2 in jaggery affects its quality in terms of imparting acidic taste to the product, triggers formation of invert sugar during storage because of absorption of water, leading to increased microbial growth and overall spoilage of the product.

There is considerable demand for good quality jaggery in the country and also, outside India. Hence, production of good quality jaggery by following good manufacturing practices can enhance its consumption and provide employment to the people. A considerable portion of refined sugars is mixed with molasses and sold as brown coloured sugar in many of the developed countries. Instead of this, good quality jaggery itself can be supplied to the consumers. There is lot of scope for value addition in terms of preparation of jaggery with diversified forms. Liquid jaggery is very popular in parts of coastal India and Northeastern states. This is extensively used in the preparation of several sweetmeats in West Bengal and surrounding areas. Some of the value-added products that could be prepared of jaggery are:

3.1 Tonic Jaggery

The sugarcane extract, concentrated to about 80° brix, is a very good source of digestible carbohydrates to serve as a carrier for several micronutrients. The consistency of the product permits incorporation of vitamins, minerals and other additions with induced taste and flavour of the product. Such a product will be specially liked by the school going children and may also be useful as a bread spread. Considerable part of jaggery can be diverted for this kind of value-added product.

3.2 Spiced Jaggery

In several parts of the country, the farmers prepare small quantity of jaggery spiced with ginger, *elaichi* and such other desirable spices, especially for home or local consumption. Since some of these spices and condiments have neutraceutical value, such a product will be a value-added food and may have very high market in India as well as abroad.

3.3 Powdered Jaggery

Most of the jaggery produced in the country is in the form of cubes and buckets. However, when the jaggery is to be utilised at the household level, it is always crushed into small grits. Hence, production of jaggery in the powder form for ready utilisation may be very useful. Such a product can find application in large quantities in the government sponsored nutrition programmes such as energy food. As on date, the difficulty for preparation of such a product is non-availability of granulators at the jaggery production sites and a suitable technology for its packaging and transportation. These difficulties could be overcome easily.

3.4 Jaggery Cubes

Sugar cubes, which disintegrate and disperse when mixed with warm liquids, are very commonly used in star hotels both in developed and developing countries. However, similar kinds of products from jaggery are not being produced so far. Jaggery cubes can replace sugar cubes because of the special taste, aroma and health benefits of jaggery.

The sugarcane crushing and preparation of jaggery are totally done manually. The element of mechanisation is mostly at the crushing stage. Because of this, if the production of jaggery is increased, there will be proportional increase in the rural employment. Indirectly, it may result in increased cultivation of sugarcane which will also increase rural employment and incomes. The by-products of sugar and jaggery production units such as bagasse, the ash and also the muddy particles containing molasses are also not being efficiently used for much of value addition. The bagasse being a very good source of cellulose can find utilisation in paper and cardboard industry. The molasses containing muddy particles may be used for production of biogas along with the ingredients used for *gobar* gas.

4. Livestock

4.1 Use of Wastes

There is considerable production and consumption of meat, poultry and fish products in India. But the proper use of wastes of these livestock products is generally wanting. The wastes created in processing of livestock products include the following:

- Mutton, beef, pork – Fur, viscera
- Poultry – Feather, viscera
- Fish, prawn, shrimp, lobster – Scales, exoskeleton, viscera

Fur and feather, composed of kerato-proteins, can become a good source of sulphur containing amino acids. By application of enzyme based biotechnologies, high value amino acids can be obtained for use in food and pharmaceutical industries. The exoskeleton of shrimp and prawns are made of chitin, from which chitosan production by deacetylation process can be undertaken. Chitin can be depolymerised to yield N-acetyl glucosamine, which is an important pharmaceutical ingredient. The animal and fish viscera can be processed as sources of protein for food, feed and pharmaceutical purposes. The protein hydrolysates can also be designed as excellent media for microbial fermentation processes. The technologies of small, medium and large-scale industrial processes for production of enzymes and use of enzymes, have a big potential for employment generation.

4.2 Poultry Dressing

The process of poultry dressing involves *antemortem* inspection, slaughtering, scalding, defeathering, singeing, evisceration and *postmortem* inspection. The edible internal organs are separated, washed and packed separately. The carcasses are washed, packed and chilled in crushed ice for further storage. The marketing of dressed chicken is done in fresh, chilled or frozen form. It is also feasible to market the product in cut up portions like half chicken, drumstick, thigh, back breast and wing.

5. Biofuel Production on Waste Lands

India has considerable wastelands which can be used for cultivation of non-conventional oil bearing crops or trees. Also, on the cultivated lands, particularly which are rainfed, there is tremendous scope for growing such oil bearing short-duration and/or hardy crops like castor and jatropha. The borders of the existing crop lands can be used to grow oil bearing crops without impairing the food security of the country. This can generate tremendous rural wealth and employment in rural areas. The employment potential is very high in post-harvest operations like decortication, oil-expelling, transporting of oil to a central biofuel facility, and transportation of oil-cake to the markets. These operations can be done by entrepreneurs employing semi-skilled and unskilled labour both in rural and semi-urban areas.

6. Science and Biotechnology Applications

There are several areas where educated rural youth can be trained in science and biotechnology applications, and employed in agricultural support services as entrepreneurs, which in turn can provide employment to others. Some of the potential areas have been identified here.

6.1 Improvement of Culture Products: Utilisation of Microbial Cultures in Agriculture

Popularisation of such technologies as organic farming, organic fertilisers, growth regulators and biopesticides which are eco-friendly, can be done by training the agripreneurs who could, in turn, develop enterprises for such technological interventions and applications. In the non-farm sector, units can be set up for development and production of cultures, inoculum, and adoption of microbial biotechnology and scale-up methods. Industrial production of microbial cultures can be taken up by small-scale industrial enterprises.

6.2 Production of High-Yielding Planting Material

Adoption of tissue culture method for micropropagation, like banana and potato propagation, is another potential avenue that needs to be encouraged. This would require training in tissue culture for adopting simple and efficient technology with less dependence on electricity. It could also be an entrepreneurial activity to cater to the requirements of elite propagules essential for the region for large-scale cultivation. Industrial production of uniform propagules for providing high-yielding and assured varieties for enhancement of yields will create opportunities for setting up of SSI units based on the requirement of the crops in a particular region.

6.3 Floriculture

There is a need for identification of the local varieties which are cultivable in a geographical region and their multiplication using modern farming methodologies. This provides opportunities for rural women in cultivation and in cut flowers business with a potential to cater to the urban markets. Interface of packaging technology for extension of shelf life would be useful. This venture provides opportunities for plant tissue culture based industrial activity, which can make India as a global leader in meeting the world requirement of selected flowering plants and foliages.

6.4 High Quality Banana Planting Material by Tissue Culture

Banana varieties such as Cavendish, 'Rasabale' and others can be mass multiplied by tissue culture technology. The method involves establishment of plant tissue culture, large-scale production of plantlets, growing tissue culture plants in poly-house and field cultivation. The steps involved are selection of high-yielding mother clones, disinfection of explants, removal of bud/meristem, establishment of shoot cultures on nutrient medium with suitable hormone in recommended containers and incubation at specified conditions, multiplication of shoots in nutrient medium with specific hormones, transfer of selected shoots for shoot multiplication or rooting, hardening of tissue culture plants in poly houses and transportation for field cultivation.

6.5 Biocontrol of Microbial Pathogens and Arthropod Pests at Farm and Post-Harvest Levels

Environment management employing blends of useful microorganisms for pollution control, deodorisation and sewage treatment is gaining increased attention. These concepts could be extended to the agricultural practices as well as food processing as a means of preservation and enhancement of nutritive value and acceptability. Biotechnological processes using genetically modified and proven, friendly consortia of microorganisms can be employed. The industrial scale production of such microorganisms and creation of their blends for commercial purposes also provide good potential for employment of skilled rural youth.

Summing Up

Clearly, there is immense potential of employment generation through technological interventions for primary value addition to the agricultural products like foodgrains, fruits, vegetables, sugarcane and livestock products/by-products. The employment potential is also very high in production of biofuels and application of science and biotechnologies in agricultural activities. In addition to the areas identified in the paper, there are several other agricultural crops like barley, potato, coconut and medicinal plants which provide considerable potential of employment creation in primary value addition activities. As simple as cleaning, washing, sorting and packaging of fruits, vegetables and other farm products at the farm or village level, can provide tremendous employment and also save huge physical losses of the product that occur during storage, handling and transportation to the market. Obviously, the employment opportunities exist for unskilled, semi-skilled as well as educated youths. However, there are two preconditions for full utilisation of these opportunities. First, the technological interventions will need to be dovetailed with institutional support system including, technology transfer, credit delivery and training of entrepreneurs in the form of specific government programmes and schemes. And second, there will be a need for new enterprises (individuals, self-help groups or cooperatives) to procure the products from individual farmers and add value to them. While some such primary processing enterprises may be established in small villages, a majority will necessary become viable in large villages, towns or semi-urban areas. Complementary policies to encourage such enterprises will be quite critical for their viability and success.

2

Agriculture-Industry Interface

Value-Added Farm Products

Introduction

India is today the second largest producer of food in the world and has all the potential of becoming number one if the emerging problems after the green, white and blue revolution are properly addressed. Agriculture production has shown a growth of about three per cent per annum, and today, India is the number one producer of milk, and second largest producer of fruits and vegetables in the world, with a buffer stock of over 60 million tonnes of wheat and rice. Due to poor handling of the produce, post-harvest losses have been high, resulting in a significant gap between gross production and the net availability to the consumer. The profits on agricultural commodities have greatly diminished. Since nineties, the cost of agricultural inputs has increased faster than the market price of the outputs. As a result, farmers are about 15-20 per cent worse off, even after taking into account the gains in productivity.

The problem of improvement in agriculture needs to be tackled from two different angles, First, to increase productivity of agriculture and delivery system and Second, to increase the farmer's earning through efficient and effective value addition.

Value addition to raw food material in India is only 7 per cent while it is 23, 45 and 188 per cent in China, Philippines and UK, respectively (as per National Food Processing Policy, Draft Document, 2000). Studies also reveal that more than four dozen value-added products are produced from derivatives of paddy in a small country like Japan, which produces only 2 per cent of the total world production of paddy. We process less than 2 per cent of fruits and

---

Policy Paper 16, November 2002.

vegetables as compared to 30 per cent in Thailand and 80 per cent in Malaysia (PIS Feature, 2001).

In India, the difference between price paid by consumers for value-added products and farmer's realisation has been increasing rapidly. Further, the food processing industry is plagued by high-risk profile, poor infrastructure and outdated technologies and taxation laws. There is also a lack of backward linkage between farmers and processors. This leads to non-uniformity and inconsistent supply of raw material, longer chain of intermediaries and lack of adequate economic benefits to farmers.

Value addition is often understood in the context of adding value to the product. A new dimension from the consumer point of view is added to the existing understanding of value i.e. how a consumer perceives the value delivered to him through a bundle of product services. This new approach of value addition through the consumer's mind needs special attention. All the activities now must be seen from the consumer point of view. In other words, consumer orientation is required in all spheres of agricultural sector. Keeping this approach in mind, there are three ways in which value addition to farm produce is possible:

Level 1 - *Post-harvest level/primary processing:* proper cleaning, grading and packaging e.g. vegetables, potatoes, fruits, etc.

Level 2 - *Secondary processing:* basic processing, packaging and branding e.g. packed *atta, suji,* rice, etc.

Level 3 - *High end processing:* supply chain management, modem processing technology, packaging of processed foods, branding, marketing e.g. potato chips, breakfast food, noodles, macaroni, etc.

Need for a Change

The slow down in the Indian economy is of serious concern to every one, in the government, in industry, and in all the sectors of the economy. Today, we are at less than 6 per cent growth level, which is below the planned target. Even more worrying is the fact, that the growth rate has been tending down for the last three years. Agricultural growth has dropped to 0.9 per cent from the average of 3.9 per cent in 1980s and 3.3 per cent in the nineties. In the Indian context, agricultural growth has a strong multiplier effect across the economy. Many of the models show that the modest incremental growth of 3 per cent in agriculture would lead to another 2.6 per cent growth for manufacturing taking the overall GDP growth up by 1.7 per cent—closer to 8 per cent mark and above 10th Plan ambitions.

Time has come when agriculture has to be run as an agribusiness rather than subsistence agriculture. The concepts of profit and loss are getting extended to the farms. Profits have to be achieved through judicious selection and application of inputs, higher productivity and improved quality.

Creation of a food chain starting from the farm gate to retail outlets is inevitable. In fact, the chain should start from agricultural research institutions and laboratories to help farmers in scientific cultivation, particularly to produce quality raw material for specific end-use, proper post-harvest management, storage, transportation and marketing. There is also need for application of more and more mechanisation and automation in post-harvest management and processing operations; utilisation of agriwastes generated during on-farm processing; agro-industrial systems development; and bio-process technology for conversion of raw agricultural produce to useful products.

World markets are undergoing a major transformation. Today, major markets are product-driven and less commodity-driven. To boost economic returns from farming, we must find ways for farmers to earn a greater share of the product sale revenue after adding value to their own produce. Strategies and organisational models to be adopted will vary from crop to crop and region to region and for meeting different technological and social needs. In order to discuss these issues, the National Academy of Agricultural Sciences organised a Round Table[1] on September 14, 2002.

The Round Table focussed upon discussion on a few selected commodities where (i) we have abundant production and consequently, the storage problems, (ii) technologies for processing and/or value addition are available readily or can be generated with little modification on existing ones and (iii) national or international demand for such products exists. The group debated and identified

---

1. A Round Table organised on September 14, 2002, at New Delhi under the Convenership of Prof. Akshey Kumar Gupta. The Round Table was attended by about 25 participants representing scientific institutions, planning bodies, state agricultural universities, industry and farmers.

some suitable working models for backward linkage with farmers and forward linkage with marketing groups so as to benefit all the partners.

The value chain, as existing in India was studied and compared to those existing in some other countries. Goals were identified, and strategies were suggested for creating a favourable environment to achieve development of value-added products.

To prepare a model for initial action, the discussion was restricted to the following commodity chains only:

1. Wheat and rice
2. Fruits and vegetables—Mango, Citrus and Potato
3. Milk and dairy products
4. River and marine products

Closer farmer-processor relationship was the core of discussion. It was emphasised that this relationship needs to be a win-win type of relationship, with reduction of intermediaries. The producer should also be permitted to trade and the laws need to be amended for this.

Action needed for providing effective financial support, favourable government policies and laws, and linkages among producers, industry, R&D institutions and other partners, were broadly suggested.

Recommendations

1. General

- To meet the emerging challenges, agriculture must diversify in favour of high-value enterprises. The emphasis should be on production of high value commodities e.g., fruits, vegetables and fish with enhanced quality and specific nutritional and processing characteristics, than increasing production *per se* as in the past. Pricing policies also need to be changed, linking it with the quality of the produce or a product is the basis for fixing per unit price, just as fat content in milk; higher protein quality/quantity in wheat; better aroma or cooking quality in rice and shelf life of fruits and vegetables.
- Since marketing of products is more remunerative than raw commodities, farmer-processor linkages are needed to add

value as per demands of the consumers. There is a great scope of developing some of our traditional food items from cereals, fruits, milk and fish. Appropriate and cost-effective packaging technology for these items is needed to ensure safety and prolonged shelf life.

- Agriculture is fast becoming demand driven from the earlier supply driven situation. For the foodgrains like wheat and rice, government support in the form of assured purchase as done in past will not be the same in years to come. Farmers will have to grow specific varieties needed for processing or add value to their produce. Policy and legislation must be reformed to allow processors to purchase their produce requirement directly from the farmers.
- Intermediaries in the food chain, lock value and add to cost of the raw materials sometimes by even 80 to 100 per cent. Effective linkages need to be built between farmers and processors on a mutually beneficial contractual agreement, particularly when a large number of small farmers are to be involved in contract farming.
- Self-help or common interest groups on the model of cooperatives should be encouraged to enhance the bargaining power of the farmers and negotiate effectively with the industry, just as is being done in Kerala and Punjab.
- Model of the Town and Village Enterprises (TVEs[2]) of China is excellent for involving surplus rural labour in industrial activity by providing them alternative work at their doorstep. This should be adopted with suitable modifications, for primary processing of the agricultural produce and effectively linked to urban units for secondary processing, product development and marketing.

---

2. The Town and Village Enterprises (TVEs) of China represent a middle ground between private ownership and the state ownership. It has led to the emergence of rural entrepreneurs and enterprises at such a large scale and a rapid rate, which has not been experienced by any other country. TVEs are today a vehicle to increase rural income, absorb a large amount of surplus rural labour and check their migration to urban areas. Their growth has been critical to the success of transforming China to a market economy.

TVEs are industrial enterprises owned by local governments and citizens and generally produce consumer goods for domestic and foreign markets. They operate on a competitive-cooperative basis. Because the local governments can retain the revenues from TVEs, they have been managed efficiently and the loss-making ones are closed down. This created a domestic competitive environment. The TVEs are financed by local financial institutions and some even by foreign investors (World Bank, 1996).

- Location of food-processing units should be strategically placed depending upon the raw material availability, labour, product utilisation and domestic and/or export marketing. It should be nurtured to evolve on a natural course after initial nucleation, as done for IT industry.
- Many times indigenous technologies are better suited for application than the imported ones. There is an urgent need to have a scientific database of these technologies so that one can compare the ones that are offered to us by other countries. There are nearly 200 major indigenous technologies for food processing which have been listed by CFTRI. NABARD or other financial institutions should finance this activity.
- There is an urgent need to have commodity-based management systems to advise the governments and R&D institutions to take steps proactively, based on continuous tracking of the demand, supply, consumer needs and prices both in domestic and international markets. The intelligent information collecting system should be networked with all user agencies and farmers in the country, using latest IT technologies and infrastructure.
- Processing of agricultural raw material, generates a sizable amount of utilisable by-products, commonly termed as 'waste'. Experimental protocols for converting these into usable co-products are available. These need to be developed into commercially viable technologies. These ventures would be a success if main product manufacturing is undertaken in consonance with the by-product processing. Government regulations and policies must be suitably amended to facilitate this.

2. *Commodity-based Recommendations*

2.1 Wheat and Rice

- The varieties of wheat and rice grown, now have been developed primarily for increased yield and lack the quality characters required for making products other than *chapatis*. R&D programmes should focus on developing varieties for specific end-use and processing qualities.
- The varieties grown at present, however, are more suited for making some traditional food items. The technologies for

making these products are available with SAUs and labs of CSIR and ICAR. Their marketing potential should be explored and efforts made to make them into commercially viable propositions.

- Processing units should be linked to areas/zones that should grow specific varieties for specific products. Contract/Contact farming can meet this end. FAQ laws must be amended and production/procurement policies must provide economic incentives to farmers to produce quality grain.
- Current post-harvest handling of grains results in high losses and leads to deterioration of quality. Storage and handling technologies and infrastructure should be designed in a manner that losses are minimised and the produce retains its original quality. Post-harvest care involving grading, storage and phyto-sanitation of the produce should meet the needs for export market.
- Comprehensive processing units for developing products as well as by-products should be set up. For example, wheat flour mills can earn added benefits by undertaking wheat germ and oil production in consonance. Rice processing units should simultaneously handle de-husking and processing of bran for food, feed, oil and biochemicals, as is being done in many other countries. One such unit has recently come up in Punjab.
- Separate boards for wheat and rice should be established which should focus on all the activities from R&D to the production centres for quality grain, and from post-harvest handling to processing and marketing including image building in the national and international markets. Strategy formulation at the state and national level, development of infrastructure and facility for micro-level operators and intervention, wherein needed, should also be the role of these boards.

2.2 Fruits and Vegetables (Mango, Citrus and Potato)

- R&D programmes should focus on end-use of the produce, both for international market and the processing needs. For example, mangos of attractive skin colour and size are needed for export and potato varieties with large uniform tubers, and those with delayed sprouting will be preferred for processing and reduced storage losses.

- Mango pulp units should be set up in villages or clusters, and should ensure packing under SPS conditions. Processing units to make useful co-products like oil from mango kernels and beta carotenoids form peel should be set up alongside the pulp making units.
- Protocols for making traditional mango products e.g. *chutney, achar, aampapar, murabba* need to be standardised for manufacturing under SPS conditions and promoted to produce quality products. These must also be packed appropriately and attractively.
- In cases where debittering of citrus juice is not feasible, it should be blended with other juices after appropriate R&D to meet the taste of consumers and prolong the shelf life.
- Potato varieties *Chipsona-I* and *Chopsona-II* match, or are even better than presently used international varieties for making chips and fries. These should be promoted among farmers and farmer-linked processing units.
- Since MH is known to inhibit sprouting of potato in storage, its import should be made easy.
- Technology for converting potato powder to high fructose syrup should be exploited at a commercially viable scale.
- Peel, pulp and seed from different fruits are invariably discarded as 'waste'. These can be converted into valuable co-products. Technologies for production of oil and natural colour from citrus peel; fibre from banana leaves; oil and a rare protein from mango kernel, have been developed. These have to be made as economically viable propositions.

2.3 Milk and Dairy Products

- Even though over 80 region-specific dairy products are known yet no scientific documentation of these products is available. Agencies and institutions concerned, must urgently take up this work. Some of these traditional products, backed by R&D support for processing and packaging could be elevated to commercially viable ventures for national and international markets (just as is being done for *Shrikhand* and *Rasogulla)*.

- Attention must be paid to develop nutritive foods using whey that is rich in protein. One possibility is to supplement it with nutra-ceuticals, followed by gelling and suitably packaging to enhance its shelf life.
- Milk quality suffers from high levels of pesticide residues and heavy metals (which come through feed and fodder) and antibiotics mixed in feed or given for treating mastitis. This should be checked and monitored at the time of procurement to develop internationally competitive and pesticide-free products.
- There is an urgent need to develop low cost technologies for mechanisation to hermetically pack milk products as per needs of end users and involving the manufacturers of packaging material and equipment in an integrated manner.
- There are natural anti-bacterial agents in milk of camel/goat/ sheep, which has a shelf life of 8-10 days. R&D efforts should be initiated for blending cowl buffalo milk with these milks, wherever possible. This will enhance shelf life and reduce handling costs.

2.4 Fish and Marine Products

- Traditional preparations like fish/prawn pickle, fish *papad* and fish curry, battered and breaded products, fish mince *(Surimi)* based products like sausages, cutlets, patties, balls, pastes etc., need to be promoted in ready-to-eat packs. The packaging should be such that ensures quality and hygiene as per international standards, with proper certification.
- Since India has a higher fish processing capacity than the total catch capacity, the idle workforce could be effectively engaged in making products based on *Surimi* from low value/by-catch fish which otherwise fetch poor returns to the farmers.
- Sea-weed farming should be commercialised and sea weeds used for preparation of nutritive foods and feed items.
- Post-harvest handling of fish and prawn catches must be done as per international standards of sanitation. This is essential if we are to compete in the global trade. It will require extensive cold-chain and appropriate storage infrastructure till the catch is processed or consumed.

- Fisheries waste generated by the processing industry should be used for developing usable co-products like chitin and chitosan from prawn shells, sutures from fish guts, the technology for which is available.

2.5 Public Policies

- Policy and legislation must be amended to enable the processors to procure cereals like wheat and rice directly from the farmers. This will be cost saving and also ensure the authenticity and traceability of material.
- Policy reforms should be made so that cooperative units work as competitive units, self-supporting on their own strength, after a one-time initial financial support, on the lines of the TVEs of China. State government should act as a facilitator of cooperatives and not as controller, as is the practice now. These should be self-managed units, with appropriately defined independent dispute-handling procedures.
- Multiple taxation on value-added products hinders development. The taxation laws should be simplified soon, as also envisaged in the draft National Food Processing Policy.
- Special laboratories must be established in public and private sector for export certification of food consignments for the levels of antibiotics, pesticide residues, pathogenic organisms, heavy metals, impurities etc. as per Codex standards.
- There should be time-bound commitment from the central and state governments to develop much needed infrastructure like cold storages, warehouses, sorting, grading and packing at the farm level. Proper incentives should be provided for these ventures.

2.6 Banking Support

- Technology generating institutions must help in developing the techno-economic project profiles, leading to bankable proposals for the prospective entrepreneurs.
- Self-help women groups have successfully taken up some part-time activities with support from NABARD and other institutions. These groups should be assisted in making traditional foods as per Codex standards, with proper training and financial back up.

- Simplified credit facility should be provided to farmers and their cooperatives that wish to undertake value addition to their produce. First-loss-risk-cover should be provided to these farmer-entrepreneurs.

2.7 Management Models

- Processing units, adding value to agricultural produce, should lay emphasis on developing effective backward linkages with farmers, to procure quality raw material.
- Contract farming should be promoted in a manner that processing firms are not able to exploit an unequal relationship with growers. Most contracts in the past have been tilted against farmers and also some farmers have diverted the produce to open markets.
- The New Generation Cooperatives (NGC) model is ideal since it binds both farmers and processors to honour commitments and agreements. This should be applied in our cooperative enterprises. Contract farming involving small farmers should be promoted rather than involving a few big farmers.
- Small cooperative processing units should be promoted in rural areas. These could undertake primary processing, grading and cleaning of produce for adding value. The pattern of Town and Village Enterprises (TVEs) of China, could be used, with necessary modifications.
- Rural processing and value-addition groups should be promoted. This will allow small and marginal farmers to be part-time farmers and work in rural enterprises for enhanced income and returns on their produce.

3

Dichotomy between Grain Surplus and Widespread Endemic Hunger

We have sufficient stock of foodgrains. No one need go hungry in this country. Nevertheless, it is true that many millions of our countrymen still go hungry to bed every night. Malnourishment, especially among women and children, is widespread. We are determined to change this situation...

While we need to substantially increase our food production, we are faced with the paradoxical problem of surplus food stocks. There is also the related problem of substantial quantities of food being wasted. This has happened because of inadequate attention in the past to its storage, preservation, processing, and proper distribution.

— *Atal Bihari Vajpayee*[1]

Introduction

India has the largest number of women, children and men in the world going to bed undernourished. Although, there are numerous safety net measures of Government of India, for economically and socially underprivileged sections of the society, the FAO has listed India as one of the countries that has not been able to achieve the target of the World Food Summit of 1996, in terms of reducing the number of undernourished. This calls for serious introspection of both our strategies for fighting hunger as well as of our ethical commitment to the cause of hunger-free India. The prevalence of endemic hunger should no longer be accepted as unavoidable.

Agricultural Progress serves as the best safety net against poverty, hunger and unemployment. Therefore, sustainable agriculture is the foundation for sustainable human security. The capacity to support even the existing human and animal populations has been exceeded

1. Inaugural Address by the Hon'ble Prime Minister of India on the occasion of release of the *Food Insecurity Atlas of Rural India*, United Nations World Food Programme, M.S. Swaminathan Research Foundation, Chennai, 2001.

Policy Paper 20, August 2003.

in many parts of the developing world. Hence, the future of food security depends upon population stabilisation, the conservation and care of arable land, through attention to soil health and replenishment of fertility, and the conservation and careful management of all water sources, so that more crop can be produced per drop of water.

To quote the famous Harvard biologist, E.O. Wilson, "The problem before us is how to feed billions of new mouths over the next several decades and save the rest of life at the same time, without being trapped in a Faustian bargain that threatens freedom and security. No one knows the exact solution to this dilemma. The benefit must come from an evergreen revolution. The aim of this new thrust is to lift food production well above the level obtained by the green revolution of the 1960s, using technology and regulatory policy more advanced and even safer than those now in existence" (E.O. Wilson: 2002, *The Future of Life*).

The Growing Paradox: Grain Mountains and Hungry Millions

India has fortunately, through concerted efforts of scientists and farmers, surmounted the state of food deficit and come to a stage of food surplus. In fact, it is now faced with the paradox of a huge buffer stock of foodgrains, while also housing the largest population of undernourished in the world. As per the *Economic Survey of India*, 2002-03, we had a buffer stock of 63 million tonnes of food grains in July 2002, as against a requirement of 24 million tonnes; and on January 1, 2003, we had a buffer stock of 48.2 million tonnes of food grains as against a requirement of 16.8 million tonnes. The Survey Report further goes on to state that given a distorted policy of minimum support price not aligned with the cost of production, the Food Corporation of India has become the sole buyer, seller and stockist of foodgrains. Shortage of storage space has pushed up the costs due to wastage and deterioration of stocks. Food subsidy increased from Rs. 6066 crores in 1996-97 to nearly Rs. 17600 crores in 2001-02. Continued rise in the MSP and the consequent rise in economic cost without rise in issue price have led to a continuous rise in food subsidy.

The Public Distribution System (PDS) with a network of 4.74 lakh fair price shops is the largest network of its kind in the world. It is operated under the joint responsibility of the central and state governments. In June 1997, the Targeted PDS was introduced which brought in a two-tier subsidised pricing structure for Above Poverty Line (APL) and Below Poverty Line (BPL) families. In view of the surplus stocks, the monthly allocation was increased to 35 kilogrammes per family for both APL and BPL families. However, low off take from the system due to various reasons, *viz*. poor quality, poor delivery and lack of purchasing power, to cite a few, call for a relook at the whole system. The Long Term Grain Policy of the government has, in fact, emphasised on reintroducing the universal PDS.

Thanks to the Green Revolution, we have attained self-sufficiency in food grain production, but there is no room for complacency on that front in the face of continued population growth and growth in agricultural production reaching a plateau and even declining. If everyone was able to consume the required quantity of cereals as per the recommended daily allowance, we are just about producing enough. Availability of foodgrains in the year 1999, was only 428.8 grams per capita per day as against the requirement of 420 grams per capita per day, recommended by the Indian Council of Medical Research. The per capita availability has declined to 428.8 grams per capita per day in 1999, from 468.5 grams per capita per day in 1991.

This calls for serious attention to the production front and measures to diversify the production base and the food basket.

In fact, not only agricultural production, but factor productivity, investment in irrigation, post-harvest technology and rural infrastructure are all declining in India. Prices of many agricultural commodities have collapsed. Since about 70 per cent of our population depends on agriculture and allied activities (fisheries, forestry, animal husbandry, agro-processing) for their livelihood, this has had a major impact in terms of purchasing power. Indian agriculture is at the crossroads today. As an immediate measure for strengthening food security on the production front, systematic efforts have to be initiated to identify and remove the constraints responsible for the prevailing yield gaps. Food and ecological security are closely linked. We have to link conservation, cultivation and consumption in a mutually reinforcing manner. Future agricultural production programmes have to be based on a strategy that defends the gains already made, extends the gains with the use of yield enhancement technologies in rain-fed, semi-arid and hill areas, makes new gains through farming system intensification, diversification and value addition, and institutional support by way of infrastructure and market linkages. Land and water care, water harvesting, restoration of degraded and wasted lands, all need focussed attention. Our agriculture extension services need to be reorganised to bring a viable mix of traditional and frontier technologies to the farmers and can be a source of jobs for the agricultural graduates that we are churning out.

The Food Insecurity Atlas of Rural India has analysed the problem of food insecurity in India in great detail. It examines food insecurity from the perspective of the three major dimensions of food availability, food access and food absorption.

Availability of food—a function of production, vulnerability to natural disasters and environmental sustainability (which will have an impact on production in the long run).

Access to food—a function of purchasing power/access to sustainable livelihoods, socioeconomic and gender discrimination and existence of infrastructure facilities like roads, and measured in terms of indicators like the percentage of population below the poverty line, percentage of population dependent on labour income, calorie intake of the lowest deciles, percentage of literacy and sex ratio.

Absorption of food in the body—determined by access to safe drinking water and non-food factors like environmental hygiene, primary health care and primary education, measured in terms of indicators like life expectancy, mortality rate, percentage of population suffering from chronic energy deficiency, stunting, wasting, etc.

Integrated attention to these components of food security is crucial. The problem in India today in the face of huge food grain stocks, is largely one of access both due to poor infrastructure, low off take from the government delivery system and lack of purchasing power among a large majority. The food is there but it is not reaching the people. To ensure that the food reaches all, there is the basic issue of addressing the questions of access and affordability. It calls for a thrust on job-led economic growth. We are facing a paradoxical situation, where we are sitting on mountains of grain while millions are going hungry due to lack of access or purchasing power.

The National Academy of Agricultural Sciences (NAAS) organised a Workshop to deliberate on this paradoxical situation.[2] The recommendations of the Workshop are detailed below.

Recommendations

1. Food for Sustainable Development Initiative

There are now unique opportunities for launching a Food for Sustainable Development Initiative, in the form of a 'grain for green' movement. Such a programme could accord priority to the:

- Restoration of hydrological and biodiversity 'hot spots', particularly in mountain ecosystems.
- Coastal agro-aqua farms (planting of Salicornia, mangroves, Casuarina, palms, etc. along with coastal agriculture and aquaculture).
- Water harvesting, watershed development, wasteland reclamation, and anti-desertification measures.
- Recycling of solid and liquid wastes, and composting.
- Agroforestry and other sustainable land use systems in the fields of resource poor farmers.

The Food for Sustainable Development Initiative could be managed at the local level by Community Food Banks (CFB) operated by women's self-help groups. Such CFBs can be designed in a manner that they can concurrently address issues relating to chronic, hidden and transient hunger. The merit of CFBs is low transaction cost and transparency. They can also help to widen the food security basket, thereby saving what could become 'lost crops'. Wherever animal husbandry including poultry farming, is important to provide additional income and nutrition to families living in poverty, CFBs can also operate feed and fodder banks, if needed.

Conferring the right to food and thereby an opportunity for a productive and healthy life on those who go to bed undernourished, now is the fundamental duty of the state as well as of the well to do

---

2. NAAS Workshop (July 11, 2002) "Dichotomy between Grain Surplus and Widespread Endemic Hunger", under the convenorship of Dr. M.S. Swaminathan. Invited speakers drawn from organisations like Food Corporation of India, the World Food Programme, FAO, Hunger Project, TISCO, IGIDR, IRRI, JNU, MSSRF and IARI presented their views which were deliberated upon.

sections of the population. Thanks to both the spread of democratic systems of governance at the grassroot level and technological advances, we now have a unique opportunity to foster a community centred and controlled nutrition security system. Such decentralised system of community management will help to improve delivery of entitlements, reduce transaction and transport costs, eliminate corruption and cater to the twin needs of introducing a life cycle approach to nutrition security, and meeting the challenge of seasonal fluctuations in nutritional status. If the CFBs are operated by women, this will help to bridge the gender divide in the area of nutrition.

The CFBs may be organised with the following four major streams of responsibilities.

- *Entitlements:* The benefits of all government and bilateral and multilateral projects intended for overcoming under- and mal-nutrition can be delivered in a coordinated and interactive manner (as for example those intended for overcoming the deficiencies of macro- and micro-nutrients).
- *Ecology:* Food for sustainable development with particular reference to the establishment of water banks, land care, control of desertification and promotion of afforestation. Thus, grains can be used to strengthen local level water security
- *Ethics:* This group of activities will relate to nutritional support to old and infirm persons, pregnant and nursing mothers and infants and preschool children.
- *Emergencies:* This activity will relate to the immediate relief operations following major natural catastrophes like droughts, floods, cyclones and earthquakes, as well as to meet the challenge of seasonal slides in livelihood opportunities due to, natural causes as well as human conflicts/and mini-wars.

Each of the above four streams of activities can be managed by four separate self-help groups of local women and men. This will help to generate a self-help revolution in combating hunger. The overall guidance and supervision may be provided by a multi-stakeholder Community Food Bank Council. A CFB for 1000 persons may need annually about 250 tonnes of foodgrains.

2. *Adopting a Whole Life Cycle Approach to Nutrition Security*

2.1 Pregnant Mothers

Overcoming maternal and foetal under- and mal-nutrition is an urgent task, since nearly 30 per cent of the children born in countries in South Asia are characterised by low birth weight (LBW), with the consequent risk of impaired brain development. LBW is a proxy indicator of the low status of women in the society, particularly of their health and nutrition status during their entire life cycle.

2.2 Nursing Mothers

Appropriate schemes will be necessary to provide support to enable mothers to breast feed their babies for at least six months, as recommended by WHO. Policies at work places, including the provision of appropriate support services should be conducive to achieving this goal.

2.3 Infants (0-2 years)

Special efforts will have to be made to reach this age group through their mothers, since they are the most unreached at present. Eighty per cent of brain development is completed before the age of two. The first four months in a child's life is particularly critical, since the child is totally dependant on its mother for food and survival.

2.4 Preschool Children (2-6 years)

A well-designed integrated child development service will help to cater to the nutritional and health care needs of this age group.

2.5 Youth (6 to 18 years)

A nutrition based noon meal programme in all schools (public and private and rural and urban) will help to improve the nutritional status of this group. However, a significant percentage of children belonging to this age group are not able to go to school due to economic reasons. Such school 'push-outs' or child labourers, need special attention.

2.6 Adults (18 to 60 years)

The nutrition safety net to cater to this category should consist of both entitlement programmes like Food Stamps and Public Distribution System (PDS), as well as a Food for Sustainable

Development programme (also called 'Food for Work' programme). The Food for Sustainable Development programme can accord priority to the use of foodgrains as wages for the purpose of establishing water harvesting structures (Water Banks) and for the rehabilitation of degraded lands and ecosystems. In designing a nutrition compact for this age group, persons working in the organised and unorganised sectors will have to be dealt with separately. Also, the intervention programmes will have to be different for men and women taking into account the multiple burden on a woman's daily life.

2.7 Old and Infirm Persons

This group will have to be provided with appropriate nutritional support, as part of the ethical obligations of society towards the handicapped.

The above whole life-cycle approach to nutrition security will help to ensure that the nutritional needs of everyone in the community and of every stage in an individual's life, are satisfied. Such an integrated approach is being adopted under the 'Malnutrition-free Tamil Nadu' programme, as shown below.

Pregnant Mothers	:	Special support to prevent the incidence of low birth weight children.
Nursing Mothers	:	Appropriate support for six months.
Infants (0 to 2 years)	:	Programme designed to reach the infant through the mother.
Child (2 to 6 years)	:	Integrated child development service.
School Children (6 to 18 years)	:	Nutritious noon meal in schools.
Adults (18 to 60 years)	:	Food for sustainable development.
Old and Infirm Persons (over 60 years)	:	Special programmes.

Note: Over seven million tonnes of foodgrains (wheat and rice) were allotted for the above programmes during 2001 and 2002 in India.

3. *Developing and Spreading a Holistic Action Plan to Achieve Sustainable Nutrition Security at the Level of Each Individual*

The major components of such an integrated action plan are the following:

- *Identification:* Request the local families themselves to identify those who are nutritionally insecure. Trained community volunteers of the kind mobilised in Thailand will be useful for this purpose. Invariably, the ultra-poor who do not have any productive asset, fall under this category.

- *Education and Information Empowerment:* Empower those who are not aware of their entitlements about the nutritional safety nets available to them and also undertake nutrition education. An entitlements database can be developed for each area and household entitlement cards can be issued, indicating how to access nutritional, health care and educational programmes. The educational programmes should also lay stress on culinary habits in relation to the conservation of essential nutrients in cooked food.
- *Overcome Protein-calorie under Nutrition:* The various steps indicated under the whole life-cycle approach will have to be adopted. The problems of child labour and of persons working in the unorganised sector will need specific attention.
- *Eliminate Hidden Hunger Caused by the Deficiency of Micronutrients in the Diet:* Introduce an integrated approach including the consumption of vegetables and fruits, millets, grain legumes and leafy vegetables and the provision of fortified foods like iron and iodine fortified salt and oral dose of Vitamin A. The basic approach should be a food based one, with emphasis on home and community nutrition gardens, wherever this is socially and economically feasible.
- *Drinking Water, Hygiene and Primary Health Care:* Attend to the provision of safe drinking water and to the improvement of environmental hygiene. Also, improve the primary health care and primary education systems. Non-food factors like health care and education are equally important for nutrition security.
- *Sustainable Livelihoods:* Improve economic access to food through market-linked micro-enterprises supported by micro-credit. Also, create an economic stake in the conservation of natural and common property resources. Ensure that agreements under the World Trade Organization (WTO) provide a level-playing field for products coming from decentralised small-scale production (production by masses or farmers' farming) as compared to those emerging from mass production or factory farming often supported by massive inputs of subsidy, capital and technology. Promote job-led economic growth and not jobless growth.

- *Pay Special Attention to Pregnant and Nursing Mothers and Pre-school Children:* Measure progress through monitoring maternal mortality rate (MMR), infant mortality rate (IMR), incidence of LBW children and male-female sex ratio. Iron-folate supplements during prenatal care should be accompanied by steps to overcome protein-energy deprivation. A carefully designed maternity and childcare code will help to speedily bring down MMR, IMR, LBW and stunting. Sex ratio is a good index of the mind-set of a society in relation to the girl child.

4. *Institutional Structures for Extending the Extrapolation Domain of Successful Experiences and Efforts*

Without appropriate institutional structures, isolated success stories will remain just talking points. Based on local cultural traditions and socioeconomic conditions, institutional structures should be fostered which can take new technologies to the unreached and which can give the power of scale to small producers at the production, post-harvest and marketing phases of farming. The following are some of the structures which have given promising results in India. They are based on a 'win-win' situation for all, as otherwise the efforts will not be sustained over time.

.	*Sector*	*Institutional Mechanism*
1.	Dairy	Cooperatives
2.	Poultry	Egg and broiler coordination councils
3.	Integrated on-farm and off-farm employment	Biovillages
4.	Power of scale to small producers	Small farmers' agri-business consortium
5.	Technological upgrading of production and post-harvest sectors	Agri-clinics Agri-business centres
6.	Group action for micro-enterprises supported by micro-credit	Self-help groups
7.	Timely and affordable credit	Kisan (Farmer) Credit Cards, integrated informal and formal banking system
8.	Ensuring minimum support price	Food Corporation of India and state corporations; buy-back arrangements with the private sector

Without socially compatible and socially owned institutional structures, the extrapolation domain of successful experiences and development efforts will remain limited. Community involvement will ensure low transaction costs and a high percentage of success and help to convert unique examples into more universal ones.

5. *National Alliance against Hunger: Launching a National Food Guarantee Scheme*

The huge stock of foodgrains provides a unique opportunity for launching a national alliance against hunger, with the alliance partners reaching the remotest village and hamlet based on Mahatma Gandhi's principle—"To the hungry, God is bread; this God should be present in every house and hut of the country." The strategy of a hunger-free nation could consist of the following three inter-related groups of activities :

- Ensuring a whole life-cycle approach to nutrition security at the level of each individual by providing the needed horizontal linkages among ongoing vertically structured programmes (often operated by different national, bilateral and UN agencies).
- Organisation of a Food Guarantee Scheme on the model of the Employment Guarantee Scheme (EGS) of Maharashtra. This will essentially be a 'Food for Work' programme having the following features—highest priority in this project, which should cover both rural and urban areas, needs to be given to water harvesting, watershed and wasteland development, eco-restoration of hydrologic and biodiversity 'hot spots' and waste recycling (composting) and bioenvironmental management of mosquitoes.
- Promoting the establishment of Community Food Banks at the local level, to serve as the focal point for according concurrent attention to ending poverty induced endemic hunger, micro-nutrient deficiency induced hidden hunger and transient hunger caused by human conflicts and natural calamities.

The national alliance against hunger can provide policy guidance and undertake resource mobilisation (financial, technical and managerial) for achieving the goal of 'food for all'. While the above approach would help to alleviate hunger today, we can avoid hunger tomorrow only by sustaining advances in agricultural production through an evergreen revolution approach using environment-friendly ecotechnologies.

Immediate Action Points

Although, the Food Corporation of India has huge foodgrain stocks, the country still has the largest number of undernourished in the world. This calls for serious introspection of both our strategies for fighting hunger as well as of our ethical commitment to the cause of hunger-free India.

- This paradoxical situation can be effectively addressed by introducing the social security systems for the able-bodied poor in the form of a National Food Guarantee Scheme on the model of Maharashtra's employment guarantee scheme. The public distribution system should be strengthened and should be designed to reach the unreached. On the basis of ICMR norms, about 158 million persons, belonging to about 32 million households, fall under the category of ultra-poor and need immediate assistance to help them to lead a healthy and productive life. They are best identified by *Gram Sabhas* and local bodies. They can be issued with Food Entitlement Coupons, which should entitle them to be provided with work under the Food Guarantee Scheme. The total requirement for such an open-ended employment-cum-food security project may need at the maximum about 10 million tonnes of foodgrains per year. Meeting this need is well within our national capacity.
- Consumption inequality decreases as income increases. Diversification of diets also happens with increased purchasing power. Unfortunately, inequality of income distribution is growing. A job-led economic growth strategy will help to reduce protein-energy malnutrition, and at the same time, stimulate farming systems' diversification.
- A whole life-cycle approach should be introduced by providing a horizontal dimension to the numerous on-going vertically structured programmes supported by the central and state governments as well as by bilateral and multilateral agencies. At the local level, the management of a life-cycle based nutrition security system, beginning with pregnant women and extending up to old and infirm persons, is best left to the one

million elected women members of the *Panchayats*.

- Proactive approach to implementation of recommendations submitted by the Committee on Long Term Grain Policy, with regard to MSP, universal PDS.
- Decentralised procurement and decentralised storage will help to minimise transport and transaction costs. Uttar Pradesh, Madhya Pradesh, West Bengal and Tamil Nadu have accepted the principle of a decentralised procurement system. The storage can be done through a national grid of Community Food Banks (CFBs) managed by self-help groups. The construction of CFBs can be done at the local level under the 'Food for Work' programme.
- We must strive to produce more in a manner that high yields can be obtained in perpetuity without associated ecological or social harm. Farming systems' intensification, diversification and value addition are extremely important to generate the needed on-farm and non-farm employment. Productivity improvement in both irrigated and rain-fed areas will help to increase marketable surplus at the household level and thereby the cash income. Livestock husbandry and livelihood security tend to be closely correlated in the case of poorer households. There is greater equity in livestock ownership as compared to land. Hence, support services should be organised for small scale livestock farming families in the form of fodder and feed banks and healthcare and insurance facilities.
- There is an urgent need for spreading quality literacy including Codex Alimentarius standards and sanitary and phytosanitary measures. In addition, there is need for greater research in the field of breeding crop varieties having the quality characteristics needed for food processing and exports. If productivity and quality are improved, farm income will go up substantially.
- Sustainability of food security measures should be ensured. PDS at the local level should provide reasonable income to those operating the ration shops. Community Food Banks may receive their initial food supply from government but subsequent replenishment should come from local communities.

- There is a need for Organisation of a National Consortium for Sustainable Food Security. NAAS can act as a catalyst to promote the organisation of a National Consortium for Sustainable Food Security consisting of representatives of FCI, CII, FICCI, ASSOCHAM, NHDB, NDDB, APEDA, TISCO, Hindustan Lever and appropriate agricultural universities and ICAR institutes for fostering sustainable food security both in the hunger 'hot spots' of India as well as in other countries in Asia and Africa.

There is no time to relax, both, on the production and consumption fronts. The stock of 63 million tonnes should not lull us into complacency. Concurrent attention to production and consumption is important. Enhancing production through an evergreen revolution is a priority task. However, eliminating widespread endemic hunger is a prime responsibility and ethical obligation. A universal and user-sensitive Public Distribution System, Food Guarantee Scheme, Community Food Banks and various other food entitlement projects need to be implemented in an integrated manner, so that the goal of hunger-free India can be achieved. Nutrition status will, however, continue to fall, unless the purchasing power of the poor is increased. Hence, livelihoods for all should be the bottom line of all national development and import and export policies.

4

Diversification of Agriculture for Human Nutrition

Introduction

The current agricultural production scenario reflects some disturbing tendencies from the point of view of human nutrition, particularly in rural areas where production and consumption are directly linked, and for the poor everywhere. Decline in areas of coarse cereals and pulses and other so-called 'low-value' crops which provide access to better nutrition for the poor, illustrates this concern. On the other hand, substitution of these crops by those with higher productivity has improved calorie availability and incomes of farmers who can increasingly afford better nutrition. Over the last 10-15 years, there has been a remarkable spurt in production of horticultural crops, livestock and fisheries, driven largely by buoyant domestic demand. Average per capita consumption of these commodities has increased.

Data on nutritional status in the country in relation to norms developed by nutrition scientists, on the other hand, reveal the appalling status of nutrition, in general and of vulnerable sections of the population in particular. The National Academy of Agricultural Sciences organised a symposium,[1] to deliberate on these issues.

Some salient observations, policy issues and recommendations emerging out of the deliberations are summarised below.

Food and Nutritional Security

Thanks to the Green Revolution, per capita availability of foodgrains has increased. With 200 million tonnes of foodgrains,

---

1. Symposium on Diversification of Agriculture for Human Nutrition on December 16-17, 1999 at the National Institute of Nutrition, Hyderabad.

Policy Paper 7, May 2001.

calorie and protein deficiencies have largely been overcome. The country has achieved basic food security in terms of foodgrains. The averages, however, conceal some important differences. For example, distribution of food is highly unequal and the poor, who constitute nearly one-third of our population, continue to be food-deprived. Interregional and intrahousehold deficiencies are also significant.

Also, almost all the increase in foodgrain production has come from wheat and rice. The production of coarse grains (millets) and pulses has declined or remained stagnant, implying a decline in per capita availability of these commodities over time. Coarse grains have been important in the diets of the poor. These have relatively higher nutritive value in terms of proteins, vitamins and minerals compared to rice. Declines here, thus, affect food security of the poor qualitatively. Similar are the implications relating to pulses which are the major source of protein in vegetarian diets. Both pulses and coarse grains are important crops for dryland and fragile environments where poverty levels are high. Legumes have been the traditional restorer of soil fertility and declining area poses a threat to sustainability. Finally, the spectacular increase in foodgrain production and a reasonable overall agricultural performance has failed to eradicate malnutrition.

Nutritional security, thus, goes beyond food security as we understand the term. The latter means ensuring adequate availability of foodgrains to provide calorie and, may be, protein needs of the people, the former implies adequate supply of micronutrients such as vitamins and minerals as well. As we move up the development ladder, this becomes more relevant indicator of food security. To ensure nutritional security, increased availability of diverse types of food, such as millets, pulses, fruits and vegetables, foods of animal origin (milk, eggs, meat, fish), besides cereals, is essential. While fruits and vegetables are rich sources of micronutrients, animal-based foods abound in quality proteins as well. Vegetables and fruits also contain some health-giving phytochemicals which are powerful antioxidants and detoxifying agents which protect against degenerative diseases. Marine fish is a rich source of long chain n-3 fatty acids which have important physiological role.

Diversification Scenario

Cropping patterns have traditionally been dominated by food needs. Commercial crops were confined to some regions and on relatively larger farms. The system we inherited at the time of independence became unsustainable as rapid population growth outstripped our capacity to produce food. Fruits, vegetables, milk, meat, fish, became luxury foods, whose demand was confined to the very small rich class in rural and urban areas. A food insecure nation, despite devoting bulk of its agricultural resources to food production, became chronic importer of food. Even extension of cultivation to marginal and sub-marginal lands did not help.

The Green Revolution transformed this scene. In less than a decade we were able to achieve reasonable food security. High growth in productivity of cereals spurred agricultural growth and incomes. Rising incomes prompted shifts in consumption patterns and demand for non-cereal food became buoyant. By mid-eighties expansion of area under cereals ceased. Producers too began to look for alternatives and the process of diversification set in. It became the *mantra* for agricultural development in the nineties. The following paragraphs document these developments briefly:

- In the *horticulture* sector, there has been 2-3 fold increase in production of fruits and vegetables in the last 10-15 years. Their area has gone up, particularly in rainfed, hilly, and coastal regions prompting fears of diversion of land from foodgrains. This has so far not materialised. Among fruits, banana, papaya, citrus and mango have contributed to the significant increase. All these are nutritionally rich fruits and their increased production has resulted in decline in real prices and penetration in rural markets.
- *Livestock* production has been increasing at 10-15 per cent over the last decade or two. Milk, eggs and meat production have registered spectacular growth. Despite these increases, the annual per capita availability is much below the amounts recommended by ICMR. In terms of these, the present availability is 86.4 per cent for milk, 16.7 per cent for eggs and 29.6 per cent for meat.
- *Fish* is one of the best sources of protein and fisheries sector has grown substantially. Marine products' industry is the largest foreign exchange earner. Scientific acquaculture and deep sea fishing have contributed to the growth of these sectors in the nineties. There are, however, reports that these developments have been unsustainable and have not

contributed to livelihood security of small fishermen in coastal areas. The sector faces infrastructural constraints which inhibit exploitation of the potential.

- *Genetic engineering* is a powerful tool for improving the yields and quality of both plant and animal foods. Both macro- and micro-nutrient content can be enhanced with the now available and evolving biotech tools. Some promising breakthroughs in improving the protein and micronutrient content of crops, such as maize, rice, rapeseed, tomato, etc., have been achieved. However, safety aspects of GMO foods have become an area of concern and debate, globally.

Issues

Changes in Food/Nutritional Security Status

There is agreement that increased per capita availability of cereals and a number of other foods has contributed to greater food security at the national level. At the household level, consumption expenditure surveys by the National Sample Survey Organisation reveal declining consumption of cereals and increases in livestock and horticultural products, sugar and edible oils implying an average improvement in nutritional status. These, as well as, results of the surveys of the National Nutrition Monitoring Board, bear out gains in calorie and protein content of diets, essentially from improvements in non-grain food consumption, driven by income growth and poverty reduction. These surveys, however, also reveal the inequities across regions, farms of different sizes, sociocultural groups, as well as in intrafamily distribution of food. To illustrate, nearly 26 per cent of the rural farming households, mainly sub-marginal and marginal farmers, were nutritionally deprived.

Nutritionists evaluate nutritional security in terms of adequacy of macro- and micro-nutrient with reference to clinical nutritional norms. NNMB surveys indicate that, in general, the consumption of cereals and millets in rural and tribal groups was comparable to balanced diets, but the intakes were low in urban slums. The consumption of micronutrient-rich foods, such as green leafy vegetables, other vegetables and fruits falls below recommended levels, particularly among the poorer population groups. These studies also show sub-standard consumption of green leafy vegetables, milk, fats and oils, sugar and jaggery in the diets of children. The deficit in mean energy intakes among children of preschool and school age was about 25 per cent, but intake of iron, vitamin A and riboflavin was higher across the board. In case of rural pregnant and nursing women, vitamins A, C, and B complex and calcium deficits were higher than those of energy and protein. Both these kinds of surveys, suggest the need to look beyond total production, availability, and average profiles.

Economists view food consumption as driven by self-provisioning, food habits and changes therein, education, incomes, prices, and availability. These and other interacting variables determine the food choices of households as consumers. Empirical studies on recent Indian data indicate that, on the average, calorie consumption has ceased to be income-responsive, implying a switch to non-calorie food with further income growth. At lower end of the income distribution, however, overcoming calorie deficiency remains a priority. All such analyses point to the need for proper targeting of food intervention programmes. Universal public distribution kind of programmes have outlived their utility.

Diversification

Driven by self-sufficiency motive, cropping patterns in India were locked to foodgrain production for a long time—a few regions devoting some resources for commercial crops like cotton, sugarcane groundnut, etc. or plantation crops. As foodgrain production saturated markets, a trend towards diversification set in. Changes in pattern of domestic demand and, to some extent, export demand in the wake of trade liberalisation, resulted in changes in resource use and increasing diversification of enterprises. States which diversified the crop sector in a big way, have attained relatively higher growth in the net state domestic product of agricultural sector during the past two decades.

The factors that led to diversification of agriculture have varied, over time. During the first 15 years following the onset of Green Revolution, irrigation played the most important role, predominance of small holdings discouraged it. Abundant and cheap supply of electricity also fostered specialisation. Since early eighties, credit availability emerged as a significant determinant of diversification. Smaller farms continued to face rigidity in cropping patterns because of binding food production constraint. They diverted their attention to livestock enterprises. At the end of the millennium, there was consensus that diversification to higher value enterprises like, vegetables, fruits, other specialty crops, livestock products, fisheries, value-added agricultural products etc., was the new pathway for income growth in agricultural and rural sector. This would also help in bridging the quality gaps in terms of nutrition. Most of these enterprises were traditionally supplementary in nature and had become the domain of women. This bias adds to the strength of this paradigm.

Concerns and Constraints

- There is some apprehension that increasing diversification of land to non-food crops may affect basic food security adversely. Stagnation or decline in area under foodgrains would undermine our self-sufficiency in food. Some projections suggest massive imports of foodgrains, driven by rising food needs of a growing population and for animal consumption. A country of our size cannot reliably depend on the world market to meet these needs.
- This perception is countered by the scientific community. They contend that productivity levels are very low for most of the crops and animals. Exploitation of this yield gap through research can ease the pressure on land resources substantially. It was also pointed out that animal production systems in India are based on by-product and waste utilisation. Increased efficiency of this system would moderate demand for foodgrains for animal consumption.
- In fact, there have been instances of ecological degradation in the wake of growth in such enterprises. Brackishwater aquaculture, massive use of pesticides on fruits and vegetables, biodiversity erosion, salinisation and waterlogging in Punjab-Haryana and parts of Rajasthan, are examples sometimes alluded to, as consequences of diversification.
- These instances need to be carefully diagnosed because analysts also attribute these negative externalities to deficiencies in policies relating to pricing, investments, etc. The real challenge would be to achieve such intensification in an ecologically benign manner, through appropriate policies and technologies.

- High post-harvest losses significantly undermine the prospects of diversification, particularly through high-value perishables, like horticultural and livestock produce. Losses through inefficient handling, transport, storage etc., rise exponentially as scale of output expands. There are increasing instances of total loss in the fields as farmers cannot even cover harvest costs due to crash in prices at harvest time. We need to strengthen processes, institutions and R&D efforts to tackle these constraints. The priority accorded to agroprocessing industries is a welcome policy initiative in this direction.
- Also important are infrastructural constraints. Most of the commodities in the new diversification basket were traditionally confined to local markets in unprocessed form and in small volumes. Facilities like cool chains, refrigerated transport, modern abattoirs, processing plants and other infrastructure were not relevant. Now the market is global and weak support in infrastructure restricts growth prospects. Similar constraints emerge in institutional infrastructure. Extension, market, credit, information—all have focused on food and commercial enterprises of major importance. In the present context, these create bottlenecks.
- There is evidence to show that because of binding food needs, small farmers are able to diversify only to a limited extent. This implies that as this process accelerates, such farms will lose ground further in terms of income growth. There are special programmes like IRDP which provide support for diversification in respect of small and poor farmers, but more needs to be done. Increase in productivity of their basic staples would provide a significant leeway for incorporation of other enterprises. Small producers also face scale related constraints—the size of their output is small and income-enhancing options like access to superior markets, value addition, storage, etc. become non-viable. New institutional arrangements are necessary if this handicap is to be overcome.
- There is also an apprehension that shift from traditional foods many of which are rich in vitamins and micronutrients, would result in decline in nutritional quality of diets. Substitution of millets by wheat and rice, for example, does not augur well for poor consumers who cannot afford supplementary alternatives.

Fall in consumption of pulses is also a cause for concern. This nutritionally regressive substitution in production patterns arises from technology and price changes which affect profitability and incomes of producers. Lack of nutritional awareness leads to undervaluation of nutritive food by consumers and in the marketplace. Nutrition scientists highlight such dimensions of commercially-driven diversification and plead for investment in nutrition education.

Conclusions and Recommendations

Shifting agricultural resources to higher-valued options is the new strategy for agricultural development. Buoyancy in domestic demand for such commodities has generated congenial incentive environment for such transition and the process has begun. Good export prospects reinforce this trend. Non-conventional crops like aromatic and medicinal plants, floriculture, etc., figure importantly in this strategy but the major impetus comes from horticulture, livestock, dairy poultry, fisheries, etc., which have traditionally been minor constituents of Indian diets. The nutritional implications are obvious. Growth in incomes has spurred demand for these commodities even as foodgrain consumption stabilises, and producers have responded to such market signals.

General inadequacy of Indian diets in terms of micronutrients and vitamins is well established. For the poor, access to even macronutrients is constrained. As overall well-being improves with future growth in incomes, special attention will have to be paid to nutritional aspects. Salient recommendations relating to diversification and nutrition emanating from the symposium are summarised below.

1. Policy Imperatives

- Thrust on raising productivity of foodgrains must remain a central feature of agricultural policy. Only through this route can the twin objectives of self-reliance and rapid rural income growth be realised. It will enable unlocking of resources which would otherwise remain tied to less remunerative enterprises.
- Diversification of production base of Indian agriculture requires massive investments in rural and other infrastructure. Apart from roads, electricity, irrigation, greater emphasis on storage, specialised handling and transport, assembling,

wholesale and retail markets, effective market intelligence, etc. will be needed.

- The emerging economic regime requires dismantling all distortions in input-output pricing. These impart incorrect price signals and farmers are distracted from efficient production patterns. Massive irrigation subsidy and its effect on cropping patterns is an illustration. The structure of tariffs is also a case in point.
- A reorientation of the institutional support for agriculture will be necessary for exploitation of new opportunities by small producers who constitute more than 80 per cent of the farming households. Input-output marketing, value addition and processing, credit, insurance, R&D, extension, etc. need to shift from foodgrain and large-farm-based approaches to a more holistic paradigm. Leasing and tenancy reforms will be necessary.
- Agroprocessing investments must move to the countryside where production is concentrated. While technology and quality considerations may necessitate foreign investments in this sector, mechanisms will have to be developed to ensure effective small farm participation.
- A task of this magnitude and complexity will necessitate a dominant role for the private sector. The public sector will need to withdraw from some areas and strengthen others, like R&D, information, natural resource management, regulatory processes and so on. A set of policies to provide incentives to the private sector will be necessary.
- A large population will continue to be economically and nutritionally deprived. People below the poverty line, women and children, particularly in rural areas, and urban slums, will need strong safety nets. Weaknesses in existing programmes have been well identified and are being addressed. These need to be pursued more vigorously.
- These challenges are beyond the competence and resources of governments. It will be necessary to involve people in planning and executing decentralised initiatives. Non-governmental organisations, self-help groups, cooperatives, *panchayats* will need to play a greater role and these must be strengthened.

2. *Nutrition*

- Nutrition education must be made part of regular curricula in schools. Sustained drives using mass media, particularly in rural areas and urban slums are necessary to create greater awareness.
- Programmes like homestead gardening, urban gardening, household preservation and enrichment of food, etc. must be actively supported. Health and hygiene, sanitation, etc. make significant contributions to nutritional well-being and should be accorded greater priority.
- Food enrichment, fortification strategies need to be supported. Assessment and incorporation of indigenous ingredients offer considerable opportunities and should be exploited.
- National nutrition monitoring effort must be further strengthened and focused on target themes and populations.

3. *R&D*

- Continuous increase in productivity of agricultural enterprises—crops, animals, fish, is essential. This would ease the subsistence pressure on natural resources, relating them to commercial enterprises. The research system must continue to accord high priority to food crops, particularly those which are of importance to the rural poor and tribal farmers.
- Advances in modern sciences, particularly biotechnology, offer exciting opportunities for incorporation of marketable and nutritional qualities in food crops of various kinds. Even as research on genetically modified organisms is accelerated, proper testing and safeguard procedures need to be put in place. The point to note is that this area of research can tackle several constraints inhibiting yield, quality, and nutrition.
- More resources should be allocated for nutrition research. There are basic as well as applied research issues relating to indigenous food, nutraceuticals, formulations, food safety, standards, etc. which need to investigated. There is enormous variability in food habits, tastes and preference products etc. across the country and these must be captured and analysed.
- Unmindful pursuit of market opportunities often exacerbates pressure on natural resources and ecology. Safeguarding future

production potential and ecological balance should be high priority for research.

- Wide diversity in growing conditions implies a wide range of options for diversification and income growth. This is a big strength for Indian agriculture, but this necessitates decentralised research approaches which maximise comparative advantage of different regions. A careful regional prioritisation of research is called for.
- Preventing post-harvest losses has emerged as a critical element, and so has value addition and processing. Known technological options in these areas are highly capital intensive and not really appropriate for small scale operations which are characteristic of the Indian rural scene. The research system faces this unique challenge of developing efficient small scale technologies which will benefit small scale rural producers and entrepreneurs.
- Finally, success of the diversification strategy would demand research on a number of socioeconomic parameters like, market structure, conduct and performance, input-output demand, comparative cost and returns, price analysis, organising producers' and entrepreneurs' private sector role, etc. While the research system is gearing up to meet production research challenges, this area must also receive attention.

5

Sustainable Fisheries and Aquaculture for Nutritional Security

Introduction

Feeding seven billion people on earth through agriculture in an environmentally sustainable manner has posed a major problem. Based on the current demographic trend, the world population is projected to reach 9.4 billion by 2050 and Indian population 1.5 billion. Though living resources are self-renewable, they have to be utilised rationally on a sustainable basis in harmony with the environment. Hence, scientists all over the world are exploring the possibility of using water as a major source of food production. Seawater is a rich and cheap medium for aquafarming. Moreover, the mechanism of production of organic matter in the ocean is more efficient than that on land. In terms of productivity, slow-growing angiospermous plants cannot match with rapidly multiplying planktonic algae.

Aquatic resources of India are vast and diversified. The marine jurisdiction alone spans 2.02 million km^2 (EEZ) and the coastline stretches to a length of 8,129 km. Among lagoons and backwaters, the Chilka, and Vembanad alone cover an area of 800 km^2, and estuaries spread over 300,000 ha. The potential brackishwater area available for aquaculture is 1.4 m ha. Replenishment and creation of water bodies through Southwest and Northeast monsoon in India are nature's gifts. The river system of the country comprises 14 major rivers (catchments > 20,000 km^2), 44 medium rivers (catchments 2,000 to 20,000 km^2) and innumerable small rivers and desert streams that have a drainage of about 2,000 km^2. Different river systems of the country, having a combined length of 20,000 km, provide one of the richest fish genetic resources in the world. India has 19,134 small

Policy Paper 8, May 2001.

reservoirs with a total water surface area of 1,485,557 ha; likewise, the medium (180) and large (56) reservoirs spread over 527,541 and 1,140,268 ha, respectively. Thus, the country has 19,370 reservoirs covering a total area of 3,153,366 ha. Hence, scientific utilisation of the vast and diverse aquatic resources should immensely benefit us, especially in the food production sector.

Fish production in our country has increased by more than five times, and the contribution of fisheries to the GDP of India has also increased three times, a growth, arguably, one of the highest among the food production sectors.

The success in induced breeding of carp in 1957 and subsequent technologies on induced breeding and larval rearing for a number of species of carps, paved the way for the current annual freshwater carp production of > 1 million metric tonnes.

The Central Marine Fisheries Research Institute (CMFRI, Kochi) developed the hatchery technology of penaeid shrimps during 1973-1978 followed by commercial scale production of PLs started by MPEDA in the late eighties helped the country to export farmed shrimps worth > $ 1 billion in 1999-2000.

Hatchery and mariculture technologies for the pearl oyster, edible oyster, mussels and clams have also been developed in the country. Of these, pearl culture and mussel culture have made significant impacts among the entrepreneurs and fishermen.

Fisheries scientists have designed a high opening trawlnet with the help of gear experts in India. The design has revolutionised the capture fisheries' sector. In two decades, all the trawlnets (150,000 in number, in 1998) in the country are of high opening type. They are responsible for an annual fish harvest of > 1.2 million mt. These developments have paved the way for what is hailed as the 'Blue Revolution' or 'Aquaplosion' in India.

Marine fisheries production increased from 0.5 million mt in 1950, to 2.7 million mt in 1999. However, the scope for its further increase is limited. During the corresponding period, the inland fisheries production also increased from 0.2 to 2.5 million mt, despite declining riverine fish production. A progressive increase in inland fisheries production has been possible owing to the development of eco-friendly, culture-based fisheries. There is scope for its further improvement.

Per capita fish availability to an average Indian is about 9 kg, less than the world average (12 kg), and the quantity (11 kg) recommended by the WHO for nutritional security. In India, aquaculture activity was ventured first by the farmers of West Bengal and later, in Andhra Pradesh. With 30 billion USD, investment on global aquaculture by 150 countries, which practice aquaculture, and with more than 10 per cent annual growth rate, contribution to fish production through aquaculture is in the range of 36 million mt, worth $ 50.4 billion and Asia's share to this production is 91 per cent. In India, about 100,000 ha have been brought under aquaculture during the last decade to produce about 82,000 t shrimps by small and medium farmers, with the necessary hatcheries and processing plants at a total investment of about Rs. 70,000 million. Unfortunately, our coastal aquaculture has begun to suffer even at the very early stage due to its failure in diversifying candidate species and system and having no regulations for sustainable development of coastal fish farming. Following the directions of the Supreme Court, the Government of India has constituted the 'Aquaculture Authority', whose main function is to give approval to farmers to operate traditional and improved traditional systems of aquaculture with improved technology for increased fish production.

Asia has been the centre of fishing and aquacultural activity. Among the Asian countries, India ranks second in culture and third in capture fisheries. More than 6 million fishermen and fish farmers of India depend on fisheries for their livelihood. About 5.6 million mt of fish and edible invertebrates worth Rs. 2,22,000 million are now captured or cultured. Our annual fisheries export is 0.4 million mt worth 1.2 billion USD. Thus, the fisheries constitute a highly productive sector, a source of valuable food and employment, and a net contributor to the national income.

The export earnings substantially increased from Rs. 25 million in the 1950s to Rs. 52,000 million in 2000. There has been a shift in export from low value fish and fish products towards the high value products. For instance, export of finfishes decreased from 99 per cent of the value during 1940-1946 to 12 per cent during 1990-1994, but that of high value shrimps and cephalopods increased to 72 per cent and 12 per cent, respectively. Also, the export pattern shifted form dried, salt cured and canned products to frozen products/live organisms.

These issues were examined carefully at a national seminar organised by the National Academy of Agricultural Sciences.[1] Recommendations, emerging out of the deliberations held at this seminar are summarised below.

Policy Recommendations

1. Policy and Legal Issues

- The instrument on fisheries' policy was first devised in 1898 and has undergone only peripheral modifications since then. It is, therefore, necessary that the Government of India formulates a new fisheries' policy taking into consideration the need for increasing sustainable fish production. Incidentally, the MPEDA Act, 1972 and a few other Acts of the central and state governments are obsolete in the present context of new developments; they all need revision with proper amendments.
- India is one of the wettest countries in the world. The average rainfall of the country is 117 cm per annum. The potential of the Indian river basin is 1,869 billion cubic metre, equivalent to 66,000 tmc but the utilised water flow, excluding that of groundwater is 24,367 tmc only. Hence, the unutilised water resource remains as high as 41,637 tmc. Even in penninsular India, which is relatively dry, the level of unutilised water is quite high, *viz.* 8,637 tmc. Water is the most precious commodity, essential to sustain domestic, agricultural (including aquaculture) and industrial activities, but its per capita availability will progressively decrease from 2,200 m^3 in 1998 to 1,500 m^3 by 2025. Hence, there is an urgent need that the National Water Research Council takes into consideration the requirement of fisheries sector to promote fisheries and aquaculture development and conservation of aquatic resources. All efforts should be made to utilise the sheltered bays for mariculture. This will require an enabling conducive policy and creation of basic facilities and essential inputs.
- In recent years, increasing attention is being paid to the environmental problems of the coastal areas. India too has exercised its powers under Section 3 (1) and 20 (v) of the

---

1. Seminar on Sustainable Fisheries and Aquaculture for Nutritional Security at Chennai from 29th November to 2nd December 2000.

Environment (Protection) Act, 1986, through a legal instrument called the Coastal Regulation Zone (CRZ) Notification, February, 1991. The CRZ Notification of up to 500 m from the high tide line with the landward boundary has inherent weakness. For a country like India, with diverse physiography, development needs and urban concentrations, a common regulation for the different maritime states is not realistic and desirable. Hence, there is need to evolve scientific policies and legal framework for supporting coastal and open-sea aquaculture, particularly on the following critical aspects: carrying capacity of coastal and marine ecosystems of different areas, quantity and impact of groundwater abstraction for shrimp culture, extent of conversion of agricultural lands and impact of untreated wastewater from shrimp farms.

The present Environmental Protection Act, 1986 and Rules February 1991, may have to accommodate the coastal aquaculture in the CRZ and declare aquaculture as a permitted activity requiring water front facilities. Coastal aquaculture should be integrated with other activities of the coastal region.

2. Financial Support

- There is a justifiable claim for increasing financial support for further development of the fisheries sector. Firstly, the projected increase for the production targets for the next decade in India, is in the range of 22-28 per cent for cereals but 100 per cent for fish. Secondly, the contribution of the fisheries' sector to the GDP of India has increased from 0.7 per cent in 1980-81 to 1.4 per cent in 1999-2000, compared to stagnation of the agricultural sector to the total GDP at 28.7 per cent. Consequently, the contribution of fisheries to the GDP of the agricultural sector has increased substantially from 1.97 per cent to 5.35 per cent. Yet, the government investment in the fisheries sector is stagnating at a mere 0.35 per cent of the total plan outlay through the five year plans, compared to the substantial increase from 1.74 per cent to 5.49 per cent in the agricultural sector. Therefore, the fisheries sector deserves better recognition and greater investment for its growth and sustenance.

3. Capture Fisheries

- The scope for increasing capture fisheries' production from the coastal waters is limited. To sustain its present level, which contributes 50 per cent of the total fish production in India, we will have to concentrate on resource enhancement programmes such as sea-ranching and eco-friendly mariculture. Effective implementation of fishing regulations also will have to be given priority. A rational approach to match the fishing capacity with maximum sustainable yield has to be evolved. There is also a need for a scientific introspection on practicality and efficacy of the Marine Fishing Regulation Acts. It is necessary to create a Fishing Regulatory Authority with powers of implementing sustainable fishing. This authority should be bestowed with legal powers.
- Contributions from inland fisheries to India's total fish production increased from 25 per cent in 1950, to 46 per cent in the 1990s. Despite declining riverine fish production, there is substantial increase due to development of culture-based reservoir fisheries. But the Indian reservoirs produce much less than their potential; their productivity can be increased to three million mt. There is need for an action plan to develop appropriate model for determining management parameters, such as stocking density, stocking size, size at capture and species selection. Though they are highly productive, we know very little about our wetlands and beels; their development in the Northeast is strategically important. Management norms for the upland lakes are virtually non-existent. A master plan is required for the development of fisheries in wetlands, beels and upland lakes.
- Suitable strategies for exploiting oceanic tunas and Antarctic krill need to be developed soon.

4. Aquaculture

- Aquaculture is the fastest growing enterprise within the agricultural sector. Global aquaculture production increased from 15.5 to 36.0 million tonnes during the years 1988-1997, indicating an annual increase of 11 per cent. The level of increase in aquaculture production is likely to remain high for quite sometime in the future, while other water- and land-based food production systems are reaching their limits. However, aquaculture is more complex than agriculture/animal husbandry, owing to diversity of taxonomic groups and unfamiliarity of the environment/medium to the cultivator. China harvests 67 per cent of the world aquaculture production, while India's contribution is stagnating at 6.2 per cent. Unlike China, diversification of aquaculture in India is relatively poor. Even though production is less than half that of India, Japan (4.37 $/kg) and Indonesia (2.75 $/kg) fetch higher value for their products compared to 1.11 $/kg of India. Hence, diversification of candidate species for aquaculture in India has to be judiciously manoeuvred. Of 297 species cultured, 53 per cent are fishes, 15 per cent crustaceans, 27 per cent molluscs and 5 per cent seaweeds. India claims to have mastered the technology of aquaculturing 15 species, whereas as many as 39 and 52 species are profitably cultured by China and Korea, respectively. The need for diversifying taxonomic species and aquaculture systems is obvious to increase cumulative aquaculture production, enhance value for produced fish. Sustainable aquaculture system can be developed in harmony with the physico-chemical and biological environment as well as the socio-economic environment involving other sub-sectors/sectors, namely fishing, agriculture, forestry, tourism, public health and housing among others, ensuring protection of all the stakeholders involved. Aquaculture thus, would be part of a total sphere development, as envisaged in integrated rural area development or coastal area management.

- Being basically a small scale enterprise, freshwater aquaculture provides for the domestic food security of rural India. This sector has undergone due intensification and diversification. Likewise, brackishwater aquaculture is a highly promising sector and about 1.4 m ha of potential brackishwater area is available for farming in the country, of which less than 10 per cent area is under cultivation to produce 82,000 tonnes. Its productivity is low (800 kg/ha), in comparison to that of Thailand (2,500 kg/ha). There are 226 shrimp hatcheries with a capacity of 10.5 billion seedlings. There is an urgent need to have a uniform leasing policy. To boost aquaculture of inland, brackishwater and marine organisms, a scientific policy of leasing water bodies will have to be formulated to utilise them rationally and legally. There is also need to provide the required infrastructural facilities such as seeds, finance by the NABARD and aquaculture insurance scheme. Since, recycling freshwater will become increasingly important in the years to come, policy guidelines must be framed for mandatory provision of depuration ponds in farms undertaking sewage-fed fish culture.

- Total feed requirement for carp farms of Andhra Pradesh alone is 200,000 t but only 35,000 t is produced by our feed mills. Hence, there is a gigantic need to increase commercial production of inexpensive, balanced feeds for finfish and shellfish. To encourage establishment of larger fish-feed industries, the aquaculture feed industry must be granted a status of parity with poultry and cattle-feed industries, and given a waiver and/or tax concessions, as enjoyed by other feed industries. Regulation devised for other feed industries may also be imposed on the fish-feed industries to ensure quality control of the products. When formulating standards for quality monitoring, the following must be considered: (i) particle and pellet size, (ii) fibre content, (iii) water stability, and (iv) the permitted levels of steroids, antibiotics, pigments. Specialised fish-feed plants have to be encouraged to provide sectoral development of aquaculture feeds.
- A major constraint to the development of aquaculture is the loss caused by microbial diseases which inflict heavy mortality. The loss due to such disease problems in Asia has been estimated to be several hundred million dollars. To contain microbial diseases, early detection of pathogens is very important. Molecular techniques for early and rapid detection of pathogens are being developed in India. Diagnostic laboratories with facilities for rapid detection of pathogens by molecular methods, such as PCR should be set up in all the fisheries colleges of India. Farmers should have easy access to these laboratories to get the desired diagnosis, medicine and suggestions at subsidised cost, as is being done for livestock animals.
- Intense anthropogenic activity and heavy load of pollutants have driven several aquatic organisms, especially commercially

important fishes to the status of threatened/endangered species. There is a need to prioritise species and aquatic habitats for conservation. Financial support for research on cryopreservation of fish sperm and cell-lines has become of paramount importance. Training and capacity building for fish taxonomy and genetics merit support on priority basis. With regard to exotics and related issues, more research attention has to be given to mitigate their negative impacts.

- Greatest emphasis and financial support must be given to the post-harvest technology. On the global basis, around 27 per cent of the harvested fish are being discarded as bycatch. The proportions of discards are also high in India. Larger trawlers should be provided with additional storage facility for the bycatch of low value fish, or they should have an installed plant, that would process about one mt bycatch per day for fish meal and fish oil. This installation must be encouraged and subsidised.
- A survey conducted at different fish landing centres, clearly indicated that five per cent of the landed fish is of unacceptable quality. Hence, fishermen should be encouraged to carry ice on board to preserve the harvest. There is a need to create an awareness among fishermen and processors about the need for hygienic handling of the harvested fish. The traditional bamboo basket must be replaced by insulated container. The fish can be carried to the urban consuming centre in good condition within four hours after harvesting. Transport by air, where required, should be encouraged and subsidised.
- Fish is a highly perishable commodity; the processes responsible for its decomposition are autolysis, bacterial spoilage and oxidative rancidity. These processes can be retarded effectively by chilling with ice. Radurisation is an irradiation process for extension of shelf life of fresh fishery products in ice or under refrigeration by reducing the number of spoilage bacteria. The major benefit of the application of radiation in fishery products is in the reduction of post-harvest losses and improving the hygienic quality. Recognising the technoeconomic advantages, 40 countries including India have permitted such radiation treatment. It is recommended that

irradiation facility be established by the government at least in two strategically important landing centres in each maritime state, each installation at the minimum cost of Rs. 75 million.

- Long-chain polyunsaturated fatty acids present in fish and fish oils have several positive health benefits during different stages of human life—conception, growth and development, and prevention of diet-related chronic diseases. Iron deficiency is the most common micronutrient disorder. This problem is most severe in India, where 88 per cent of pregnant women are anaemic. Eating fish 1-2 times/week before conception, during pregnancy, lactation and breast feeding will significantly contribute to the growth and development of babies and ensure good health of both mothers and babies. The public should be made aware of the health benefits of fish-eating habits.

5. Fisheries Education and Research Publications

- Qualified and trained manpower is a critical input for sustainable fisheries development. Fisheries education had a late start in India. Till the sixties 'fisheries' education was confined only to a special paper in postgraduate programmes, in disciplines like zoology in some universities. Under the State Agricultural University (SAU) system fisheries education the first College of Fisheries was established at Mangalore under the auspices of the University of Agricultural Sciences, Bangalore, in 1969.
- A four-year degree course in fishery science is offered in 11 colleges to train about 300 students. Two-year-postgraduate and doctoral degree courses are offered only at 8 and 6 institutions, respectively. About 150 and 50 candidates secure their M.F.Sc. and Ph.D. degrees from these respective institutions annually. Unfortunately, 28 per cent of the sanctioned faculty positions lie vacant in these colleges. The Education Division of the ICAR must devise a method to make the vacant positions filled by the respective colleges. As over 18 per cent of the sanctioned scientific and technical positions are vacant in the eight national fisheries institutions, many qualified fisheries graduates remain underemployed or unemployed; the government should give consideration to

these nascent institutions to grow by filling the vacancies. There is a need for colleges of fisheries to establish linkages with other national laboratories and institutions of the conventional universities.

- Although, fisheries scientists have accomplished a commendable task of increasing fish production in the country, both the number and quality of their publications suffer at the international level due to poor visibility. An important reason for this is that our fisheries scientists publish mostly in non-indexed journals. An all-out effort must be undertaken to improve the image of the premier fisheries research journal of the ICAR, namely, *Indian Journal of Fisheries*.

6

Hi-tech Horticulture in India

Introduction

India is endowed with a wide variety of agroclimatic conditions and enjoys an enviable position in the horticultural map of the world. Almost all types of horticultural crops can be grown in one region or the other. Horticultural crops occupy 8 per cent of the gross cropped area of the country contributing 24.5 per cent of the gross value of agricultural output and 54.55 per cent of export earnings in agriculture (1998-99).

India is the second largest producer of both fruits and vegetables. Total production of fruits has been estimated at 44.04 million tonnes from 3.72 million ha. Vegetables occupy an area of 5.78 million ha with a production of 87.53 million tonnes. Our share in world production in fruits and vegetables is 10 and 13.38 per cent, respectively.

India is the largest producer of coconut, cashew, tea and spices, with a total production of 14,9248 million nuts, 0.46 million tonnes, 805.61 million kg and 2.90 million tonnes, respectively.

India has made noticeable advance in production of flowers. Floriculture is estimated to cover an area of 73,970 ha with production of 4,59,163 mt of loose and 61,21,523 lakh cut flowers. The area undercut flowers has increased in recent years, so has the product range.

India exported horticultural produce and product worth Rs. 90315.20 million during 1998-99. Tea, fetched the highest exchange of Rs. 21918 million followed by spices, coffee and cashew. During this period, fresh fruits and vegetables earned foreign exchange of Rs. 5360.20 and 7056.80 million, respectively.

Policy Paper 13, October 2001.

Horticulture today, is not merely a means of diversification but forms an integral part of food and nutritional security, as also an essential ingredient of economic security. Adoption of horticulture, both by small and marginal farmers, has brought prosperity in many regions of the country, of which, Maharashtra, Karnataka, Andhra Pradesh and Kerala are prime examples.

Research and development in horticulture has received impressive support in the last 15 years. As a result, the research infrastructure has increased manifold with the setting up of several new institutes and national research centres on several export-oriented crops. The development activities received a boost with approximately 40-fold increase in budget from the 7th and 8th Plan. At present, 10 per cent of the total budget of ICAR and 18 per cent of the total budget of DOAC is earmarked for the horticulture sector in the 9th Plan.

So far, Indian horticulture has been insulated from forces of outside world. Time-bound removal of quantitative curbs on imports and other barriers to access to domestic market under WTO, of which India is a signatory, will require Indian horticultural produce/product to be competitive both in the domestic and export market. This would call for use of hi-tech horticultural technologies. Such technologies can be defined as 'those, which are modern, less-environment dependent, capital intensive and have the capacity to improve productivity and quality'. These will include, use of genetically modified crop varieties, micropropagation, integrated nutrient and water management, integrated pest management, protected cultivation, organic farming, use of modern immuno-diagnostic techniques for quick detection of viral diseases and hi-tech post-harvest technologies, including cold chain.

Ensuring the nutritional and aesthetic security of our nation with a population of 120 crores by the end of 2005, would be a major challenge for the horticulture sector. With shrinking land for agri-horticultural activity, the sensible option before the nation is to increase further the production levels per unit area. Intensive cultivation in hi-tech protected environments with hi-tech production inputs will ensure higher productivity levels.

With the opening of global market and removal of quantitative restrictions under the WTO agreement, export-import scenario is likely to change at much faster pace. Horticulture stands as the major beneficiary in such a scenario to contribute to the much needed forex reserves.

Sensing this importance, the National Academy of Agricultural Sciences (NAAS) organised a National Seminar on Hi-tech Horticulture.[1] The present trends, issues and constraints were discussed under seven theme areas and recommendations made.

1. Opportunities for New Biology for Germplasm and Crop Improvement

1.1 Issues in Germplasm Identification, Conservation and Exchange in Horticultural Crops

- Of the total germplasm collections held in various gene banks, horticultural crops comprise only 15 per cent. Hence, there is an urgent need for intensification of collecting and conservation activities, in these crops. There is need for duplicates of collections as back up for gene bank collection and intensifying *in situ* conservation activities. For doing so, the CBD mandate for *in situ* conservation offers appropriate guidelines.
- Conservation activities should involve participation of farmers in addition to the government agencies. Appropriate methods of conservation and emergency strategies, especially for threatened wild species need to be developed. The approaches for *ex situ* conservation of different species were reviewed. While technology-intensive methods like *in vitro* conservation and cryopreservation would help conserve genetic diversity, a low cost conservation technology like ultra dry storage, is the need of the hour.
- Cryopreservation of pollen has great significance, particularly in perennial tree crops. For *ex situ* conservation of perennial tree species, techniques like bonsai culture may be explored. There is need for screening of germplasm for pest resistance and for characterising important wild species and land races using molecular techniques such as RAPD analysis and DNA finger printing.

---

1. The National Seminar was held for three days at Bangalore from 26th to 28th June, 2000. The seminar was jointly organised and hosted by the Horticultural Society of India, New Delhi and Indian Institute of Horticultural Research, Bangalore.

1.2 *New Biology for Developing Improved Cultivars of Horticultural Crops*

Molecular characterisation is very important for IPR, patenting and safeguarding against gene piracy. Molecular marker-aided-selection and genetic engineering will have to be employed in future genetic improvement of horticultural crops. Breeding of horticultural crops should have broad genetic base. Isolation of somaclonal and protoclonal variation and *in vitro* mutagenesis need to be explored. Use of molecular methods of transferring male sterility need to be explored for development of hybrids. In near future, integration of conventional and molecular breeding will have to be resorted to, for genetic improvement of horticultural crops. Special areas, in which genetic enhancement in quality horticultural crops requires attention are: dwarfing, herbicide resistance, pest resistance, increasing shelf life, improvement in processing nutritional quality.

1.3 *Hi-tech Propagation of Horticultural Crops*

- Micropropagation has emerged as an important tool for quick production of large number of plants. It is already being exploited in crops like banana, capsicum, tomato, chilli and several ornamental plants. There is an enormous scope in India for micropropagation of crops especially ornamental crops, in years to come. Their protocols, therefore, need to be further refined for higher efficiency. There is also need to develop protocols for micropropagation in crops like mango, cashew, litchi, walnut and date palm which have a great commercial potential in India. Thrust on quality and strict avoidance of contamination hold the key for the success of Indian tissue culture industry. Efforts are, therefore, required to achieve high standards to meet international requirements. Further, innovative management practices like efficient use of techniques, such as, low cost polythene bags for raising plantlets, cheaper laminar flow facilities with more capacity, minimum contamination and faster sterilisation need to be developed and standardised.

1.4 *Greenhouse Technology for Plant Propagation*

- Presently greenhouse technology is practiced by very few in the country. Scope, therefore, exists for large-scale expansion of nursery activity under favourable microclimate created in cost-effective low cost structures. Efforts also need to be intensified to develop suitable technology for production of healthy and well-developed seedlings in plastic perforated trays for Indian conditions. New growing media also need to be standardised for optimum plant growth. Refined techniques for hardening of seedlings, including root pruning, have to be standardised. Greenhouse design, structure and technology for producing high quality planting material need to be fine-tuned to suit Indian conditions.

- Commercial grafting techniques with computer control are available in Japan for mass production of certain plants. Such machines need to be developed indigenously or need to be imported by a nodal agency in India, to standardise and perfect the technology for further dissemination.

1.5 Hybrid Seed and Seedling Production Technology for High Value Vegetable Crops Under Protected Environment

- Climatic factors, like temperature, relative humidity and sunlight, influence plant growth and pollen production. Optimum climatic requirements need to be worked out for all the crops suited to Indian conditions to exploit properly the greenhouse structures.
- Nearly 40 per cent of the total cost of hybrid seed production is incurred on labour employed for emasculation and pollination. Use of male-sterile lines, gynoecious lines and growth regulators facilitate hybrid seed production by cutting cost of hand emasculation. Efforts are, therefore, needed to isolate/develop male-sterile/gynoecious lines in important vegetable crops for hybrid seed production.
- While potato varieties suitable for production of TPS (True Potato Seed), have been evolved, there is need to standardise package of practices for TPS cultivation. In future, TPS production programmes' emphasis should be laid on selection for early maturing and improved tuber size and yield by improved agrotechniques which will greatly help to bring down risk and shall enhance efficiency.

2. Hi-tech Production of Horticultural Crops

2.1 *Developing High Density Planting (HDP) in Horticultural Crops*

- High density planting has a potential to increase yields and reduce cost of production of several horticultural crops. It has already been exploited successfully in kinnow, orange, pineapple, banana and to some extent, in apple and mango. There are several ways of achieving this objective such as selection of dwarf cvs and scions, use of dwarfing rootstocks/ interstocks, pruning and growth regulators for canopy management. A better option would be to achieve dwarfing through genetic means.

2.2 *Developing Nutritional Need Diagnosis and Leaf Nutrient Norms*

- Leaf analysis, particularly in perennial crops, offers a means of assessing nutritional requirements. Leaf nutrient standards for a large number of crops have been developed and are in use in developed countries. Work on these lines is also underway in India. This needs to be intensified to ensure economy in fertiliser use through optimised application. To aid practical application, however, there is a need to correlate nutritional composition of leaf with fruit yield on the basis of extensive nutritional surveys of orchards and comprehensive field experiments.

2.3 *Micro-irrigation and Fertigation in Horticultural Crops*

- Micro-irrigation has proved its efficacy in saving water and improving yield but its large-scale adoption is limited to a few southern states especially for high value crops. High initial investment has been one of the critical factors. Therefore, there is need to develop a holistic approach to the problem through effective integration of research, development and industry, so that the gap between potential and current status could be bridged. There is need to refine technology, improve design and articulate mechanisms for efficient utilisation of water through micro-irrigation. Availability of water-soluble fertilisers for fertigation and related policy matters, also need to be addressed on priority.

2.4 *Organic Farming in Horticulture*

- In their quest for higher yields, farmers are using heavy doses of fertilisers, herbicides, fungicides, pesticides, growth regulators. This results in not only increased cost of production but also drastically changes our environment. In recent years, there has been emphasis on organic farming to obtain pesticide residue-free fruits, vegetables, spices and other horticultural commodities. However, scientific technology with sound alternatives is not available. Keeping in view the benefits, which this technology offers alongwith higher returns in international markets for such products, systematic programmes need to be taken up. To ascertain and guarantee the consumer/importer that the produce has been genuinely raised organically, the production has to follow basic organic standards. An Indian Organic Food Production Act, has to be enacted early to safeguard consumers' interest and check exploitation of farmers.
- Suitable analytical laboratories need to be established, both by the Government of India and state governments, to assess and evaluate the quality of inputs. Research and development on upgrading compost quality, biomass production, biofertiliser and biocontrol measures should be the mandates of all agricultural research centres in the country.

3. Protected Cultivation

3.1 *Design and Development of Greenhouses for Different Agroclimatic Conditions*

- At present the greenhouse technology is imported either from the Netherlands or Israel, resulting in high cost of structures. Economically and ecologically sustainable polyhouse structures suited to different agro-climatic conditions need to be evolved indigenously.

3.2 *Greenhouse Technology for Production of Cut Flowers*

- Rose continues to be the prime floriculture product in the international market. India has proved its capability in growing and exporting roses despite several bottlenecks. Indigenous production technology based on ethnic growing media including coir-pith and post-harvest management to overcome impediments needs to be developed and fine-tuned to suit our climate. Therefore, research efforts need to be intensified on roses. To sustain and maintain superiority a diverse product mix is essential. As such, efforts need to be made for identifying suitable products for viable diversification.

3.3 Scope of Protected Cultivation of Vegetables in High Altitudes

- At high altitudes, especially Ladakh, polyhouses with low cost technology have been employed successfully to grow vegetables round the year and also to take up seed production of vegetables round the year in spite of adverse climate. There is need to promote these polyhouses in other arid temperate regions of the country.

4. Hi-tech Plant Protection

4.1 Integrated Pest Management (IPM)

- Horticultural crops are adversely affected by a variety of pests. These pests result in annual loss of over 36 per cent. For the success of hi-tech horticulture, integrated management of these pests is essential. Research efforts are required to develop models for application of IPM practices for important crops in different locations. With changing horticultural practices new pests like thrips, white flies, mites, phytoplasma, viruses and viroids are emerging. Some of these are exotic pests. To protect horticultural crops from such pests, greater vigilance and stringment quarantine measures are required.
- With increasing awareness of the ill effects of environmental pollution, biological control of pests will be the most favoured and desirable approach for minimising losses caused by pests, like insects, fungi and bacteria. Transgenic crops alongwith biological control strategies will be the main armoury in fighting the pest menace in this country. Some initiatives in these directions have been taken in the country, but considering the range of horticultural crops and agroclimatic zones requiring high-tech approaches, a much greater effort is required for utilising the full potential of biological control and transgenic crops.

- Due to enormity of pest problems, need-based application of pesticides will have to continue as part of IPM. Present-day indiscriminate use of pesticides, which results in undesirable levels of residues in the produce and contamination of water and soil, will have to be stopped through appropriate policy measures. Success of such a policy will depend on the availability of 'safe' pesticides like botanicals. The country has good information about availability of basic materials like *neem*, but more emphasis is required for developing economical and environmentally compatible technologies for commercial production of safe pesticides.

4.2 Diagnostic Molecular Techniques

- Use of healthy and quality planting material is the key to improving production of horticultural plants—especially those propagated vegetatively. Success achieved in improving production of potato as a result of well-thought and meticulously implemented 'Certified Potato Seed' programme is a good example. Now that sensitive molecular diagnostic techniques are available, it is time to start certification programme for all horticultural crops grown in the country for export as well as internal consumption. A beginning has been made with the establishment of a 'Network National Facility for Virus Testing and Quality Control of Tissue Culture Raised Plants' by the DBT. A similar facility is also required for certification of horticultural exports for freedom from undesirable chemical residues, toxins, antibiotics and incipient pest infestation in order to prepare us to meet SPS (phytosanitary) regulations. This can be achieved through establishment of a network of micro and macro level diagnostic laboratories around the country.

5. New Developments in Post-Harvest Management

5.1 Marketing Opportunities and Infrastructural Support

- Implementation of nutritional security programmes, increase in purchasing power, shift in food habits, growing international trade and removal of quantitative restrictions (QR) are bound to influence the domestic market in India, both for fresh and processed horticultural products. Quality improvement and exploitation of both competitive and comparative advantages will largely determine future growth in the global trade.

- Redesigning the supply chain from seed bed to consumers' plate is the need of the hour. Technological advancement made so far and global competitive pressure should determine the path of restructuring the supply chains. The huge post-harvest losses particularly in perishable horticultural produce in tropical environment and unorganised market system requires priority developmental attention. Public-private partnership and massive investment in infrastructural development in post-harvest handling and modernisation of marketing systems of perishable products are essential.
- By checking losses through reducing multiple handling of fresh produce and adding value at producer's end by sorting, grading, waxing, precooling and improving storage, packaging and transportation systems, better economic return to the growers can be ensured. A network of farmers' service centres is recommended to be set up in major production areas to galvanise the whole production and supply system. Similarly, establishment of primary processing networks and a grid of centres of HRD will further consolidate the gains in modernising horticultural marketing system in India.
- Strategic marketing requires strengthening several areas e.g., sourcing for export and distant markets; contract farming may ensure both supply and better quality produce to maintain supply chain, modernising markets through central auction, reducing multiple handling and price setting mechanism. Direct involvement of growers' cooperatives, greater participation of stakeholders, retaining regulatory powers of the government may greatly help in re-organising the system. Market information database, market intelligence marketing system and use of information technology, are essential components for modernising horticultural produce marketing in the country.

5.2 *Mechanising and Post-Harvest Handling*

- Separation through conventional and modern techniques and gadgets based on photometry, acoustic response, short wave radiation, machine vision and laser technology hold promise to meet quality standards for markets and operations down-the-line including value addition. Appropriate field heat removal techniques, *viz.*, cold air system, hydro-cooling, vacuum cooling or slush system, need to be adopted at the field level *(in situ)*.

5.3 Processing and Value Addition

- Minimally processed for quick cooking of vegetables and their products, which are prepared with intermediate moisture and hurdle technology, have great potential in the civil sector besides the defense sector. These convenient foods can be stored without refrigeration and can be marketed in both domestic and international markets. Entrepreneurship in this area needs to be promoted. Designer foods developed by Defence Food Research Laboratory can go as calorie specific functional foods in the civil sector in the field of nutrition.

6. Logistics and Policies for Hi-tech Horticulture

6.1 Human Resource Development

- Since, hi-tech technology is new to the country, our curricula need to be fine-tuned to impart latest knowledge to both undergraduate and postgraduate students. Suitable nodal centres need to be developed to impart training in various aspects of organised farming and post-harvest management.

6.2 HACCP Standards for Horticulture

- With regard to the quality and safety parameters, the Indian standards are in agreement with those where we are not in concurrence with the international standards which need to be bridged. There is an urgent need to educate all organisations, institutions, both government and non-government, and the producers, regarding the HACCP standards and sanitary and phytosanitary measures.
- There is a need to develop special curriculum and vocational training for specific type of agribusiness. There is also urgent need for HRD for crop executives and exposure trips to other countries for the scientists/extension workers and personnel of financial institutions and industries.

6.3 Mechanisation and Automation in Horticultural Crop Products

- Mechanisation and automation has not yet been fully exploited for the benefit of Indian agriculture. These offer several benefits, e.g. improvement in efficiency and quality of work alongwith reduction of drudgery. Some of these innovations promote employment generation. In horticulture crops these efforts should help in vegetative propagation, planting and pruning of plants; harvesting, grading and post-harvest management, including value addition.

7. Policy Issues

- There is need for simplification of policies and issues such as plugging of loopholes in the land acquisition law. Priority treatment of hi-tech floriculture as an industry needs to be given. Availability and importing of quality inputs such as water-soluble fertiliser, pesticides, throughout the year, across the counter should be ensured. The high cost of finance, sudden increase in quarantine duty and duty on plastics which have been hampering the growth of the industry, need to be reviewed.
- Information technology has the potential to shatter all the barriers reaching across the continents. A suitable mechanism to safeguard the interest of the Indian growers and consumers needs to be evolved. Since, this concept is new to Indian markets, suitable training needs to be imparted to all those involved in the cyber trading.
- In view of the fragmented landholdings in all the states of the country, contract farming/cooperative/consolidated farming needs to be encouraged so that modern technologies of cultivation can be adopted for increase of productivity and quality.
- In view of low productivity and poor quality, there is urgent need for effective transfer of technology within the country to increase production, productivity and to improve quality. Certification of nurseries should receive high priority.
- Domestic market has to be strengthened in order to increase exports. Without organised domestic market base, we cannot promote exports. Models like NODS (F&V units) and Hopcom

need to be created in all the metropolitan cities in the country.

- High interest rates for export finance also make our exports non-competitive. In India, pre-shipment export credit is available at the rate of 10 per cent, that too, after observing so many formalities, whereas, the international exports in other countries get finance at the rate of 4 per cent. Such anomalies need to be removed.
- Immediate emphasis should be on creation of post-harvest infrastructural facilities, like, collection centres, grading centres, washing facilities, packing, refrigerated vans, pre-cooling units, cold storages, intermediate cold storages, processing units, etc. A complete cold chain facility needs to be created.

7

Peri-urban Vegetable Cultivation in the NCR Delhi

Introduction

Peri-urban agriculture is recognised for its potential role in increasing food security, employment and income generation, poverty alleviation, community resource development, waste management and environmental sustainability. A large part of the world population is living in peri-urban areas. The rich social and economic variability of peri-urban areas offers opportunities for innovative science, living condition improvements and agricultural and ecosystem management requiring novel policies, and action to promote harmony both between society and nature and between groups of people.

The population of Delhi is around 1.4 crores which is causing phenomenal pressure on land, infrastructure and civic amenities. Availability of rail, road and air transport, cold storage, processing units, export houses and well established market network, will be helpful for taking up vegetable cultivation in peri-urban areas of Delhi in an organised way. Also, a huge quantity of solid waste generated during handling and marketing of fresh vegetable produce in NCR, Delhi, which is creating health and environmental hazards, can be used or recycled to produce vermicompost, etc., for use in organic vegetable production. Peri-urban vegetable cultivation can provide farmers the possibility to cultivate a small piece of land, and obtain an income to meet their essential and basic needs. In recent years, around big cities, green belts are being developed which can provide a very intensive and profitable network of small farms specialised in production of perishable vegetables for consumption by the urban consumers. This is likely to result in a social symbiosis between farmers and city dwellers with mutual benefits and advantages. This

---

Policy Paper 26, July 2004.

will require involvement of a large number of institutions to address many issues related to the peri-urban vegetable cultivation in a viable participatory approach to make an impact and deliver anticipated results to the peri-urban population.

Status

Delhi has five Community Development Blocks comprising 209 villages, of which, 199 villages are inhabited while the remaining 10 villages are uninhabited. Now-a-days about half of the NCR area is urban. The rural area of the NCR is divided into five Rural Development Blocks: Alipur (north), Kanjhawla (northwest), Najafgarh (southwest), Mehrauli (south) and Shahdara (east). Out of a total area of 1,47,448 ha, net sown area is 48,357 ha, *diara* land 7,000 ha, fallow land 12,886 ha, and cultivable waste land 10,850 ha. In addition, NCR Delhi has 74,248 ha non-agricultural land, 11,708 ha other uncultivated land, 1,561 ha forest land, 1,267 ha under tree crops and grooves. The total percentage of agricultural land comes to 41.7 per cent, 35 per cent of the total agricultural land is under vegetable cultivation. The annual collection of vegetables from Delhi farms is around 22 lakh tonnes. This is from around the periphery of the city from where men folk commute daily to the city and women attend to the needs of peri-urban vegetable cultivation. Major vegetables grown in and around Delhi include cauliflower, cabbage, carrot, spinach, mustard (leaves), cucurbits, okra, brinjal and tomato. In addition, culinary herbs such as fenugreek and coriander are also cultivated.

In NCR Delhi, 42 per cent landholdings are of up to 2 ha while rest (58 per cent) are of up to 3 ha. This size of holding suits best for development processes and technology adoption. Approximately, 1 per cent of the urban and 17 per cent of the rural population is engaged in agriculture and adoption of vegetable cultivation is on a rise replacing cereal crops, due to its being more remunerative than other crops. As such, up to 35 per cent of land in some villages has gone under vegetable cultivation. Almost, all kinds of vegetables (tropical, subtropical and temperate) can be grown in Delhi farms, which is again an advantageous position. An important factor affecting both, land use and cropping patterns, is the access to transport networks for linkages with urban markets, labour and capital.

Resource Constraints

- Most of the farmers are ignorant about the quality of the seed material, its usefulness and availability. The improved high yielding varieties and hybrids have also not reached them due to either poor extension services or high cost. The quantity of such seed material available is also meagre. There is very little or sometimes no dissemination of advanced/improved production technologies including high-tech vegetable cultivation and off-season vegetable cultivation, probably due to lack of trained manpower/system.
- There is no awareness about low volume high value vegetable crops like broccoli, Chinese cabbage, asparagus, celery, parsley, leek, baby corn, cherry tomato, etc. and as such their demand by starred hotels and elite class is also only partially met with.
- Farmers are mostly ignorant of right stage of harvesting, proper cleaning, grading and packing of the produce, which results in low price for their produce. Sometimes, improper labelling of different grades also results in confused marketing and poor sale price.
- Though a good network of wholesale markets exists, less suitable transportation facilities and unplanned production, creates market glut, resulting in poor payment to the farmers. Lack of sufficient cold storage facilities to store surpluses for next day marketing also amount to a major constraint. Non-existent or meagre local marketing system also results in low sale price as well as market glut.
- There has been inadequate attention towards post-harvest management to prevent losses and to ensure supply during odd periods. Infrastructure facilities like precooling units, cold stores, refrigerated transportation system, packinghouses, modernised market places are either absent or negligible. Market information system is also either non-existent or not available in big *mandis* resulting in irregular production, uncertain availability of the produce and unorganised market prices.

Resource Availability

- Indian Agricultural Research Institute (IARI), the premier institute of agricultural research, education, technology development and extension is situated in the state and can be looked at, for all sorts of technical guidance.
- A large cultivable area in plain and also a sizable spread of Yamuna riverbed is under cultivation. Another 10,850 ha of cultivable waste land, 11,708 ha of uncultivated land and 12,886 ha of fallow land are available to be brought under vegetable cultivation.
- Unemployed but educated rural youth and also women are available to be trained in vegetable cultivation for gainful employment and increased production.
- A fresh vegetable export zone (VEZ) can be established comprising a group of villages involved in custom production conforming to norms of export in certain low volume high value vegetables.
- Protected cultivation of off-season vegetables with high quality, particularly during rainy season and winter can be promoted with the improved technologies.
- Organic farms can be established and recycled wastes from city and APMC *mandis* (fresh vegetable sale points) consumed as organic manure. This will also reduce environmental pollution and health hazards.
- A large number of processing and preservation units are functioning, though on part time basis, because of irregular supply of raw material. Ensuring their full time functioning in an organised way will assure utilisation of market glut as well as availability of products during off-season.
- An excellent network of marketing of agricultural produce under APMC already exists to satisfy wholesale marketing requirements. There are over nine regional *mandis* in different parts of Delhi.
- Delhi climate is suitable for cultivation of all types of seasonal as well as perennial vegetable crops. Off-season cultivation is also possible as has been demonstrated through polyhouse and

polytunnel cultivation during rainy season and low temperature regimes.

- Facilities for training the farm supervisors as well as farmers exist both in the Department of Agriculture/Horticulture, Delhi, as well as at IARI, for making available trained manpower in production including seed.
- An array of vegetable varieties suitable for cultivation under both biotic and abiotic stresses, for production of high nutrition, higher biomass (quantity) and for specific end-use is available (courtesy IARI, ICAR) to suit different conditions of production and utilisation.
- Infrastructure exists for seed multiplication and distribution and also for production technology demonstrations in the form of several farms under Delhi government's Department of Agriculture/Horticulture.
- A network, to some extent, also exists under the Department of Agriculture/Horticulture of Delhi government to help in dissemination of new technologies up to farm level and for monitoring of their progress/adoption.
- IARI has developed a number of production technologies (some are listed below), which are available for popularisation and adoption:

  i. Off-season nursery production to help enhancing availability of produce in cucurbitaceous vegetables.

  ii. Off-season production technology under protected cover for high value vegetables like cucumber, summer squash, tomato and capsicum.

  iii. Almost year-round production of radish, tomato, cauliflower using suitable varieties in successive sowings.

  iv. *Kharif* onion production to reduce storage requirement, fetch higher returns, also stabilise market price.

  v. Easing and economising hybrid seed production in pumpkin, bottle gourd, ash gourd, bitter gourd, luffa, etc., to help employment generation, reduced cost of F_1 hybrid seed, increased F_1 hybrid seed availability.

  vi. Ratooning in okra and brinjal to extend the crop availability beyond the normal season under open field. This ensures availability of the produce during lean/off-periods.

vii. Off-season production technology of cucurbitaceous vegetables in riverbed and *diara* land.

- A good road network exists in the state for transporting cultivation inputs to the farm and for transport of produce from farms to the markets/consumers. Similarly, sufficient irrigation facilities in the form of canals and bore wells also exist. However, drip and sprinkler irrigation/fertigation facilities need generation.
- To help organic farming *goshalas* and dairy farms also exist in the state to provide organic manure (e.g., FYM, composts, etc.) and such other related inputs.
- Branches of different banks are available in every locality to possibly extend credit facilities to farmers through these branches.

Strategy

Delhi, having ever-increasing consumer population, requires extra procurement of horticultural produce, particularly vegetables from neighbouring states of Uttar Pradesh, Haryana and Rajasthan. The present day production of around seven lakh tonnes of vegetables is not sufficient and, as such, intensive vegetable cultivation on available land, on scientific lines, using latest technologies, should be adopted/promoted. Further, additional land, including *diara* land, has to be reclaimed and brought under vegetable cultivation. Efforts have to be made on reducing post-harvest losses and utilising/avoiding market gluts, improvement in produce management and utilisation strategy, including processing and systematic marketing. Improved marketing information system has also to be developed. Important vegetable crops which hold promise in cultivation are onion, *bhindi*, chilli, bitter gourd, tomato, potato, brinjal, cauliflower, peas, bottle gourd and cucumber. The thrusts have to be decided for programmes on long term and short-term basis. In addition, some medicinal and aromatic plants also need promotion in cultivation for use in therapeutic and pharmaceutical industry.

A brainstorming session was held in New Delhi, to deliberate on these issues.[1] The following recommendations emerged as a result of deliberations at this Session.

Recommendations

- Horticulture advisory service and extension network of the NCR Delhi, needs to be strengthened. Awareness generation about IPM, IPNM, biological pest control and off-season vegetable production technologies should be given major thrust through preparation and distribution of informative material on vegetable cultivation on scientific lines.
- Modern micro-irrigation systems (drip, sprinkler) including fertigation, arrangement of supplies for installation of systems on cost basis needs to be promoted to improve cost-benefit ratio in vegetable production.
- Protected crop and nursery production technology should be popularised so as to make it profitable/viable on commercial scale. This will also help in off-season production of vegetables.
- Production of hybrid seed and use of pest-resistant high yielding F_1 hybrids and varieties of vegetables in commercial cultivation need promotion. Hybrid seed-production techniques developed by IARI should be used to economise seed production cost.
- Export production farms, particularly for vegetables should be established.
- Infrastructural facilities for post-harvest management *viz.*, cool transport chain, precooling units, packing houses, short- and long-term cold stores, etc., should be improved.
- Human Resource Development programmes, by giving more

---

1. A brainstorming session on Peri-urban Vegetable Cultivation in NCR Delhi, for improving urban health, generating rural employment and improving income of the farmers was held on 23rd December 2003, under the convenership of Dr. S. Nagarajan, at IARI in New Delhi. The participants comprised representatives from National Horticultural Board (NHB), Agricultural Produce Marketing Committee (APMC: *Subzi Mandis*), National Horticulture Research and Development Foundation (NHRDF), Agricultural and Processed Food Products Export Development Authority (APEDA), NABARD, Fruit & Vegetable Project of Mother Dairy, Delhi State Department of Agriculture, NSC, IFFCO, KRIBHCO, ICAR Headquarters and scientists from various divisions of the IARI.

emphasis on training of farmers and farm supervisors in the field of scientific cultivation of vegetable crops, as well as multiplication of seed and planting material should be strengthened.

- Yamuna riverbed (main *diara* and upland *diara*) area for vegetable cultivation on scientific terms should be developed.
- There is need for generation of awareness and establishment of on-farm storage, processing and preservation units.
- System towards making available information regarding demand, supply, price, market outlook, consumers' preference, marketing channels and practices to help streamlining sale of the produce should be developed.
- Extension network to help dissemination of knowledge/ technologies (IARI vegetable production technologies) and monitoring their adoption; generation of written material on technology transfer and arrangement for its adoption should be strengthened.
- Commercial nursery production units should be established.
- Contract cultivation/cooperative farming should be encouraged.
- Model production farms with quality considerations for fresh vegetable export should be established.
- Micro-irrigation (drip, sprinkler), and fertigation systems need to be promoted.
- Reclamation and conversion of unutilised cultivable land, waste land, fallow land, and riverbed and *diara* land area into scientific vegetable cultivation.
- Development of market information system to regulate supply, price, and efficient sale of the produce, establishment of Farmers' Sale Points and promoting farmer based cooperative marketing.
- Establishment of organic vegetable farms for elite class and for selected vegetables.

Government of NCR Delhi, should establish viable linkages with the concerned government departments of the adjoining states of Haryana, Uttar Pradesh and Rajasthan for implementation of the policy issues in a mission-mode approach. IARI can be the hatching ground to form small projects on different component issues on the suggested lines. Financial institutions should provide credit in the form of short-term soft loans to the farmer. The technologies and improved vegetable seed could be made available by IARI. Expertise in the form of contract vegetable production and sale, can also be taken from Fruit and Vegetable Project of Mother Dairy. Expertise available in the IARI could be utilised for imparting training on various aspects of scientific vegetable cultivation in peri-urban areas. The ultimate aim should be to improve socioeconomic status of the vegetable growers of NCR Delhi, later to be adopted by other metros.

8

Empowerment of Women and Diversification in Agriculture

Introduction

Empowerment in the context of women's development is a way of defining, challenging and overcoming barriers in a woman's life through which she increases her ability to shape her life and environment. It is an active, multidimensional process which should enable women to realise their full identity and power in all spheres of life.

Since time immemorial, women have played, and continue to play, a key role in conservation of basic life support system, such as, land, water, flora and fauna. Rural women play a crucial role in agricultural development and allied fields, including crop production, livestock production, horticulture, post-harvest operations, fisheries, etc. Without total intellectual and physical participation of women, it is not possible to achieve the goals of rural uplift.

In spite of all this, it is unfortunate that because of centuries of inertia, ignorance and conservatism, the actual and potential role of women in the society has been ignored, preventing them from making their rightful contribution to social progress. It is also because of distorted and/or partial information about their involvement in agriculture and consequent devaluation of their contribution that they are denied their rightful status as active producers in agriculture and access to developmental resources and services contributing to their marginalisation. Women must be empowered by enhancing their awareness, knowledge, skills and technology use efficiency so that agricultural production multiplies at a faster pace, environmental degradation is reduced and conservation of resources is practised earnestly, thereby, facilitating overall development of the society.

Policy Paper 11, September 2001.

All this signifies the importance of having a full understanding of the role and contribution of farm women so that the extension services may accordingly be devised and geared to fully integrate them in agriculture by better serving their specific needs and interests. It is the concern for their integration in the process of agricultural development, which has today become instrumental for policy debates, research initiative and organisational efforts from women's perspective.

Sectoral Concerns and Constraints

Women have to do jobs that are time and labour intensive such as sowing, transplanting, weeding, interculture, harvesting, threshing, and post-harvest operations like, shelling, cleaning, grading and processing. They also contribute to decisionmaking processes for crop production, seed production and management, post-harvest management of agricultural and horticultural produce, biomass utilisation, livestock management, marketing and financial management. In addition, women take up small ventures to generate additional family income. Therefore, the role of women in agriculture and allied occupations, and household activities needs proper recognition.

Women constitute a major component of agricultural workforce. They have, however, lagged behind in use of improved crop production and processing tools and machinery. Now, a number of gender neutral and simple devices/equipment for crop production and processing have been developed. Their adoption by women will not only help improving agricultural production but will also elevate the status of women through better jobs and greater role in the economy of the family.

In addition to their role in agricultural production, women are gainfully employed in agribased allied activities like dairying, animal husbandry, poultry, goatery, rabbit rearing, beekeeping, floriculture, horticulture, fruit preservation, post-harvest technology, value-added food products, etc.

The National Academy of Agricultural Sciences organised a National Seminar,[1] where the participants considered these aspects critically and made the following observations.

---

1. Seminar on Empowerment of Women in Agriculture, from March 4-6, 2001, at Rajendra Agricultural University, Patna.

Dairying

Dairying as a profession fits very well in the framework of rural economic activity. It is compatible with inherent routine chores performed by women. Women's involvement in dairying will not only provide them employment but will also bring in social change. The National Dairy Development Board has steered a cooperative development programme designated to promote and support self-reliant autonomous, member-owned and controlled cooperatives. This has helped in recognising the very important role played by women in dairying and to raise the potential of women in contributing positively to the evolution of dairy cooperatives and women's leadership in these areas.

Beekeeping—An Ideal Enterprise for Women

Beekeeping is an ideal, economically-viable hobby. A few bee colonies (boxes) can be kept in a kitchen garden or backyard of the house. The honey produced therefrom adds to the income of the farmer.

Beekeeping can be taken up on a commercial scale now, as we have available with us the high yielding Mellifera type, commonly called Italian bee. Also, beekeeping requires less labour and attention and can be taken up as a profession by the farmwomen. There is less initial investment but returns are quick.

Beekeepers should be encouraged with good support price and easy marketing, with appropriate loan for this low cost adventure. The farm women, families and unemployed youth can be encouraged to adopt this hobby which could turn into a large scale profitable business.

Mushroom Cultivation—A Women Friendly Profession

Mushroom growing is one agricultural activity in which women can play a vital role without sacrificing their household responsibilities. Mushroom cultivation is simple, low cost, suitable for rural areas, is labour intensive and can provide employment in both the semi-urban and rural areas. Mushroom cultivation will improve the socioeconomic condition of farmers, families and solve employment problems of both literate and illiterate, especially women.

Rural Based Low Cost Nutritional Foods as an Enterprise for Women

A large number of low cost nutritional products grouped under infant food, health beverages, nutritious snacks, convenience foods (even for people on fast), speciality foods (for patients under stress condition) and home preserves have been developed which require simple processing methods and have the ability to fulfil the nutritional requirement of our population, especially the vulnerable segment of our society. Babycorn, quality protein maize, nutritional grains, etc., are the gift for people under stress conditions and can be used to develop a number of products. These products are a great source of rural entrepreneurship, particularly suited to women, due to low capital requirements.

Coconut Sector

Coconut plantation plays an important role in the rural economy by generating income and employment. It is a major sector, which is women dominated. The role of women in coir sector is as high as 80 per cent and 60 per cent in coconut processing and broom making, respectively. Involvement of women in coconut based handicrafts is up to 40 per cent. Women generally work as groups and hence, there is need to organise women's societies with a view to empower women socially and economically.

Lac Cultivation

Lac cultivation is one of the supplementary sources of income in most of the tribal areas of the country, mainly Jharkhand, Chhattisgarh, Orissa and West Bengal. There is large involvement of women in lac cultivation, processing and lac-based small scale cottage industry like sealing wax, bangles and ornaments making, etc., where women give more than 60 per cent physical input. Women in lac-processing factories share around 55 per cent of the total skilled labour input.

Household Appliances

Improved *chulha*, biogas, solar cookers, solar heaters, photovoltaic energy operated appliances, etc., have great scope of adoption by women. This must be encouraged and if necessary, subsidised.

Agroprocessing for Additional Income

Agriculture as a source of cheap raw material cannot survive alone. There has to be some kind of processing to add value to farm produce, which will lead to better income and employment of women. Many simple farm-produce-processing technologies have been developed using minimum equipment and small investments. Women should be trained for handling these equipments/gadgets.

Rural Entrepreneurship

There is need to design programmes which could gainfully utilise the services and skills of women in relation to their involvement in agribased allied activities. This would also help in generating self-employment and rural entrepreneurship. Some of the ventures would include agriservice centres' sale of quality seeds and other agricultural inputs; advisory services and consultancy; village-level marketing; multipurpose warehouses and controlled atmosphere storage; hiring of implements; micropropagation; hatcheries for fishery; production of biofertilisers; biocontrol agents; beekeeping; livestock feeding and health management and artificial insemination; food processing and testing units; post-harvest management units; social agroforestry; agri-horticulture; cultivation and management of medicinal and aromatic plants; utilisation of crop residues; mushroom cultivation; poultry; piggery; goatery; agribased handicrafts; child health care centres; use of renewable and unconventional sources of energy; low cost household equipments and appliances etc.

Entrepreneurship skills can help in building confidence and self-awareness through income generation. Various activities can be taken up by rural women for income generation while being at home and by organising themselves in 'self-help groups'.

Financial Institutions

Various economic activities, investment outlet enterprises are needed to create employment opportunities including self-employment avenues and income generation, particularly for downtrodden women in rural areas. For financial support to these ventures, a rural network is required to provide credit on appropriate and easier terms. Regional rural banks should continue to play an important role in empowerment of rural women through various credit schemes.

Various financial institutions have special financial schemes for women. These include KVIC, NABARD, cooperative societies and regional rural banks. They provide loan opportunities to rural women for dairying, poultry, beekeeping, food processing and preservation, mushroom cultivation, spices processing, potato chips, *dalia* and other cottage industries. These agencies provide loans to women up to five lakhs without any security.

Other facilities include, 13.5 per cent rate of interest up to a loan of Rs. 2 lakhs; relaxation of registration and respective fees related to land mortgages involved in security for a loan in rural areas up to the limit of Rs. 3 lakhs; convenient and easy installments in case of long term loans and provision of immediate cash under a limit as working capital; loans for housing and consumption loans to raise status of rural women; and special credit schemes for enterprises related to computers and other electronic equipment at easier terms.

The Group finally made the following recommendations.

Recommendations

Empowerment of women, demands recognition of their rightful place in the decisionmaking process. Women should have knowledge and resources, greater autonomy in decisionmaking, greater ability to plan their lives, greater control over the circumstances that influence their lives and free them from shackles imposed by prevailing customs, beliefs and practices. They should become active members in development process rather than mere beneficiaries of programmes and services.

- Essential to empowerment is acquisition of knowledge and skills in modern technologies such as soil testing; use of high yielding varieties; scientific use of seeds bio- and chemical-fertilisers, pesticides, water, etc. Compilation and documentation of relevant technologies and a network for technology dissemination is essential.
- Small entrepreneurial outlets involving agribased allied professions should be created to ensure self-employment for rural women.
- Planners, social welfare agencies and women's development organisation should provide a rightful place to women. It is necessary to expand networking of support services so that

women are freed from some of their gender imposed roles.

- The NARS, ICAR and SAUs have developed several agro-technologies directed at reducing drudgery of farm women. Use of these technologies must be ensured by dissemination of information. Women should also be freed from socio-cultural bindings. Progress towards latter can be improved by obtaining feedback and suitably modifying the procedures in this respect.
- Technologies which are labour saving, drudgery reducing, income generating and productivity increasing should be given wide publicity and their use encouraged through pragmatic extension.
- Infrastructural facilities like road, electricity, cool chain, etc. must be developed in rural areas for increasing entrepreneurial opportunities.
- Women must be provided opportunities to have control over production resources that would lead to better life for their families and their children.
- Women mostly work in groups. There is need to organise women's societies for various activities involving group engagement. A strong and effective women's cooperative movement is desirable to boost economic upliftment process of rural women and families.
- Women's dairy cooperatives' success in Bihar, should be extended to other parts of the country for which a working group of technocrats from milk unions and universities should collaborate to develop integrated approach for empowering women in dairy sector.
- Women should have access to financial resources of the family. They should assume larger role in making decisions related to setting up of small cottage and agribased enterprises.
- Collaboration between university and banking sector for implementation of the findings of research alongwith ease in procedure for granting bank loans to rural women, should be facilitated.
- Various economic activities, investment outlets and enterprises are needed to create employment opportunities which must be

accelerated for the upliftment of rural women. This would need creation of a network to facilitate provision of credit at appropriate and easier terms through rural banks.

- Entrepreneurship Development Centres should be created in agricultural universities to integrate entrepreneurial skill development among rural folk, particularly women for building self-confidence and self-awareness in them through income generation. Women should be made direct recipients of technology to provide access to sources of income.
- Trained rural women should form a mandatory component for all agencies and organisations engaged in rural development. Suitable training methods should be evolved and materials developed, keeping in mind the needs of women farmers and their level of education and awareness. It is important for the trainers to have full understanding of the role and contributions of farmwomen, so that extension services could be devised and geared to serve specific needs and interests of women in agriculture.
- Feedback and follow up mechanisms for rural women's programmes should be strengthened and the National Agricultural Research System should play a major role in this endeavour.
- Media should be encouraged to play a strong role in projecting success stories of rural women entrepreneurs and women farmers.
- Procedures should be devised and implemented for recognising outstanding rural women entrepreneurs and farmers at state and national level.

9

Gender, Science and Agriculture: Ten Years after Beijing Conference

Introduction

The United Nations Fourth World Conference on women, held at Beijing was unique. The Beijing Declaration and Platform for Action, adopted unanimously at this conference of women (4-15 September, 1995) by representatives from 189 countries, reflected a new international commitment to the goals of equality, development and peace for all women, everywhere. The Platform was further reaffirmed, as the governments also recognised the need to ensure a gender perspective in their policies and programmes.

The Platform, divided into six chapters, identified 12 'critical areas of concern' to represent the main obstacles to women's advancement. It defined strategic objectives and spelled out actions to be taken over the next five years by the governments, the international community, non-governmental organisations and the private sector for the removal of the existing obstacles.

Over 60 per cent of people live on less than one dollar a day in the rural areas, and more than 70 per cent of the labour force in low-income countries work in agriculture. Women are an overwhelming majority of the one billion people living in abject poverty and illiteracy. Furthermore, by 2050 the world population will be nine billion up from six billion, today. Food demand will double and water use will increase by 50 per cent. If we are to eradicate poverty and have any chance of feeding future generations, we must improve agricultural productivity and rural incomes. Improved agricultural technologies promote economic opportunities for women who do most of the agricultural work in the developing countries.

Nearly 30 per cent of the world's population suffers from some form of malnutrition. A balanced diet, with reference to protein calories and hidden hunger caused by deficiency of micronutrients like iron, zinc, iodine and vitamin A in diet, is of great concern. Those who do not get enough energy or key nutrients, cannot sustain healthy and active lives. It greatly reduces the immunity, which in turn results in devastating ailments like HIV/AIDS. According to a FAO report, 95 per cent of people living with HIV/AIDS are in the developing countries including Africa. HIV infection rates are 3 to 5 times higher in young women than in young men. Once a largely urban problem, HIV/AIDS has moved to the rural areas. For over 30 years, agricultural research in India has promoted science-based solutions for the farmers. If scientists can provide knowledge and skill to these farm women for economic empowerment, it will go a long way in enhancing food production.

The environmental problems such as tsunami and earthquake have contributed to large permanent migration and could eventually displace hundreds of millions. According to a recent UN report published in a Daily, by 2010 the world will have as many as 50 million 'Environmental refugees'. In another shocking news, Human Rights Organisation, claims that children orphaned by the Kashmir earthquake are being 'adopted' by terrorist groups that hope to train them to fight in the *jehad*.

In India, the government had taken steps for the welfare of women much before the Beijing conference in 1995. The National Commission for Women was set up under the Parliament Act, 1990. The Delhi State Commission for Women was set up under an Act of the Legislative Assembly of the National Capital Territory of Delhi, passed in 1994. All governments and/or other organisations have a women cell chaired by a senior woman official and also a separate women harassment cell for the welfare of the women employees. Recently, on November 14, 2005, government announced an award of Rs. 2000 per month for postgraduate education to single girl child of a parents.

The Tenth Five Year Plan according to the first Report of National Commission for Farmers (NCF) entitled "Serving Farmers and Saving Farming" calls for a paradigm shift from food security at the national level to nutrition security at the level of every child, woman and man. Women farmers and farm labourers must get access to credit, knowledge, skill, technological and marketing empowerment. NCF has recommended National Board for "New Deal for Women in

Agriculture" headed by the Union Minister for Food & Agriculture, Government of India.

With Horticulture Board coming into force, post-harvest management requires adopting the well known concept of 'Packing House' in each production zone, equipped with facilities for collection, clearing, grading, packaging, precooling, storage and transport. Training of the Farm-Women in desired skills for these various jobs will help them in attaining economic empowerment.

The trade of agroproducts requires international quality measures to be adopted. Sanitary and phytosanitary (SPS) measures are highly inadequate. Several food consignments have been rejected in recent past due to stringent Codex Alimentarius food safety standards imposed by the importing developed countries. There is need for urgently launching a quality and food safety literacy movement in the villages and for strengthening the SPS infrastructure. There is a recommendation to establish a Food Safety Council of India, chaired by the Union Ministry of Agriculture. Both women and men can be useful in managing all aspects of crops, livestock, fisheries and herbals.

The pursuit of new scientific knowledge and its use in service to the society requires the talent, perspective and insight that can only be assured by broadening participation in the science and technological workforce. Despite advances made in the proportion of women choosing to pursue science careers, women continue to be significantly under-represented in almost all science fields and constitute only approximately 25 per cent of the science workforce at large. In India, women scientists are employed in institutions like CSIR, DAE, DBT, DST, ICMR and ICAR. In IARI, out of 400 scientific staff, women scientists are 85 (22 per cent) and about 65 are technical staff (20 per cent). Several technologies developed by the scientists must reach farm women to derive the maximum benefit.

Academic institutions of higher learning play a pivotal role in preparing the science workforce. Faculty members and academic as well as administrative leadership at these institutions serve as intellectual, professional, personal and organisational role models that shape the expectations of many prospective scientists. Currently, women constitute less than 21 per cent of science women faculty in last few years in the colleges and universities, and hold even smaller percentage of high-ranked positions. Persistent under-representation of senior women faculty members is likely to affect women student's critical relationship with mentors, full participation as members of research and education teams, and self-identification as potential researchers. This situation leads to a minimising effect on the number of women choosing to pursue science careers. A number of factors have been hypothesised to account for the lower proportion of women in the senior ranks of science faculties, e.g. differential effects of conflicts between work and family demands, unequal access to resources such as space and supporting facilities, to name but a few. The cumulative effect of such diverse factors has been to create formidable barriers to the advancement of women in academic science.

About the Conference

In order to improve the situation of all women, specially of the most disadvantaged groups and to take stock of the situation in the advancement of women in the country 10 years after the Beijing World Women's Conference 1995, a two-day consultation on "Ten Years after Beijing: Gender, Sciences and Technology" was organised by the National Academy of Agricultural Sciences and the National Commission on Farmers (NCF).[1]

Beside inaugural and valedictory sessions, the consultation had five technical sessions: (i) Science for Women and Women in Science, (ii) Agriculture, Health, Environment and Food Security, (iii) Civil Society Initiatives—Learning from Successes, (iv) Access to Technology and Skill Empowerment, and (v) The Beijng Platform for Action and the UN Millennium Development Goals: Next 10 Years. Prof. M.S. Swaminathan, President, NAAS and Chairman, NCF, in his introductory address gave an overview of the situation like engendering various programmes and suggested the formation of a National Action and Policy Research Network for Women in Agriculture.

---

1. Organised on November 18-19, 2005, with the Convener, Dr. Gita Kulshrestha, Head, Division of Agriculture Chemicals, IARI and Fellow NAAS at the National Academy of Agricultural Sciences (NAAS) premises. The concept paper on the consultation was circulated in advance to the participants. There were about 75 participants from all over the country, largely women professionals from agricultural universities, home science colleges, government departments, national and international funding agencies and NGOs.

Each technical session had key presentation alongwith panelists, followed by general discussion. At the valedictory session, HRH Princess Maha Chakri Sirindhorn of Thailand was the Chief Guest. The recommendation emanating from the deliberations in various sessions were finalised.

Recommendations

It was well recognised that a people-centric sustainable development should ensure women's equal access to economic resources, land, credit, science and technology, vocational training, information, communication and markets. S and T has to be developed, disseminated and harnessed in a gender-sensitive manner with equal participation by women not only in S and T development, but also by women who are the end users. There is a greater need for self-sensitisation, confidence building and psychological management, among women. To ensure this, the following areas are required to be strengthened.

1. Women in Science

In order to promote and ensure full and equal participation of women in all S and T activities by taking into account the multiple burden on their time and creating suitable support systems to reduce this strain, there is need to:

- Enhance the retention of girls in the leaky pipe-line, especially at the tertiary level, and the mid-career women by bringing in flexibility in the procedures and personnel policies to meet their special needs.
- Assess and draw up appropriate schemes to enhance their professional participation and increase the number of fellowships, opportunities for participation in conferences/ workshops.
- Increase their representation and participation in decision-making and professional bodies.
- Institute many recognition and reward systems for meritorious work.
- Prepare an electronic directory of women scientists and technical personnel as a ready reckoner for facilitating and

enhancing their participation in scientific meetings and decisionmaking committees.

- Promote entrepreneurship among interested women by drawing up innovative schemes which help in the incubation and scaling up of innovative ideas to address the specific needs of women.

There is need for a special venture capital fund for enabling women entrepreneurs to take up a career of self-employment by converting new technologies into market-driven products which also support at the incubation level.

2. Science and Technology for Women

Although, tremendous developments have taken place in the areas of S and T, there are still critical gaps in dissemination of their benefits to the unreached women in the following areas:

2.1 Health, Food and Nutritional Security

- There is a need to engender the proposed National Rural Health Mission.
- Enhance the strategies to address anaemia and hidden hunger caused by micronutrient deficiencies, particularly, iron, iodine, zinc, and vitamin A and B-complex vitamins.
- Enhance food and nutrition security by paying attention to availability, access and absorption, and also to food sovereignty.
- Promote conservation and revitalisation of the traditional landraces and ethnic foods inclusive of wild edible foods to diversify the food basket.
- Create awareness of and ensure access to preventive measures (i.e. vaccines) and develop low cost diagnostic kits for early detection of Cancer, HIV/AIDS, Diabetes, Tuberculosis, etc.
- Promote construction of toilets by making it mandatory while giving house loans.
- Minimise occupational hazards by installing appropriate environment and other safety measures at workplace.

2.2 Agriculture

- Recognise that women play a critical role in the conservation-cultivation-consumption-commerce chain.
- Land and productive assets are to be held jointly, so as to enable women in agriculture to have access to institutional credit, including Kisan Credit Cards.
- Land reforms should be implemented in a gender-sensitive framework.
- Banks should foster credit and insurance literacy programmes in order to enable rural women to take advantage of credit and insurance programmes.
- There is need for institutions, concerned with gender roles in agriculture, to get together as partners in a longitudinal study of the specific needs of women working in agriculture (i.e., crop and animal husbandry, forestry, inland and marine fisheries, as well as in agroprocessing) with reference to technology, services (including credit and market linkages) and support services (like creches and day-care centres for children). This should be supported by public policies in terms of mainstreaming gender in all agricultural and rural developmental programmes.
- NAAS could facilitate the design of a project for initiating a National Action and Policy Research Network for Women in Agriculture by convening a Project Design Workshop involving interested partners from across various agroecological regions of the country with a long-term commitment in this area.
- Such a National Action and Policy Research Network could deal with all links of the food chain such as conservation of agrobiodiversity and natural resources, cultivation, consumption and commerce.

2.3 Common Property Resources (CPR) and Natural Resource Management (NRM)

- There should be efforts for greater awareness generation to support poor women/marginalised groups in their struggle to protect their existing rights and use of CPR like land, water and other physical resources.

- Provide appropriate technologies and support services to enhance their use in a participatory and productive manner (JFM, Group farming, M&A plants; water use/management and augmentation; in harnessing renewable energy like biomass, biogas, solar and wind).
- Engender the various Acts, especially the BD, PPV&FR and launch a literacy drive to understand them.
- Enhance the participation of women in establishment and management of the community seed, food, water, fodder and feed banks to address the local needs.

2.4 Skill Empowerment and Livelihoods

- Identify appropriate technologies for multiple livelihoods that will enhance incomes and improve the way of life for the poor and vulnerable women.
- Draw up well-coordinated programmes to reach the rural women through well-networked models for technology transfer, demonstration, dissemination and adoption.
- Establish training and mentoring centres for farm women and farm labourers as well as the rural women at households, and SHGs. Capacity building centres on the model of KVKs should be established soon.
- Establish rural technology parks which will also help in adoption of appropriate technologies to address the local problems like reduction in drudgery and occupational hazards, appropriate tools, identification of income generating activities, etc. Such parks are useful in assessing the real needs and providing research agenda to S and T institutions.
- Broad sectors where such intervention is needed, could be horticulture, cultivation of medicinal and aromatic plants; village level agroprocessing and value addition centres, organic farming, etc.
- Identify and promote micro-enterprises based on assured and remunerative markets, low transaction cost and economic viability.
- Easy access to credit and markets to enhance multiple sustainable livelihoods; develop microcredit programmes into 'Livelihood Finance' systems which involve provision of credit

together with appropriate services in the areas of technology, training and trade.

- Financial institutions to address issues like flexibility, sensitivity of the needs and status of the poor women, so that they have an easy access to credit.
- ICT based information service like e-governance, e-agriculture, e-education, e-medicine, e-commerce, etc. should be promoted under the Mission 2007—Every Village, a Knowledge Centre initiative.
- The knowledge centres should have provision for appropriate contacts and follow-up programmes to help those seeking special assistance.

2.5 Generic

- A need to engender the curriculum at the school level and technical education (medicine/engineering/agriculture) level for gender-sensitivity among future scientists and development personnel, and to strengthen science and technology education and communication skills among the community development workers.
- Need to engender all technology mission mode programmes and to conduct periodic gender audit.
- Generate disaggregated gendered data regarding the impact of S and T on the livelihoods of rural women.
- Harmonisation of all activities and schemes coming under different ministries/departments, so that women benefit optimally.
- Establish a network of women scientists and institutions interested in engendering the development through S and T based interventions to develop an end-to-end approach for the various agroclimatic zones.
- Greater focus may be given to extension services in all areas of technology, and a strong cadre of extension workers built up, who may be given frequent exposure for sharing these ideas with the members in the group.
- Ensure adequate resource allocation for women under various programmes.

- Implement equal wages for equal work.
- Create a strong group of men advocates who can help in promoting gender equity in the present development paradigm.
- S and T applications should foster job-led economic growth and not jobless growth. The opportunities for rural women to earn their daily livelihood through mini-enterprises and vending methods of marketing should be safeguarded by introducing engendered employment impact analysis.

To conclude, science and technology should help strengthen the health, nutrition and livelihood security of women. Science and technology should also help in reducing infant mortality rate (IMR) and maternal mortality rate (MMR) and assist in achieving the goal of 'Literacy for All' through the introduction of computer-aided learning techniques.

Section II

Soil Health Enhancement and Fertiliser Use

10

Fertiliser Policy Issues (2000-2025)

Introduction

By the turn of the century, India's population is expected to cross the one billion mark. Pragmatic assessments reveal that between 220 and 240 million tonnes of foodgrains will be required by 2000. With the current food production around 190 million tonnes, India has the compulsive need to raise its foodgrains targets at a rate of more than 6 million tonnes per annum. This obviously is a daunting task. However, in view of our past achievements, and resilience of the system, there is no reason why this phenomenal acceleration in food production cannot be achieved. Technological adoption, using required inputs, especially seeds, fertilisers and water, would be the major determinant of food growth in future. Among these, the most costly input would be fertilisers, whose judicious and need-based use would trigger the process of accelerated growth in Indian agriculture in the coming decades. As such, need for a national policy on fertiliser use, especially considering the emerging paradigms of productivity, profitability and sustainability becomes highly relevant at this juncture. In view of this, the National Academy of Agricultural Sciences (NAAS) based on the discussions of a symposium organised by it in 1996 developed the following policy issues on fertiliser use for the years 2000-2025.

An Assessment of Future Needs

In view of shrinking land and water resources, the singular option for India to meet future challenges on food, fibre, fuel and other needs is through increase in productivity. The only way to achieve this goal is through the scientific and need-based use of fertilisers.

Policy Paper 2, March 1997.

Whether it is irrigated or rainfed agriculture, plant nutrient deficiency is the main impediment to agricultural production. The only way of alleviating this constraint is by increasing the use of fertilisers and manures. The current gap between annual drain of nutrients from the soil and inputs from external sources is 10 million tonnes, which is likely to grow further. The low organic matter content of our soils and poor prospects of recycling of crop residues, add to the importance of inorganic sources of fertilisers in Indian agriculture.

No country in the world, developed or developing, has been able to increase agricultural productivity without expanding the use of fertilisers. India cannot be an exception to this phenomenon. We believe that the future of productive and sustainable agriculture in India lies in increasing the amount and harmonious use of nutrients through inorganic, organic and biological sources and enhancing their use efficiency.

There is a great concern about the adverse effects on soil health as well as productivity due to widening ratio of N:P:K (8.7:2.5:1.0 in 1995-96) due to unsound policy decisions taken earlier, favouring prices of N and ignoring P and K. This has now been corrected to some extent. But without wholesome policies on pricing of fertilisers and of the agricultural commodities, with long-term perspective in view, the impending disastrous effects of nutrient imbalances cannot be ruled out. We feel that steps should be taken to narrow down NPK ratio to 4:2:1.

Keeping is view the conservative population estimate of 1.4 billion and minimum foodgrain need of 301 million tonnes by the year 2025, it will be necessary to use 30-35 million tonnes of NPK from fertiliser carriers and an additional 10 million tonnes from organic and biofertiliser sources. Thus, it will be essential for India to raise the consumption and production of chemical and organic sources of plant nutrients by 2025, to meet these goals. It is interesting to observe that China is already using more than 30 million tonne nutrients from fertiliser sources, whereas in India, we are consuming 13.8 million tonne nutrients with indigenous production of only 11.36 million tonnes. While in 1994-95, China used 309 kg NPK ha^{-1}, India used merely 73 kg ha^{-1} of gross cropped area.

The disparity in fertiliser use in different agroecological regions is the principal cause of the vast differences in crop yields. For example,

Assam is using 12.8 kg NPK ha^{-1} as against 167.3 kg ha^{-1} in Punjab. Strategies and policies are needed to remove this inequality in fertiliser use. Without changing this situation, Indian agriculture cannot realise its full growth potential.

Fertiliser research agenda must change its focus and orientation. Emphasis should be as much on improving fertiliser-use efficiency as on its productivity, profitability, sustainability and eco-friendliness. Though about 70 per cent of fertiliser used in India consists of nitrogenous fertilisers and 80 per cent of it is urea, its use efficiency is hardly 30-50 per cent. The losses of N through nitrification, denitrification, ammonia volatilisation, leaching, and surface runoff are very high, particularly in the rice-based cropping systems. Both research and development agencies need to intensify their efforts in popularising the technologies on increasing fertiliser-use efficiency and minimising N losses. Models of integrated nutrient management for diverse but well-defined agroecological zones and cropping systems should be developed, to provide guidance for rational and efficient fertiliser use.

Organic Manurial Resources

Past impassivity in exploitation of the potential of organic manures, composts and agricultural wastes and their synergistic effect with chemical fertilisers must give way to responsive and more vigorous strategies on their use.

Our soils have very low organic matter content. Therefore, without regular application of organic manures and recycling of crop residues, we cannot hope to maintain and sustain productivity and ensure high responses to NPK fertilisers. We need policies and strategies to prepare and distribute for use in agriculture, bulky organic sources of nutrients after enriching them with chemical fertilisers and using modern technology for reducing their bulk. This assumes greater significance in the coming years as with rapid urbanisation the bulky organic wastes are increasing and their disposal and profitable use in agriculture in rural areas is hampered because of transportability and cost constraints.

Green Manuring and Incorporation of Legumes

The green manuring practice besides improving N economy also has many other beneficial effects on soil health. Greater potential of green manuring lies in humid, sub-humid areas and in plantation and horticultural crops. It is also promising for agroforestry and under irrigated farming, specially in intensive cropping systems based on sugarcane, rice, potato, etc. It can result in saving of 40-60 kg N ha^{-1}. In dryland agriculture the legumes as a part of the crop rotation, as sole crops, intercrops and relay crops, are important both for improving productivity and sustainability of agriculture. Location-specific technologies and strategies need to be developed for different agroecological regions depending on availability of water, labour, seed and the critical period in the cropping systems. Special drive on production and distribution of seed of crops for green manuring is emphasised.

Biofertilisers

The responses to biofertilisers are environment, crop and management specific and generally vary: Rhizobium > blue green algae> azospirillum > azotobacter > P solublisers. Their use in soils of low fertility and for those areas and systems which use low amount of fertilisers is most relevant and needs encouragement. The use of BNF is for partial replacement of nitrogen fertilisers and can result in saving of 20-40 kg N ha^{-1} and this is not able to meet the full requirement of N for high yields. Moreover, unless the quality of biofertilisers and their shelf life are improved and deterioration in the quality during storage, transport and distribution is minimised, the desired results cannot be realised. Intensive research using biotechnology and other emerging frontier sciences is needed to develop more efficient biofertilisers, to supplement the fertilisers for productive agriculture.

Rhizobium cultures are most relevant to legumes and their greater benefit can be realised in environments where such crops and their varieties are newly introduced, e.g. soybean and berseem, and N status is low. The blue green algae and azolla to rice-based system, VAM-micorhizae fungi for tree-based system and phosphorus solubilisers for high phosphorus fixing soils, are most relevant.

Phosphorus and Potassium (P and K)

Soil test summaries show that 50 per cent soils are low and 48 per cent medium in available P and 13 per cent low and 53 per cent medium in available K. There is also growing evidence of increasing deficiency of P and K, aggravated by the disproportionate application of higher doses of N in relation to P and K. The recent aberration in prices of fertilisers has also changed the NPK ratio of fertiliser use from 5.9:2.4:1 in 1991-92 to 8.7: 2.5:1.0 in 1995-96 indicating less use of P and K. This unhealthy trend, needs to be reversed through development of appropriate strategies and policies to avert disastrous consequences.

Sulphur

There is a growing evidence of increasing responses to S for oilseeds, pulses and legumes and high-yielding cereals. Presently, the gap between S removal and its addition to the crops is estimated to be about 0.5 million tonnes available S equivalent and it is likely to go beyond 2 million tonnes by the year 2025. Strategies and policies need to be developed to reduce this gap and to encourage more use of S either through fertilisers containing S as component or by-products of fertilisers and sugar industry such as phosphogypsum and press mud, respectively. Pricing structure of S containing fertilisers, also needs consideration.

Micronutrients

The micronutrient deficiency in crops is growing rapidly both in extent and intensity with the Green Revolution and it has been observed that 47, 4.8, 11.5 and 4.0 per cent soils are deficient in zinc, copper, iron and manganese, respectively. The deficiency of boron and molybdenum also have been recorded in many areas. The deficiency of Zn is the most widespread, followed by that of iron, copper, manganese and boron and it is anticipated that with higher yields and more intensive agriculture the micronutrient deficiency will increase both in amount and extent. Estimates based on sufficiency approach show that the requirement for zinc will be 324, iron 130, copper 11, boron 3.9 and manganese 22 thousand tonnes by the year 2025. Except for Punjab and Haryana which used about 4000 tonnes zinc and 120 tonnes iron, during 1995-96, there are no statistics available about the amount of micronutrients being used by other status though the fact remains that many of them are using various amounts. While serious efforts are needed to collect the reliable data about micronutrient needs and use, the policies and strategies are required to ensure availability of necessary micronutrient carriers and to protect farmers' interest and environment. It also needs to be stressed that, through increased use of organic manures some of the micronutrient requirements can be met.

Soil Amendments

Gypsum for alkali soils and liming material for acid soils are undoubtedly the most needed amendments. Strategies and policies are required to encourage the use of these amendments or other alternative sources to ameliorate and enhance the productivity of these problem soils. The critical aspects such as transport facility and incentives for use of these amendments need special attention.

Soil Testing

India has a network of about 472 (mobile 94) soil-testing laboratories. However, there is sufficient room to improve the quality of advice and impact on rational use of fertilisers. A critical review of soil-testing system is warranted. Strategies and action plans are necessary to modernise and improve the quality and service provided by soil-testing laboratories. Soil testing should include plant tissue testing also to provide advisory service for general as well as for specialised farming including horticulture, floriculture and plantation crops. Selected laboratories should be developed for such exclusive advisory service in each agroecological region on payment basis. Initiation of a national networking arrangement is absolutely essential and the Indian Council of Agricultural Research (ICAR) should be encouraged to provide the leadership for quality control and advisory service backup.

Fertiliser and Soil Health

The critical analysis of the presently available information gives little evidence of any deterioration of soil health or pollution of groundwaters under balanced and integrated fertiliser use. However, possibility of any ill-effect arising from a long-term unbalanced fertiliser use, such as N alone must be guarded against.

Indian soils are of low fertility and productivity and also assuring national food security demands, balanced and integrated nutrient use as well as monitoring changes in soil health. Thus, building up a national networking arrangement to coordinate the studies on fertiliser use and soil health in different agroecological regions is of highest priority.

Fertiliser Pricing Policies

The *ad hoc* changes in pricing of fertilisers, subsidies and decontrol of P and K fertilisers caused a sudden and disastrous effect on the ratios of consumption of NPK. A long-term sound policy and mechanism to encourage fertiliser consumption growth, and balanced and efficient fertiliser use, needs to be developed.

The question of subsidy on fertilisers should be viewed from a national perspective of food security, nutritional security and national independence. A gradual change is preferable to sudden *ad hoc* change in subsidy. There is need for a critical review of pricing policy on fertilisers, agricultural commodities and of subsidy by a panel of experts, agronomists, soil scientists, plant scientists, agricultural economists, fertiliser industry and farmers' representatives.

In view of the special case of phosphates and potassic fertilisers, the suggestion is reiterated to make long-term contracts with the exporting countries for their uninterrupted supplies at affordable prices. We believe that the need for import of phosphate and potassic fertilisers will continue to increase in future and there will always be a crisis in meeting the growing fertiliser needs.

Strategies and policies have to be evolved for assured and regulated supplies of essential quantities of nutrients both from internal and external sources as the future of agriculture depends on them.

Fertiliser, being the key to national food security and agricultural development, must get the highest priority in any strategy of national planning.

11

Low and Declining Crop Response to Fertilisers

Introduction

With an annual compound population growth rate estimated at 1.97 per cent, nearly 20 million new mouths equivalent to the present population of Australia are added to India each year. Providing two meals a day to its people will always remain one of the major concerns of the government. Foodgrain demand of India is estimated at about 300 million tonnes per annum by 2020, necessitating an increase of about 91 million tonnes (mt) from the estimated 209 mt production for 2005-06. Since there is no likely prospect of any further increase in the area under cultivation over the present 142 million hectares, much of the desired increase in foodgrain production has to be attained by enhancing the productivity per unit area. The productivity of rice has to be increased from the present 2077 kg/ha to 2895 kg/ha by 2020, with an average increase of about 5 per cent per annum. The productivity of wheat has to be increased from the present 2713 kg/ha to 3918 kg/ha with an average increase of about 7.5 per cent per annum, while the productivity of pulses has to be increased from the present 637 kg/ha to 1282 kg/ha with an average increase of about 5.3 per cent per annum. On the contrary, the productivity of most foodgrain crops except wheat has shown a negative growth rate of 0.72 to 1.84 per cent per annum during the period 2000-01 to 2002-03. This poses not only a matter of great concern but also a formidable challenge.

The foodgrain production, following the Green Revolution in 1969-70, was 99.5 mt and nearly doubled by the end of the last century. The highest average annual increase of 6.1 per cent in foodgrain production was recorded during 1980s; from 110 mt in 1979-80 to 171

Policy Paper 35, May 2006.

mt in 1989-90, but the annual increase in foodgrain production during 1990s dropped to 1.5 per cent. The fact that the fertiliser was the key input in augmenting foodgrain production after the availability of the seeds of high yielding crop varieties, is evident from the increase in fertiliser ($N + P_2O_5 + K_2O$) consumption from 1.98 mt in 1969-70 to 18.07 mt by 1999-2000. Nevertheless, the average annual increase in fertiliser consumption witnessed a declining trend in these three decades, as reflected by 16.5 per cent during 1970s, 12.04 per cent during 1980s and only 5.6 per cent during 1990s.

A simple regression analysis between the foodgrain production and fertiliser consumption during 1960-61 to 1999-2000 showed that the partial factor productivity of fertilisers has been continuously declining. The data available from some centres under the Project Directorate of Cropping Systems Research (PDCSR), Modipuram also indicate a reduction in crop response to fertiliser application, specially when balanced fertilisation is not practiced. This is supported by the fact that the farmers in the rice-wheat cropping system belt (specially Punjab, Haryana and Western Uttar Pradesh) are forced to apply more and more fertiliser to obtain the same crop yields as in the preceding years. The data from the trials on the farmers' fields conducted by the PDCSR, Modipuram during 1999-2003, showed that the average response of cereals to fertiliser was 8-9 kg grain/kg fertiliser. The efficiency of fertiliser nitrogen is only 30-40 per cent in rice and 50-60 per cent in other cereals, while the efficiency of fertiliser phosphorus is 15-20 per cent in most crops. The efficiency of K is 60-80 per cent, while that for S is 8-12 per cent. As regards the micronutrients, the efficiency of most of them is below 5 per cent. Attention to this serious problem of low and declining crop response to fertiliser was drawn by Professor M.S. Swaminathan, President, NAAS in his Presidential address on June 5, 2005, New Delhi.

In view of the above disturbing agricultural scenario, NAAS organised a two-day Session on Low and Declining Response of Crops to Fertilisers.[1]

1. Session organised on February 20-21, 2006 with Dr. Rajendra Prasad, Ex-ICAR National Professor and Ex-INSA Sr. Scientist as the Convener. It was attended by 41 experts including some of the Directors of leading National Agricultural Institutes and Professors/Scientists/Officials of the State Agricultural Universities/Ministry of Agriculture/FAI, etc. The deliberations included four technical sessions with key and panelist presentations followed by discussion. Dr. R.B. Singh, Member, NCF, GoI in his inaugural address emphasised the need for soil health card, and increasing the farm profit by improving the crop response to fertilisers. Dr. J.S.P. Yadav, former Chairman, ASRB in the wrap up/concluding session listed several causes of declining/low crop response to fertilisers and summarised recommendations made thereof.

It was vividly brought out in the deliberations that the major factor responsible for the low and declining crop response to the fertilisers was the continuous nutrient mining of the Indian soils without adequate replenishment to the desired extent. It is estimated that about 28 mt of primary plant nutrients are removed annually by crops in India, while only 18 mt or even less are applied as fertiliser, leaving a net negative balance of about 10 mt of primary plant nutrients (NPK).

An analysis of the data pertaining to rice-wheat cropping system from 24 research stations in the Indo-Gangetic plains revealed that the rice yields are declining more rapidly as compared to wheat yields, and soil K depletion seemed to be a general cause, whereas P and Zn depletion emerged as a cause at some centres. Similarly, a study at ND University of Agriculture and Technology, Faizabad showed that the yield decline in both rice and wheat was the highest when N alone was applied at 120 kg/ha. As the depletion of native soil P increased with time, the response of both rice and wheat to applied P showed an increase after 10 years of cropping. Also, as the native soil K decreased over time, the response to applied K started to increase after 5 years in both rice and wheat. In wheat, it continued to increase even after 20 years, while in rice the increase in response to K stopped after 15 years. Such reports are several in number.

Furthermore, the soils are also getting continuously depleted of secondary plant nutrient S and micronutrients. The marked deficiencies of S and Zn are widespread in the country and significant responses to application of these nutrients are well documented. The deficiencies of B, Fe and Mn are also being reported from some parts of the country. Thus, inadequate and imbalanced fertilisation is a major causative factor for low and declining crop response to fertilisers. The need for integrated nutrient management including the use of organic sources such as farmyard manure, rural and urban compost, vermicompost, green manures, inclusion of legumes in the crop rotations, growing of dual purpose short duration legumes such as mungbean in the summer months, in the rice-wheat cropping system, and biofertilisers was emphasised for meeting a part of the plant nutrient needs of crops.

The major factor, responsible for the low response of crops to fertiliser nitrogen, is its low use-efficiency, especially in case of rice crop where it is only 30-40 per cent of applied N due to various N loss mechanisms, namely, surface run-off, ammonia volatilisation, leaching and denitrification. In 1995, the global estimate of nitrogen loss, from the applied fertiliser N in 1995, through ammonia volatilisation was 11.2 mt (14.45 per cent), while that through NO and N_2O through denitrification was 1.5 mt (1 per cent). India's contribution to these losses could approximately be 10 per cent of the total. Ammonia added to the atmosphere leads to the acid rain, while NO and N_2O are responsible for the depletion of ozone layer in the atmosphere. In addition, nitrates leach to the groundwater and lead to the nitrate pollution of drinking water which is injurious to health. An increase in the nitrate content due to heavy N fertilisation has been reported in Punjab by the researchers at the Punjab Agricultural University. There is, therefore, an urgent need to develop more efficient nitrogen fertilisers.

Some slow-release N fertilisers developed in the foreign countries are used on a limited scale in case of high value crops due to their high costs. Similarly, nitrification inhibitors have also been developed in some foreign countries as well in India, but their high cost and application problems with some, such as N-serve, are constraints to their use. Studies in India showed that coating urea with *neem* cake/ oil/*neem* bitterns could retard the nitrification for 2-3 weeks. A technology for coating urea with *neem* oil emulsion at the factory scale is now available in India and the Government of India has permitted National Fertilisers Limited (NFL), Shriram Fertilisers and Chemicals and Indo-Gulf Fertilisers to manufacture and market the *neem*-coated urea on an experimental basis. The results obtained with *neem*-coated urea on the farmers fields have shown a 2 to 10 per cent higher grain yield of rice as compared to uncoated urea. Taking a mean value of 5 per cent increase in nitrogen use-efficiency, a saving of five lakh tonnes of nitrogen per annum at the current level of consumption of 10 mt of N is estimated, which is equivalent to the combined annual production of fertiliser N at Panipat and Bhatinda plants of NFL or the annual production of fertiliser N at Jagdishpur plant of Indo-Gulf. There is a heavy demand of *neem*-coated urea in several states of India, and therefore, production of *neem*-coated urea needs to be enhanced.

Recommendations

Major recommendations emanating from the deliberations are given here under four heads, namely: (1) causes for low and declining crop response to fertilisers, (2) agenda for research, (3) rejuvenation of agricultural extension, and (4) policy decisions.

1. Causes for Low and Declining Crop Response to Fertilisers

1.1 Nutrient Supply and Soil fertility

- Continuous use of fertiliser N alone or with inadequate P and K application leading to mining of native soil P and K.
- Continued practice of intensive cropping systems like rice-wheat with high yielding varieties even under recommended NPK use, impoverishing soils of secondary and micro-nutrients specially S, Zn, Mn, B and Fe.
- Use of high analysis fertilisers and inadequate addition of organic manures resulting in widespread deficiencies of S and micronutrients.
- Fertiliser application mostly not based on soil-test values.
- Inappropriate time and method of fertiliser application.
- Excessive use of irrigation in rice-wheat cropping system, sugarcane and other heavily fertilised crops leading to leaching of nitrogen and other plant nutrients.
- Inadequate availability of appropriate kind of fertilisers at the right time.
- Antagonistic reaction between some plant nutrients.
- Low status of soil organic carbon.
- Subsoil impedance to plant root system restricting nutrient uptake.
- Soil degradation due to high salinity/sodicity/acidity/waterlogging, affecting nutrient availability.
- Lack of adequate and quality soil testing facilities and meagre availability of fertiliser recommendations under aberrant weather conditions.
- Environmental degradation, having negative impact on below-ground biodiversity, especially agriculturally important microorganisms.

1.2 Seed

- Non-availability of sufficient seeds of high yielding varieties of crops at affordable price and at the appropriate time.
- Lack of more efficient nutrient using genotypes.

1.3 Agronomic Practices

- Delayed sowings/plantings.
- Low seed rates resulting in poor crop stands.
- Poor weed management.
- Inefficient tillage.
- Inefficient irrigation and rainwater management.
- Large scale monoculture.
- Lack of consideration of previous cropping in the same field.
- Lack of capturing water-nutrient synergic interaction.
- Inadequate plant protection.

1.4 Weather Aberrations

- High intensity rain leading to nutrient loss.
- Abnormal high/low temperature.

2. *Agenda for Research*

- Nutrient budgeting in different crops/cropping system for different agroecological zones needs to be done.
- Site-specific fertiliser application practices (SSFAP) based on soil-test recommendations need to be developed for various crops and cropping systems in different agroecological zones of the country.
- For sustainable agriculture, integrated plant nutrient supply systems (IPNS), involving FYM, compost, vermicompost, green manures, dual purpose legumes (e.g. summer mung in rice-wheat cropping system) and biofertilisers, need to be developed for various crops and cropping systems in different agroecological zones of the country.
- SSFAP and IPNS also need to be developed for different agro-forestry systems, integrated farming, orchards and peri-urban-

agricultural systems involving vegetables and flowers.

- Fertiliser-irrigation water interaction studies need to be intensified.
- Considering the fact that about 40-50 per cent of the applied fertiliser nitrogen is not only lost from the soil-plant system, but also adds to the environmental pollution, more efficient nitrogen fertilisers, involving materials such as *neem* oil or cake or indigenously developed nitrification inhibitors need to be developed for field adoption.
- Customised fertilisers based on soil-test-crop response studies for different regions of the country are to be developed.
- Research is needed on the use of soluble fertilisers/fertiliser solutions specially in horticulture, vegetable crops and floriculture for drip irrigation system.
- Since fertiliser nitrogen is used in the largest amounts and its efficiency is low, development of accurate and site-specific recommendations for nitrogen, based on soil-test, plant-tissue analysis, use of chlorophyll meter and other techniques, if any, is urgently needed.
- Appropriate fertiliser recommendations for different crops grown in the dryland areas need to be developed *vis-à-vis* weather aberrations.
- There is need for developing a database at IASRI, New Delhi, for all the field experiments conducted on crop response to fertilisers in the country. IASRI should develop a suitable pro forma for reporting of such data, which the ICAR should circulate to all agricultural universities and ICAR/state institutions for compliance.
- The optimum doses of N, P and K need not be computed from the response curves. These need to be computed by fitting response surfaces and using soil fertility status.

3. *Rejuvenating Agricultural Extension*

- It is generally presumed that the farmers know all about fertilisers. This is a myth that needs to be broken. One of the major constraints is the farmers' lack of awareness about the concept of balanced fertilisation not only about major elements but also involving secondary and micronutrients.

- Extension machinery needs to be geared up and rejuvenated/ revived.
- Demonstrations are needed to emphasise the need for balanced fertilisation of N, P, K, S, Zn, B, Fe and other micronutrients and on Integrated Plant Nutrient Supply System (IPNS).
- Demonstrations on the role of fertilisers for increasing crop yields under dryland agriculture conditions are also needed.
- The farmers need to be educated about the role of micronutrient plant nutrients in human and animal health. This will encourage the application of micronutrients in crop production.
- Demonstrations are also needed to emphasise on the time and method of fertiliser application.
- The fertiliser industry should play an active role in conducting demonstrations in respect of the balanced fertilisation, IPN, as well as on time and method of fertiliser application.

4. Policy Decisions

A long-term pragmatic policy for all fertiliser nutrients including secondary and micronutrients is essential to achieve balanced nutrition to overcome soil fertility and health degradation, and improve crop productivity on a sustainable basis. This includes:

- Adequate number of well equipped soil-testing laboratories manned by well trained personnel to take care of secondary and micronutrient analysis.
- Soil health cards for all the categories of the farmers throughout the country for the purpose of periodical monitoring of soil fertility status and as one of the sources for obtaining agricultural credit.
- Quality control for fertilisers including micronutrient fertilisers.
- Standardisation of organic manures.
- Providing price incentives to the fertiliser manufacturers opting for the manufacture of value-added fertilisers such as *neem*-coated urea and balanced customised fertilisers.

- Strengthening collaboration between the fertiliser industry and national institutes engaged in developing soil test-based and site-specific nutrient recommendations to develop soil- and crop- specific quality fertilisers.
- Provision of a transparent regulatory framework for ensuring supply of tailormade crop and soil-specific quality fertilisers to the farmers and corresponding provisions in the Fertiliser (Control) Order, GOI.
- More proactive State Fertiliser Review Committees having wider representation of the scientists and the industry and empowered with quality control for blended and value-added fertilisers.
- Ensuring credit to the farmers at low interest rates for the purchase of fertiliser, seed, pesticides, and agricultural equipment.
- Providing subsidies to promote cultivation of legume crops for green manuring/grain to build-up soil fertility and to meet the shortage of pulse requirements in the country.
- Establishment of a national network on Integrated Fertiliser Development for Sustainable Agriculture by the ICAR/ Ministry of Agriculture.
- Establishment of Coordination Cell for coordinating the activities of different ministries of Government of India related to efficient use of fertilisers such as Ministry of Agriculture, Ministry of Fertilisers and Chemicals, Ministry of Irrigation, ICAR and National Commission for Farmers.

Conclusion

Fertiliser has been and will continue to be, the key input for achieving the estimated foodgrain production goals of the country. Since increase in foodgrain production is possible only through the increased productivity per unit land, an all out effort is needed to increase the crop response to fertilisers. Some of the suggested measures are balanced and adequate N, P, K, S, Zn, B and Fe (and any other deficient nutrient) fertilisation, Integrated Plant Nutrient Supply System (IPNS), development of quality soil-test facilities at district/ block level, timely availability of desired fertiliser materials, availability of good quality seed of the recommended crop varieties, implementation of recommended agronomic practices and availability of low interest credit to the farmers. Considering the fact, that about 40-50 per cent of the applied fertiliser nitrogen is lost by ammonia volatilisation, leaching, run-off and denitrification, development of more efficient nitrogen fertilisers such as *neem*-coated urea needs to be encouraged by providing price incentive to the fertiliser manufacturers. Customised soil and crop specific fertiliser materials need to be developed for major cropping and farming systems in different agro-eco regions. Creation of a national network on Integrated Fertiliser Development for Sustainable Agriculture, is urgently needed.

12

Policy Options for Efficient Nitrogen Use

Introduction

Nitrogen (N) is necessary for all forms of life and is a crucial component for increasing production of food to feed the continuously increasing human population. However, barring N_2, which cannot be directly used in agriculture, all reactive forms of nitrogen (urea, ammonia, nitrate and their derivates) used to produce food can threaten the environment. Reactive N species (NOx) are also formed from fossil fuel consumption by industries and vehicles, with attendant environmental implications. Therefore, the challenge now facing Indian agriculture is to further enhance the productivity of agricultural system without adversely impacting environment and ecology. This is the basic dilemma for N management policies. The increase in food production has been a hallmark of the Green Revolution achieved through the combination of better seeds with responsiveness to water and fertiliser, particularly nitrogenous fertilisers. N based fertilisers constitute a major fraction, nearly 70 per cent, of the total fertiliser material. The increase in fertiliser N-use in the last three to four decades has resulted in unprecedented increase in agricultural production in northwestern India leading to food security of the country. From the beginning of Green Revolution, N-fertiliser use has been a success story for Indian crop production. With 6 million tonnes N-fertiliser in 1989-90 to 10.4 million tonnes in 1998-99, every million tonnes N-fertiliser used resulted in 10 million tonnes of cereal production. The per hectare use of nutrients from fertilisers increased from less than a kilogram in the 1950s to more than 100 kg per hectare of gross cropped area by 2000. Accordingly, the all India annual consumption of fertilisers increased

---

Policy Paper 33, December 2005.

from 70,000 tonnes in 1950-51, to nearly 18 million tonnes at present. India is currently the third largest producer and consumer of fertilisers (after China and USA) and fertiliser usage is bound to increase with further intensification of agriculture. While interest in organic manures and biofertilisers is increasing (and rightly so), these can at best meet the total demand for fertilisers, at least in the near foreseeable future.

Expansion of fertiliser use has, however, not been uniform. While the soils in most parts of the country continue to be deficient in N, the fertiliser N-use may be as high as 300 kg ha^{-1} yr^{-1} in the intensively cropped regions of northwestern India. This has led to a peculiar situation in which demands for expansion of fertiliser N-use in some areas coexist with the concerns over the environmental hazards of excessive and inefficient N fertilisers' use in other areas. This, coupled with N contributions from industrial effluents/exhausts, animal wastes and geodeposits have led to soil acidification, widespread pollution of groundwater and eutrophication of surface waters, posing a public health problem and the ecosystem imbalance. The ozone depletion and greenhouse effects of NOx gases from various farm and non-farm sources may pose new concerns for N-C balance, especially for environment and sustainable agriculture.

There have also been genuine concerns over fertiliser-use efficiency, in general, and N-use efficiency (NUE) in particular, for economic as well as environmental reasons. Worldwide, NUE for cereal production (wheat, rice, maize, barley, sorghum, millet, oat and rye) is as low as 33 per cent. The unaccounted 67 per cent represents an annual loss of N fertiliser worth up to Rs. 72,000 crores. The manufacture of nitrogenous fertilisers involves huge amounts of foreign exchange and consumes large quantities of non-renewable energy resources such as naphtha, natural gas, coal etc. Poor utilisation of fertiliser N by plants also adds to the pressure on these finite resources. Low NUE for crops implies higher costs to producers and consumers and, therefore, reduced competitiveness. Loss of N from soil plant system results from gaseous plant emission, nitrification, denitrification, surface runoff, volatilisation and leaching beyond rooting 15 zones of crops. Many N recovery experiments conducted in the country on different crops have reported unaccounted losses of fertiliser N from 20 to 50 per cent depending on the local conditions. These losses of fertiliser N, or in general, the leakages of the reactive N from the agricultural systems into the environment are a cause of serious concern. Nitrogen on its own rarely increases yield much and for long; other nutrients, water, crop variety, weed and pest control must all be in proper proportion and available before fertiliser N can do its best. Harvesting high yields by applying only N is at best a short-lived phenomenon, as was shown in the early years of the Green Revolution. Clearly, 'N-driven systems' are not sustainable, as N becomes a 'shovel' to mine the soil of other nutrients, with the result that, soils initially well supplied in other nutrients become deficient in them and productivity declines. Researchers have shown that the production practices that have resulted in increased N-use efficiency relative to conventional or standard practices are those that will counter conditions or environments that contribute to N loss from soil-plant systems. The challenge is to make such modern farming systems accessible and affordable to the farmers, who are constantly under pressure to cut input costs.

Another aspect of N management is its global implications. Whereas the global C cycle is being perturbed by less than 10 per cent due to anthropogenic activities, the global reactive N cycle is being perturbed by over 90 per cent. Increasing agricultural N-use to meet the perpetual demand for food production, combined with the increasing release of N from dairies and industrial exhausts/effluents makes it inevitable that perturbation of N cycle will continue to be much faster than that of carbon. Although, eutrophication of aquatic ecosystems and health hazards due to nitrate/nitrite contamination are critical local concerns, gaseous emissions may eventually prove to be the most serious on a global scale. The accumulation of NOx gases have global consequences that cannot be constrained by political boundaries or explained away by local policies. Hence, regulatory approaches must strive to balance the need for uniform standards with the diverse local realities and the costs of compliance, and maximise the economic benefits that result from adoption of improved food-production practices that help meet environmental quality standards. Narrow profit margins do not allow farmers to absorb the additional costs of regulation.

Thus, the challenge now facing India is to find ways to further our agricultural and industrial development in a sustainable manner without adversely impacting, the environment and ecology, with respect to nitrogen. In order to prepare ourselves for this challenge, we need a precise understanding of the scale of the nitrogen use/

misuse/release through various agricultural, industrial, vehicular and other activities and their contribution to the pollution of water and air with special reference to various point and non-point sources and the biogeochemical N cycle. While well coordinated nation-wide surveys on N in our environment are lacking, the overall indications from the few sample studies by agricultural and environmental scientists, pollution control boards and others, point to the alarmingly high levels of nitrates in some areas of the country, especially in ground/surface waters. The rise of NOx levels in the polluted airs of our cities often makes eye-catching graphics in the media. We need to clearly integrate the information from various separate studies, account for their methodological differences and identify the underlying issues/sources of concern as well as areas that need further study. Considering that N is an essential part of our agricultural and industrial developmental paradigm, the options for mitigation of N pollution will have to be addressed at various levels, such as establishing/updating national N information systems, improvements in fertiliser formulations, promotion of biological N fixation, enhancement of the N-use efficiency of our crops and farming systems, reduced dependence on non-renewable energy sources, improvements in fossil fuel quality, fuel-use efficiency and reduction of fossil fuel use/abuse, reduction in NOx emissions from farming, industrial and vehicular sources, minimising anthrophogenic (including agri-industrial) N load in naturally overloaded areas (eg. geodeposits) and fragile ecosystems.

The above perspective led the Society for Conservation of Nature to hold a brainstorming session on Policy options for efficient N-use.[1] Their recommendations formed the basis for this paper. One of the most important realisations that emerged from this exercise is that the problems related to N in agriculture, industry and environment are multidimensional. The sheer diversity of research areas/expertise/approaches it encompasses and the various levels at which the problems need to be identified/tackled, calls for an integrated network

---

1. The Society for Conservation of Nature held a brainstorming session sponsored by the NAAS on "Policy Options for Efficient N-use" on October 4-5, 2005 at NAAS, New Delhi, with Prof. Y.P. Abrol, INSA Senior Scientist and Editor, NAAS as convener and Dr. S.M. Virmani, Foreign Secretary, NAAS as Co-convener. A background paper on N in agriculture and environment was circulated in advance to over 80 scientists and specialists from agricultural, environmental and industrial sectors, based on which over 50 papers on specific aspects were received. These were integrated into a few position papers that were presented and discussed during the session attended by more than 35 participants.

approach to harness our intellectual, financial and infrastructural resources effectively.

The session deliberated on the important issues in the total gamut of nitrogen use in crop production related to:

- Low recovery of costly fertiliser N input implies higher costs to producers and consumers and, therefore, reduced competitiveness.
- Manufacture of nitrogenous fertilisers consumes large quantities of non-renewable energy resources—naphtha, natural gas and coal, adding to the pressure on these finite resources.
- Nitrogen, not recovered by crops contributes to environmental problems including those of pollution of groundwater, eutrophication of surface water, greenhouse gaseous emissions of nitrous-oxides, ammonia etc.

Recommendations

The deliberations led to the following recommendations:

1. The rice-wheat cropping system with its high demand for fertiliser nitrogen has been the mainstay of the food security during the last three decades. Main attributes favourable for this system are the availability of good quality high-yielding seeds of rice and wheat, well researched production technology which the farmers could adopt easily and also abundant water, may be at the cost of overexploitation, coupled with procurement of rice and wheat by government agencies and market stability. Research on N-use in this system can contribute to curtailing the leakages of N to the environment and at the same time, make fertiliser N application more cost effective while keeping pace with the increasing demand for food. The N-use by crops can be enhanced by maintaining synchrony in water and plant growth so that leakages can be checked using farmer-friendly tools, such as Leaf Color Charts (LCC) for rapid diagnosis of N deficiency in plants. Because little additional arable land is available, site-specific precision agriculture approaches that minimise fertiliser use must be employed wherever feasible to produce optimum yields. This is especially necessary for high-

nutrient consuming crops including the major cereal grains—rice, wheat and maize. Quantifying the pathways of N loss is of immense utility in the development of strategies for reducing N losses and to enhance N-use by plants. Another important area is the development of N management recommendation compatible with new emerging resource conservation for rice-wheat system such as zero or minimum tillage, direct seeding and bed planting. A good comprehensive study at 4-5 locations, considering all the aspects of N pathway in soil-plant system, which can be further enlarged using simulation modelling to develop technologies for various rice-rice agro-ecoregions is a need of the time.

2. Use of urea super granules (USG) and nitrification inhibitors based on *neem* products improves N-use efficiency but these still have not gained popularity with farmers. Advantage of deep-placed USG in transplanted rice over split application of prilled urea was recognised long ago. Its lack of popularity at the farmers' level can be attributed to its commercial non-availability, need of suitable machinery for placing granules into the soil and lack of fiscal support by the Government of India. There is, however, a successful case study by IFDC in neighbouring, Bangladesh.
3. Fertiliser use in maize and cotton and other crops, in the dryland areas is low and needs to be looked into, to optimise with water availability. The basic constraint in dryland areas is water conservation technology. The evaluation of optimum fertiliser doses, considering the soil moisture status and rainfall pattern, and their application in a staggered manner in suitable split doses depending on the crop nutrient requirement at critical physiological stages, needs to be worked out. Also, the diagnosis of limiting nutrients in soil-plant system pertaining to local situations is necessary for improving the fertiliser N-use and its efficiency in dryland regions.
4. The subsidies on fertiliser should be rationalised so that there is parity in nutrient pricing to promote balanced fertiliser use. There should be a research cess on fertilisers which then should go to funding research for developing efficient use of nitrogenous fertilisers. Farmers must be able to make profit in

their ventures to continue farming. Because maximisation of profits is generally not congruent with minimising losses of reactive N from agriculture, optimisation of systems to meet these objectives (and others) will be required. Optimisation can best be achieved if the external costs of N losses from agriculture are internalised. This means that consumers must pay more for their food. Costs that should be included in the price of food are the costs for production and the farmer's profit, but also the environmental costs associated with fertiliser production and transport, transport of resources and products, and waste processing. Policies designed to promote greater N-use efficiency in agriculture should emphasise incentives to farmers rather than punitive regulations, so as to avoid export of crop and livestock production to areas with less stringent environmental guidelines.

5. Agricultural practices and nitrate in environment are very closely interlinked. Globalisation of agriculture demands consideration of the costs of input and output. There is no way of leaving N from crop production scenario of the country, where without N fertilisation grain production would have been 80 million tonnes which now stands at 220 million tonnes with N fertiliser. It is essential to conduct yield maximisation trials with N responsive high-yielding crop varieties with quantification of the extent of nitrate pollution and watershed contamination. However, till date there has been no concerted effort in the country to look into the fertiliser N-use and build-up in nitrate levels in the environment. Also, the reports about increased levels of nitrate in some pockets, whether due to fertiliser N, animal production systems or are of geogenic sources, have not been authenticated. There is need to address these issues in detail to formulate a comprehensive fertiliser N policy for enhanced agricultural production and safe environment. This calls for the nitrate monitoring in groundwaters in different agro-eco regions in a systematised manner to identify the sources contributing to the built-up of nitrates in ground waters under given crop, input and farming systems so that a comprehensive database can be established for development of proper policy *vis-à-vis* fertiliser N-use. Since it is impossible to have too many field studies, simulation models must be

modified and/or developed through intensive detailed field investigations and validated for diverse set of conditions. It will need extrapolating the results to other locations. At present, there is lack of both, authentic data and validated models, under Indian conditions. The information on N cycling at field, sub-regional, regional or national level is lacking and needs to be developed as well as collated from the available data. It is pertinent to mention that anthropogenic fluxes largely control the natural as well as the anthropogenic N cycles. Using remote sensing and GIS, it is possible to spatially describe dominant controlling factors for N cycle, such as land use, vegetation type, farming systems, soil and water characteristics and need to link the information through networking for optimisation of policy on N-use in different agro-eco systems.

6. Animal manures constitute 5-10 per cent of N input in Indian agriculture, but can be a significant source of nitrate pollution in some areas due to leakages during storage and handling. Efficient use of animal manure in Indian agriculture should cover minimisation of the collection loss, proper composting, reducing loss during storage, integrated nutrient management (INM), optimising application method, and *gobar* gas technology be studied and package of practices need to be developed. Information required for this includes, nutrient content dynamics and, for technique and precision, nutrient loss, techno-managerial issues of implementation, policy issues of incentive and punishment for adopting the efficient methods. Policy issues should include integration of crop and livestock production system, promotion of *gobar* gas plants and improving their efficiency, training of women and incentive for better manure management, research for improved compost preparation and socioeconomic evaluation of manure use.

7. Nitrate toxicity and health hazards has been much talked about but studies are available as far as the nitrate toxicity on human being is concerned, except reports documenting methaemoglobinemia in infants due to high nitrate ingestion. Reports indicate nitrate ingestion is a cause of methaemoglobinemia for all ages, recurrent diarrhoea, RRTI (recurrent respiratory tract infection) up to eight years of age

and recurrent stomatitis. An observation that a sudden rise in the occurrence of these diseases when the nitrate concentration is more then 26 ppm, suggests a need for downward revision of permissible limits of nitrate in drinking water. Apart from methaemoglobinemia, a few studies indicated nitrate as a cause of malignancy (liver and gastric cancer), but these are controversial studies and no firm conclusion have been drawn so far. Removal of nitrate from drinking water is difficult and costly. It is recommended that a simple policy change may be urgently introduced in government campaign in areas having high nitrates in potable waters. As the treatment options are not much, there is need to formulate guidelines for preventing nitrate toxicities specifically in infants, children and pregnant women in areas having high levels of nitrate in potable waters.

8. Tropical agriculture contributes significantly to greenhouse gases (GHGs), especially emission of methane (CH4) and nitrogen oxide (N_2O). To meet the future foodgrain production targets, the fertiliser consumption is expected to sustain production levels. The unique nature of rice production (flooded soil and high N input) makes it a significant contributor of CH4 and N_2O to the environment. Emission of N_2O from rice fields is directly related to N-fertiliser application. Crop residue and green manure gives as much N_2O as chemical N-2 fertiliser alone, but nitrate leaching is restricted. Improvements in NUE should diminish N_2O emission. This can be done even while increasing the use of fertilisers if their use efficiencies are tightly managed. Future research should focus on quantification of the fertiliser impact on GHG emission from different agroclimatic zones on an annualised basis, quantification of sink potentials for GHGs, reduce uncertainties in GHG and N deposition and develop simulation models for estimation and extrapolation of GHG emission. A strong database on N_2O emission from agriculture is essential for which intensive research is needed combining simulation models and GIS for determining the impact of N-inputs on N_2O fluxes. Development and use of environmentally compatible mitigation options for controlling N_2O emission from agriculture is urgently needed. Since mitigation of N_2O emission is closely linked to food

production and the progress of national economics, it is important to continuously review and monitor the sources and sinks of N_2O in agriculture. Unlike N_2O, NH4 is highly reactive and returns to the earth through deposition. We don't have any data on N deposition in India. Data from Europe and elsewhere, indicate that N deposition is significant and need to be collected in Indian conditions to remove uncertainties.

9. There is need to develop crop rotations involving legumes to tap the benefits of biological nitrogen fixation (BNF). Nitrogen use efficiency for cereals following legumes is greater than that for cereals following cereals or fallow. The role of legumes in N economy is well researched but the problem is how to increase N input and the options are, increased system efficiency or increase in the area under the system. N derived from BNF in legumes varies from 40-80 per cent and residual effect on succeeding crops is variable and depends on several factors. The more intensive systems (growing more crops in a given period of time) require greater fertiliser N inputs but are economically advantageous to farming community. More intensive dryland cropping systems involving legumes in rotation lead to increased water use efficiency and also better maintain soil quality. The research has shown a positive impact of BNF on nitrogen economy of cropping systems but the vast potential of BNF has remained unrealised at the farmers' level due to many reasons and needs to be looked into from the holistic approach on nitrogen use in agricultural production systems. Some aspects which need immediate attention are: increasing public investment in microbiology for teaching, research and training, encouraging private investment in manufacture of biofertilisers, constitution of nodal agency for registration of manufacturers, establishment of quality control laboratories and act as a watch dog and promoting products through DAVP and media. The private sector needs to play a crucial role and set an example by employing qualified microbiologists for production, assuring quality through creation of brand equity, ensuring niche marketing through entrepreneurship ventures and providing dealers' involvement as advisor and friend on product and proper application.

There is an urgent need to improve the inputs of organics and BNF and to increase the production of quality inoculants and popularise their use in Indian agriculture rapidly. Development of effective and competitive strains tolerant to high temperature, drought, acidity and other abiotic stresses are of high priority. Newer formulations of mixed biofertilisers, improvement of inoculant quality and devising effective delivery systems are crucial for making further progress in taking the BNF technology to farmers' fields.

10. Breeding and selecting crop cultivars that make more efficient use of soil and fertiliser N (including higher N fixation and N partition) while maintaining productivity and crop quality has been a long-term goal of production agriculture. Development of nitrogen-efficient cultivars could help decrease fertiliser N inputs and resulting reactive N losses to air and groundwater. These nitrogen-efficient cultivars could also be useful in regions where limited-resource farmers are unable to afford synthetic N fertilisers. Selection of N efficient genotypes that is, the varieties which can extract more N from soil at lower availability will enhance the production in these regions. Molecular and biotechnological approaches for searching for regulatory targets for manipulation of N-use efficiency be strengthened. Unravelling the details of N signal transduction to provide additional clues to improve N uptake and assimilation efficiency.

11. Our first duty, as scientists working on N, must be to get our science as right as we can. But we also have a duty to society to provide the best advice we can, presented as clearly as we can, on how to use nitrogen efficiently and how to keep it from getting into places where it is not wanted. Improving N-use efficiency in major food crops will not be easy. It will require collaboration among ecologists, agronomists, soil scientists, agricultural economists, and politicians. Great needs exist for accurate measurements of actual fertiliser N-use efficiency, N losses, and loss pathways in major cropping systems. Only in this way we can: a) identify opportunities for increased N-use efficiency by improved crop and soil management; b) quantify N-loss pathways in major food crops; and c) improve human understanding of local, regional, and

global N balances and N losses from major cropping systems. The starting point for any improvement has to be a clear understanding of the fluxes and balances of nitrogen at the farm level. Direct on-farm measurements are necessary because estimates from small plots on research stations overestimate field-scale fertiliser N-use efficiency. Focus new initiatives in research on options that will reuse or remove reactive N before it cascades through the environment. Especially promising target areas include: a) decreasing NOx emissions from fossil fuel combustion in power plants and industry by pre-combustion removal of organic nitrogen, b) converting NOx in flue gas streams into saleable fertiliser products, c) increasing N-use efficiency by crop and animal production systems, d) decreasing reactive N losses from animal wastes, and e) recycling reactive N back on the land from which feed grains and forages are produced. Fostering multidisciplinary innovations in the management of reactive N such as, i) research by agricultural economists on possibilities for internalisation of environmental costs in crop and animal agriculture production systems, ii) research by agronomists and agricultural engineers on means by which to integrate farm wastes such as crop residues and animal manures more effectively into crop nutrient management plans, and iii) on possible innovations in the manufacture of reactive N fertiliser materials that could result in their more efficient use in cropping systems. We need to find the answer of the 'N legacy question' by determining the extent to which reactive N accumulated in various ecosystem reservoirs in recent decades is retained and how these accumulations will affect the future productivity, stability, and resiliency of crops, forests, and natural ecosystems. We also need to, 1) develop multipollutant multi-effect strategies to optimally combat environmental effects resulting from human activities; 2) develop technological road maps for future infrastructures where fossil fuels can be replaced by renewable energy sources, e.g. through exploration of hydrogen-based rather than fossil-fuel-based energy production systems; 3) establish an institutional framework that supports exchange of information between researchers on the effects of reactive N in its various aspects—e.g. food production, use of fossil fuels, and the environment—and identify mechanisms, such as the World Bank/IMF/ADB to fund researchers in the developing world to investigate specific regional impacts and solutions; and 4) develop better ways to assess the uncertainty of current and future reactive N loading to water bodies.

Top Priorities for Innovations in Reactive N Research and Policy

- Developing economically sound, applicable policies and measures
- Developing incentives for improving N-use efficiency through integrated nutrient management practices
- Integration of plant and animal production systems
- Conversion of agriculture from total reliance on synthetic reactive N fertilisers to a significant input from BNF
- Development of modified N fertilisers—controlled release fertilisers, urea supergranules, *neem* coated urea etc. for lowland rice and other areas prone to reactive nitrogen leakages
- Novel molecular techniques to develop plants with higher N-use efficiency
- Accurate and site-specific test for estimating soil-mineralisable N and developing site-specific N management for different crops

General Remarks

The need to develop N budgeting guidelines at field and farm levels and further expanded to watershed scale requires urgent attention so that the effect of fertiliser N-use in particular and overall N-use in totality can be worked out for efficient management of N input in agricultural systems. With this objective in view we need a systems' approach across the various agricultural disciplines in a network mode to tackle the issue of N-use and develop strategies for increased production with minimal environmental consequences and at the same time, a sustainable system which is also economically viable for the farming community.

It is recommended that a Nitrogen Research Centre be established to address all facets of the problem in totality. It will be a step in the forward direction in coming decades, keeping in view the food requirements in future and the quality of environment *vis-à-vis* enhanced fertiliser N-use to achieve the future targets with least human health problems. This issue needs to be addressed urgently at national level.

13

Organic Farming: Approaches and Possibilities in the Context of Indian Agriculture

Introduction

During the era of Green Revolution, introduction of high-yielding varieties, extension of irrigated areas, use of high analysis NPK fertilisers and increase in cropping intensity, propelled India towards self-sufficiency in food production. In the process, relative contribution of organic manures as a source of plant nutrients *vis-à-vis* chemical fertilisers declined substantially. With increase in cost of production inputs, inorganic fertilisers became increasingly more expensive. Another issue of great concern was the sustainability of soil productivity as land began to be intensively tilled to produce higher yields under multiple and intensive cropping systems. Waterlogging and secondary salinisation have been the banes associated with excess and irrational irrigation. Groundwater table declined sharply as more and more deep bore wells were drilled. Recharging of groundwater has also been reduced due to severe deforestation. Indiscriminate use of chemical pesticides to control various insect pests and diseases over the years, has destroyed many naturally occurring effective biological control agents. An increase in resistance of insect pests to chemical pesticides has also been noticed. Health hazards associated with intensive modern agriculture, such as pesticides residues in food products and groundwater contamination are matter of concern. The occurrence of multinutrient deficiencies and overall decline in the productive capacity of the soil due to non-judicious fertiliser use, have been widely reported. Such concerns and problems posed by modern-day agriculture gave birth to new concepts in farming, such as organic farming, natural farming, biodynamic

---

Policy Paper 30, February 2005.

agriculture, do-nothing agriculture, ecofarming, etc. The essential feature of such farming practices imply, i.e., back to nature.

Organic Farming and Countrys' Food Security

The primary concern of all organised communities and civilised societies is to meet the food requirements of its people. The cultivated area, required to maintain the present level of foodgrain production in India without using the fertilisers, reaches more than the total geographical area of the country. At present, there is a gap of nearly 10 million tonnes between annual addition and removal of nutrients by crops which are met by mining nutrients from soil. A negative balance of about 8 mt of NPK is foreseen in 2020, even if we continue to use chemical fertilisers, maintaining present growth rates of production and consumption. The most optimistic estimates at present, show that only about 25-30 per cent nutrient needs of Indian agriculture can be met by utilising various organic sources. It is proved beyond doubt that on long-term basis, conjoint application of inorganic fertilisers along with various organic sources is capable of sustaining higher crop productivity, improving soil quality and soil productivity. The organic sources should be used in integration with chemical fertilisers to narrow down the gap between addition and removal of nutrients by crops as well as to sustain soil quality and to achieve higher crop productivity. The food security demand of the country requires that inorganic fertilisers be used in balanced doses.

Issues

Organic farming has the twin objective of the system being sustainable and environmentally benign. In order to achieve these two goals, it has developed some rules and standards which must be strictly adhered to. There is a very little scope for change and flexibility. Organic farming thus, does not require best use of options available but the best use of approved options. These options are usually more complex and less effective than the conventional ones. The philosophy of the proponents of organic farming presents considerable difficulty to the scientists as the organic movement generally resists the comparisons between the two systems.

The organic farming movement presents a challenge to the scientists who cannot and would not want to abandon a scientific approach. The present lower productivity of the system is a result of constraints that its practioners have put upon themselves. Difficulty faced is with evangelising statements, that the products of the organic farming system—food and fibre—are in some way better than those from conventional agriculture. The Indian farmer should get the advantage of emerging global market on organic farming which is at present around 26 billion US dollars, and is expected to grow to 102 billion US dollars in 2010. Currently, 130 countries are producing certified organic products. A vast scope for promotion of organic farming in the export market, without compromising with the national food security exists in the country, as farming by tribals and under rainfed conditions is generally organic, since very little chemical inputs are used.

Organic farming as a concept/philosophy is well tested in some of the western countries, though the same is not unknown to most of the nations. But in the Indian context, it needs to be looked into more critically, seeking answers to the following questions: What level of crop productivity is acceptable? Is it suitable for a country like India with such a large population to feed or can it fit in the niche areas? Are available organic sources of plant nutrients sufficient for organic farming in the form it is advocated? Are organic farming technologies sustainable in the long run?

Various aspects relating to organic farming were addressed and discussed in a workshop organised by the NAAS.[1] Following issues were taken up:

- Redefining the concept of organic farming in Indian context.
- Opportunities and constraints in organic agriculture.
- Organic agriculture perspective and future challenges.
- Soil quality, biodiversity and fertility in organic farming system.
- Standards in organic farming and quality markers for organic produce.

---

1. NAAS organised a workshop under the Convenership of Prof. P.K. Chhonkar, Head of the Division, Soil Science and Agricultural Chemistry, IARI on 25th-26th November, 2004 at NAAS, New Delhi. The workshop was organised in a panel discussion mode bringing together about 50 experts having different shades of opinion on organic farming, ranging from those in fertiliser/pesticide industry to practicing organic farmers & NGOs promoting organic farming, on a single platform to discuss various issues relating to organic farming. A concept paper developed by the Convener was sent to the participants in advance to set the agenda for discussion. Various issues related to organic farming were discussed in six sessions spread over two days. Each session had a lead paper followed by a presentation from each of the panelists and open house discussion.

- Setting up of certification and accreditation agencies for export.
- Setting up of testing units for soil and organic produce for chemical residues and quality control.
- Market development and assessment of organic produce for export.
- Evaluation of organic/biodynamic farming *vis-à-vis* integrated nutrient and pest management.
- Documentation and confirmation of existing Indigenous Technological Knowledge (ITK) on biodynamic farming.
- Development of biocontrol measures for various plant diseases and pests.
- Research agenda for comparing organic and conventional agriculture.
- Holistic approaches in organic farming research and development.

Recommendations

Major recommendations that emerged from the workshop are listed below:

1. Organic Farming: Prospects and Limitations

- For India with its ever increasing population, the sustainable agriculture has to be based on site-specific balanced and adequate fertilisation, and an integrated plant nutrient supply system (IPNS) involving organics, inorganics and biofertilisers. What the country needs today, is the conjunctive use of organic and inorganic sources of plant nutrients for sustainable productivity.
- Organic farming is a market demand-driven agriculture, aimed to cater to the foreign export and affluent section of the society in the country. However, in order to make a dent in the export market, we need to develop high-tech organic technology with strict quality control, meeting international quality standards prescribed for organic produce.
- Niches of the organic farming need to be identified. However, the real niches will be determined by the market infrastructure

and the international links. The practice should be considered for lesser endowed regions of the country rather than in resource endowed regions which serve as the backbone of the country's food security.

- The availability of organic manures in adequate amounts and at costs affordable by the farmers is a major problem. The increased mechanisation has further reduced the availability of manures with the farmers and this problem will become more acute in future. In such circumstances, post-harvest residues should be utilised to the fullest extent. However, to accomplish this objective, feasible technologies are needed for *in situ* recycling/rapid composting of on-farm residues and wastes, in addition to extension efforts to change the mindset of the farmers. Possibilities of using non-traditional organic sources e.g., slaughter house waste, should be exploited to partly supplement, plant nutrient needs of the organic farming systems.
- Changing cropping patterns with area under legumes is going down, shrinking area under green manures due to economic considerations, and reduced availability of loppings from forests, seriously restricts wide scale use of green manures. Inclusion of legumes in intensive cereal-cereal production systems as short duration grain or forage crops, as substitute to one of the cereals or as break crops, needs to be promoted.

2. *Organic Farming: Its Relevance to Indian Agriculture*

- In Indian context, organic farming has to be practiced without synthetic pesticides, but complete exclusion of fertilisers may not be advisable under all situations. A holistic approach involving integrated nutrient management (INM), integrated pest management (IPM), enhanced input-use efficiency, and adoption of region-specific promising cropping systems would be the best organic farming strategy for India.
- As organic farming is attracting worldwide attention, and there is a potential for export of organic agricultural produce, this opportunity has to be tapped with adequate safeguards so that the interest of small and marginal farmers is not harmed.
- Organic farming may be practiced in crops, commodities and regions where the country has comparative advantage. To

begin with, the practice of organic farming should be for low-volume high-value crops, like spices, medicinal plants etc., beside fruits and vegetables, for which R&D support is required.

- Organic farming should not be confined to the age old practice of using cattle dung, and other inputs of organic/biological origin, but an emphasis needs to be laid on the soil and crop management practices that enhance the population and efficiency of below-ground soil biodiversity to improve nutrient availability. Performance of cultural techniques for weed control and that of biopesticides for pest management need to be evaluated under field conditions, preferably under cultivators' management conditions. Besides the identification of regions suitable for the adoption of organic farming, the crops and their products should also be identified which are amenable for production through organic ways and have the potential to fetch a premium price in the international organic market.

3. *Organic Farmers and NGOs*

- Region-specific resource inventory, including animal wealth, farm residues/by products and their competitive uses, non-conventional nutrient sources of organic/biological origin etc., has to be prepared, for development of rational technology packages of organic farming.
- A strong technological back-up by scientific community should be provided in order to verify, confirm and further refine some selected ITKs like, *Agnihotra, Panchgavya,* pertaining to organic farming.
- Crop-specific and farming situation-specific package of practices for organic cultivation should be developed and after thorough on-farm validation, recommended for adoption. Such proven technology packages need to be documented in regional languages.
- Entrepreneurial potential with respect to production of organic inputs, processing and marketing of organic food should be fully exploited.

4. Trade and Certification Issues

- Certification of organic produce is an important issue that is central to organic production itself, as the prime goal of organic farming is and will be to fetch premium price in domestic and international market for the producers.
- For desired economic gains out of organic farming, the upcoming global market on organic produce has to be exploited, for which strict phytosanitary measures have to be followed.
- A strong research back up has to be put in place to develop and improve national standards for organic farming. The policy documents brought out so far by APEDA and DAC, including the report of Task Force on Organic Farming etc., should be considered in developing such scientifically sound standards, mainly for the crops and commodities that have export potential. The national standards should be the same for domestic market and for export.
- Weak links in the certification systems have to be identified and researchable areas flagged.
- Whereas organic certification will continue to be a process certification, strong research set-up/laboratories are required to monitor the quality of organic produce so as to prevent the sale of substandard material in the name of organic produce, and to save the interests of producers and consumers.
- Organic produce fetches premium price owing to better quality, and for credibility in the market it ought to be quality goods. Therefore, there should be adequate provision for their grading, packaging, storage and transportation. Marketing outlets on the lines of milk unions, may be established.
- There is a need to organise Producers' Cooperative Marketing Societies and establish credible marketing channels for steady flow of organic foods and materials in accordance with the demand, as also to safeguard the interest of small farmers opting for organic agriculture.

5. The Quality Aspects of Organic Food

- There is an urgent need to compare the quality of organically produced food with conventionally produced food. There appears to be a widespread perception amongst consumers, that organically produced foods are of superior nutritional quality. However, to prove or disapprove this contention, very limited research has been conducted, and whatever meagre scientific data is available, is often out-dated or based on inadequate study designs, lacking proper controls. In view of this, the following points merit consideration.

- No clear-cut evidence is available to support consumer perceptions regarding potential health benefits of organic foods. An in-depth research on quality aspects is required to arrive at any valid acceptable conclusion.
- Valid nutritional quality comparisons between organic and conventional food requires that plants be cultivated in similar soils, under identical climatic conditions, be sampled at the same time, pretreated similarly, and analysed by validated methods.
- Well-designed controlled studies in animal models and human subjects are needed.
- It is also necessary to undertake well-controlled studies to evaluate sensory properties, shelf life, and nutrient load of organic produce *vis-à-vis* produce from the conventional farming techniques.
- More importantly, organically and conventionally produced food should also be analysed for pesticide residues and microbiological safety i.e., presence of pathogenic organisms which could pose health hazards.

6. Research Agenda

- Develop package of practices for integrated pest management for organic farming in different agroclimatic regions for specific crops involving components like bioinoculants, pheromones etc., among others.
- The issue of microbial contamination of food arising from the use of manures has to be addressed, and measures suggested to mitigate it.
- It is frequently documented that fertilisers and pesticides applied at recommended rates have had no adverse effect on soil biological activity, and that integrated farming systems are

best for nutrient management, yield sustainability and soil biodiversity conservation. There is thus, a need to develop modern organic farming system integrating the best available options.

- There is a need to establish referral laboratories for analysis of pesticides, heavy metals and mycotoxins in the produce with appropriate accreditation to help organic farming movement. The maximum residue limits in organic food must be set in accordance with the CODEX standards.
- The Green Revolution technologies have been alleged to have caused depletion of soil organic carbon. The critical values of soil organic matter that can support the sustainable crop production under organic management have to be worked out.
- The patterns of rate and amount of nutrients released from various organic sources and their goodness of fit with the nutrient requirement of the crops at different growth stages need to be worked out. This information could be used in evolving appropriate nutrient management schedule, so as to ensure optimal nutrient supply to the crop at active physiological stages having peak nutrient demands.
- Allelopathic effects of various plant species need to be tapped, particularly for weed and pest management.

Section III

Water Resource Management

14

Emerging Issues in Water Management

The Question of Ownership

Introduction

India has made considerable progress in increasing the production of food and other agricultural commodities but the sustainability of India's agricultural development, food security and poverty reduction strategies are being questioned in several quarters due to an imminent water crises. Agricultural sector accounts for more than 80 per cent of total water used in India. However, only around 40 per cent of the net sown area is reportedly endowed with water, albeit limited, for irrigation. Further, with the growing population, the per capita availability of water is continuously going down and is likely to go below the scarcity level by 2025. Prudent use of available water, therefore, assumes critical importance for human and livestock sustenance.

Groundwater accounts for around 58 per cent of the total irrigated area in the country, whereas the share of groundwater in meeting the drinking water needs, is much higher at around 80 per cent. Nevertheless, the surface sources are quite important as these help in recharging the groundwater in several areas.

Wasteful use of canal water has created the problems of water logging in head and middle reaches of canal command areas, whereas tail end farmers continue to suffer due to lack of access to this water. The situation is further aggravated by low irrigation system efficiency. Overdrawal of groundwater is leading to depletion of aquifers. Tanks and wells are increasingly going dry due to unscientific water harnessing, and water conflicts are increasing. Rivers are also drying up. The demand for water in urban areas is increasing and efforts to

---

Policy Paper 32, June 2005.

transport water from rural to urban areas is being resisted by rural communities.

Importance of conservation and efficient use of water figures prominently in plan documents, policy statements and discourses. But emphasis and preoccupation continue to be overwhelmingly in favour of massive investments to augment the supply of water. Very little has been done in changing the ways supply systems are organised and managed as well as in demand side management including the reformulation of policies that influence the attitude and practices at the users' end. Agricultural policies are now emphasising high value non-cereal crops, which are mostly water-intensive crops. The use of water per unit output of rice and milk is very high but these are important for our food and nutritional security. Nevertheless, there are signs that attitudes are beginning to change. Some of the attempted changes are indeed significant in scope and intent but these are grossly inadequate and their implementation too weak to make a significant dent on the overall situation.

While in principle, there is a general consensus that, (a) water resource sector needs a holistic view; (b) all stakeholders need to be involved in its management; and (c) water being no more a free good, there ought to be some defined principle of water allocation. There is, thus, considerable hiatus between principles and practices. The gaps between principles and practices are mainly due to factors associated with absence of clear-cut property rights and lack of governance or misgovernance in water sector. It is now widely recognised that the epicentre of the problem lies in the current social, economic, legal and governance framework of water resource development, distribution and utilisation.

Role of the State and Governance

The imminent water crisis in India is being ascribed to, (a) the growing gap between the scarcity value of water and realised value under the current pattern of water use and management; (b) weak governance and misgovernance of water resources by the state; and (c) the absence of clear-cut property regime or ownership of water. Traditionally, water was perceived to be a free good available from nature. This perception has not materially changed. Existing institutional, legal and governance framework relating to acquisition, distribution and utilisation of water has not helped in changing this perception.

The role of the 'state' in sustainable management of water resources is and will continue to be quite critical. While the sovereign right of the state on any natural resource like water is undisputed, the role a government should assume to exercise this right has become questionable and needs to be dispassionately defined through consensus among stakeholders. It has become repeatedly clear that the governments have, by and large, not been successful in both supply and demand side management of water resources.

The role of the state as planner and manager of present and future demand can be assessed by water policy, water laws and water administrative machinery. The objectives and goals of water policy, howsoever meticulously formulated, cannot be achieved without appropriate water laws and efficient implementation mechanisms. The National and State Water Policies do not address all the relevant issues in proper perspective. The policy documents, though recognise the emerging water crisis but fail to provide well-defined instruments to improve the situation of increasing groundwater depletion and inefficient management of surface irrigation systems.

World over, the 'state' or governments have been guided by three different but interrelated legal doctrines *viz.*: (a) doctrine of riparian rights; (b) doctrine of prior appropriation; and (c) doctrine of public trust. In the Indian context, laws relating to water are traced to the Indian Easement Act, 1882. In the case of surface water, this follows the doctrine of riparian rights. During the British rule, the government regulated surface water for irrigation under various irrigation statutes. The Indian Easement Act, 1882 also provided unlimited right on groundwater to the owner of overlying land. There was no provision for 'prior appropriation' or 'reasonable use'.

The present legal system related to water in India represents a complex amalgam of colonial legislation, customary laws and post-independent constitutional and legal provisions. The legal framework of irrigation in India implicitly presumes the absolute right of the State over water resources. As per Indian Constitution, water is in the State list (Entry 17-List II States), subject to the provisions of Entry 56 of List I (Union List). This implies that the Union government can also assume responsibility in the matters related to interstate rivers, to be specifically declared by the Parliament through a specific law. The Centre lacks jurisdiction to enact legislation for regulation and control of groundwater. However, three important developments require a mention. First, the Centre formulated a Model Groundwater (Control and Regulation) Bill, 1970 and circulated to the states for adoption. The Bill postulated some kind of water permit system linked to land rights. However, the Bill did not find takers and was reportedly not adopted by any state. Second, the National Commission on Agriculture (1976), again emphasised the need for groundwater legislation. The Commission suggested some kind of *warabandi* system in groundwater also. And third, a much-improved version of the Model Bill came in 1992. The Model Bill of 1992, *inter alia,* suggested for mandatory installation of water meters by all the groundwater permit holders including the farmers. Only eight states have so far enacted legislation to regulate groundwater. Groundwater authorities have been set up in the states but very few powers as provided in the Model Bill are vested with state groundwater boards/departments. Further, as the states have failed to check the depletion of groundwater, the Supreme Court of India intervened and, under the Environment Protection Act, created and empowered Central Groundwater Authority to assess and declare overexploited areas as 'dark zones' prohibiting further extraction of groundwater in such areas. Though this was a good initiative but it meant superseding the powers of the State by the Centre, which is perhaps against the constitutional provisions. The Centre has established Central Groundwater Board, which now has 15023 hydrographic stations to monitor the groundwater situation in different parts of the country. However, by and large, the groundwater continues to be unregulated. Only indirect administrative measures like restricting the institutional financing of groundwater extraction is being used. But these measures further increase the inequalities, as those with sufficient resources are able to dig wells without any check. Nevertheless, in the matter of protection of water quality and settlement of interstate river water disputes (Water Disputes Act, 1956), some effective legal provisions are in place. In 1983, a National Water Resource Council was also set up.

The wasteful use of surface water, overdrawl of groundwater and consequent ecological damages can be attributed to the absence, failure, or even the inability of the governments, (a) to change the legal framework that is outdated; (b) to bring in new framework where it does not exist; and (c) establish and enforce effectively the legal framework that exists now. Every public irrigation system has rules regarding allocation, scheduling and penalties for violations, but these are flouted with impunity. Even the legal framework for groundwater could not be clearly defined. Whatever legal provisions exist, they could not be effectively implemented. The law allows landowners unrestricted right of extraction of groundwater from their plots. There are, of course, regulations on effective spacing between wells and on construction of new wells in overexploited areas but there is no monitoring. The policies relating to pricing of water and electricity for pumping water are perceived to have ignored the degree of scarcity and consequently, water use efficiency. By and large, there is enough evidence of state failure and ineffective governance in managing the precious water resources.

The Issues Relating to the Ownership of Water

Improving efficiency and equity in land resources, was attempted by undertaking land reforms. Since ownership of land is relatively well defined, it was possible to do so. Lot of research efforts went into understanding of gains and losses from tenancy reforms and conferment of land rights. However, in the case of water resources, the ownership and use rights vary across water sources, usage and states. The water use is guided by multiple property regimes *viz.* common property, common pool resource, private resource and state-owned regime. The issue is further complicated when ownership and use are combination of these regimes. Above all, there is the perception of water users about the resource, which ultimately guide its usage pattern and efficiency of use or sustainability of the resource. For example, even in drought-prone areas, people perceive groundwater as unlimited and use it to maximise short-run profits. In the case of a village pond, while water is considered a common pool resource and there are unwritten customs of sharing it, the water going to recharge the well from the pond becomes a private property. The owner of such a well-even in a period of scarcity may not agree to share water with the community. Efforts of a community in conserving water, which help in recharging individual wells, result in building the private poverty which may not be equitably shared with the other members of the community. Recharging of wells, by public funded countrywide watershed development programmes, is a good example of public investment becoming a private property.

The world over, as countries move from a state of plenty to that of scarcity, there is an increasing tendency of centralised control of water. Several countries have declared water as a state property. However, they allow decentralised independent decisionmaking. But on the other side, there are countries (UK, USA) which have opted for complete privatisation. But here also, there are prioritised (earlier use) and correlative (land related) rights and these rights are recognised as property rights and are as such transferable.

Issues relating to ownership of water are not only complex but also different than for other resources. The use, control and ownership of water is linked to the ownership of land and irrigation structures. Thus, water ownership issue cannot be discussed in isolation. Further, ownership issue relating to surface water and groundwater needs to be dealt with differently. Watershed or inter-basin management programmes have suffered mainly on account of not being able to address the issue of ownership regime. Howsoever difficult the issue may be, establishment of some sort of water rights and responsibility system, specifying the withdrawal or entitlement of water, is critical for prudent use of water resources.

There are five possible property regimes in India and water has presence in all of them.

(a) Individual private property (groundwater)

(b) State or public property (surface water resources)

(c) Common property (tanks with PRIs or communities)

(d) Common pool resource—access to identified group but no one has a right (village tanks)

(e) No man's property (open access water bodies)

The problems with state ownership and management are well known. As regards the commons, these are bound to degrade unless there is a body of people responsible for their use and maintenance. If groundwater remains under common property regime, it is bound to be overexploited because if an individual cooperates with others, he looses and if he does not cooperate, he does not loose anything. Water thefts from government-owned tanks are also not uncommon.

While some ownership or property regime is necessary, complete private ownership of water as a substitute for state or community controlled management and operation is also being questioned. Several non-governmental and civil society organisations have initiated action for social movements to influence the stakeholders including governments and the water users. There are quite a few global initiatives also. People's World Water Forum argues that no one can have a right to steal a natural resource. The attempts by the governments to raise water tariffs are being interpreted by civil society organisations as a step towards corporatisation of water resources. Globally, privatisation of water has been a mixed but mostly discouraging experience. Privatisation of water utilities in Jakarta, Manila and Cochabamba (Bolivia) left both firms and clients dissatisfied. Nevertheless, in Penang and China, these are reported to be successful.

Efforts to privatise some rivers have also failed. There is also a lack of clarity in the privatisation model. The privatisation could take several forms *viz.* (a) Build-Operate and Transfer (BOT); (b) privatisation of development only; (c) privatisation of management only; (d) complete transfer of ownership encompassing (a), (b) and (c) above.

Some argue that while water for drinking and domestic use can qualify to be a fundamental human right, water for other economic uses need not have the same ethical qualification. Transferability and exchangeability of water rights are critical to capture and reflect the scarcity and use value of water through price signals and thus, guide water allocation to attain economic efficiency.

Those who argue for the sole control of local communities (macro, meso or micro watershed basis) on water resources have raised several questions on the New National Water Policy adopted by the Government of India in April 2002, superseding the earlier policies. It is their contention that India has enough water resources to meet the needs of drinking water and sustainable irrigated agriculture, but it requires local communities' control on management of water. It is only through community involvement that cropping patterns of different regions can be dovetailed to hydrogeological and agro-climatic biodiversity of micro regions.

Water is a natural renewable resource and is the ecological basis of all life. As its sustainability and equitable allocation depend on cooperation among community members, people should own or decide who should manage it. However, the ultimate custodian of people's resource could be the state.

There is also a question with the common law related to the riparian owner. Every riparian owner is entitled to continued water flow of a natural stream in its natural condition, without any obstruction and pollution. This solves the problem of other riparian owners but does not establish claims of those living in the interiors.

What is, therefore, needed is a well-defined combination of state, community and private ownership and their clearly defined roles/ responsibilities in water management. Their relative roles may vary according to basin, sub-basin and micro watershed and from region to region and project to project.

Given the complexity and wide ranging issues involved in current water use and ownership regimes, NAAS organised a Symposium[1] on Emerging Issues in Water Management–The Question of Ownership. The Symposium was divided into the following four technical sessions, each representing a major sub-theme:

- Water Ownership and Property Regime including Legal Framework.
- Economic Policy Impacting Water Use, Water Management and Water Markets in Rural Areas.
- Community Management and People's Perception on Water Related Issues.
- Science & Technological Aspects of Water Management in Agriculture.

In each technical session, there was a keynote presentation on one of the four identified sub-themes, followed by an open discussion. The recommendations given below are an outcome of the deliberations in the Symposium.

Recommendations

1. While the 'state' should continue to have a sovereign right over a natural resource like water, the role of the government in exercising this right needs to be carefully defined.
2. Water being a precious natural and common property

---

1. The Symposium was organised at the Institute of Development Studies (IDSJ) on January 15, 2005 at Jaipur under the Convenership of Prof. Shabd S. Acharya, Honorary Professor of IDSJ, with Prof. Manohar S. Rathore, Senior Fellow of IDSJ, as Co-convener. The Symposium brought together around 40 experts from social science research institutes, agricultural universities, general universities, international water management institute, ICRISAT, development administration, and non-governmental organisations working on natural resources and water management aspects. The concept paper developed by the Convener and Co-Convener was sent to the participants in advance to set the tone for discussions.

resource, vital to the sustenance of human, animal and plant life, policies, legal framework and administrative system dealing with water resource management must take in to account the principles of access, justice and social equity.

3. In the case of water, all three aspects *viz.* development, regulation, and management, are of considerable importance and water policy should address all of these.
4. Water, as a resource should not be made a private property. However, only use rights on equitable basis should be given. There is a need for treating drinking water in rights mode and irrigation water in the economic mode. Water rights need to be delinked from land rights.
5. A judicious combination of government and community control and private water rights, the proportion varying from macro, meso and micro level and from region to region, is perhaps desirable. Such combinations can be under public trust framework when government and community have a control over water as trustees and people have well-defined usufructuary rights, i.e. right to use and profit from another's property on the condition that it remains uninjured.
6. Private water rights and community water rights can be accommodated within the public trust arrangement where the state or community has a ownership right enabling it to exercise social control so as to promote ecological security and social equity, while individual members have private and transferable user rights so as to promote economic efficiency in water use.
7. The water rights system can simultaneously address three goals of ecological security, economic efficiency and social equity. While the ecology and equity could be taken care of at the stage of allocation of water rights, efficiency could be achieved at the stage of water utilisation through the transferability of water rights. Moreover, since the water rights system, specially the one based on the public trust framework allows social control and decisionmaking at the stage of allocation of rights and decentralised private decisionmaking at the stage of water utilisation, it provides an institutional synthesis with greater conflict resolution potential.

8. The need for recognising the common property resource character of water can be met by transferring the control of water to local communities or user groups, equipped with technical information on aquifers and their storage, recharge and transmission characteristics.
9. The enforcement and administration of established rights system should be left to the local trustees who can be PRIs, WUAs or outlet level communities.
10. There is a need to purge out the lingering colonial vestiges in the realm of water laws and make them relevant to the current socioeconomic requirements of the country. There is a need for comprehensive legal reforms bringing together central and state laws, rules and orders, customary laws and court decisions pertaining to water that may necessitate a constitutional amendment.
11. Water related legal reforms should incorporate, (a) economic incentives for water conservation; (b) conferment of private transferable water rights to water users; (c) allocation of transferable water rights to the landless; (d) allocation of water rights under public trust framework be demarcated appropriately at the level of state, community and individual users; and (e) unambiguous and appropriate definition of the community at the macro, meso and micro basin/watershed levels.
12. The doctrine of riparian rights needs to be re-interpreted. The public claim on water as public and common property resource is reflected in legal concepts as 'reasonable', 'beneficial', 'non-wasteful' and 'public trust'. Water laws should take these principles into account. The law should also incorporate notions of accountability and provide for expeditious remedies for violation of rights. The law should also recognise hydrological unity of water resources and provide for conjunctive use of surface and groundwater resulting in integration of administration of surface water rights and groundwater rights.
13. Groundwater is too valuable a natural resource to be left unmanaged. In the absence of well-defined property rights to it and with rapidly growing water markets and widespread use

of modern water extraction technologies, groundwater is prone to exploitation and is in fact, being overexploited in many arid, semi-arid, hard rock areas, and even in a state like Punjab. For avoiding its overexploitation and depletion, it is necessary to change the status of groundwater from a common pool resource to a cooperative or joint property of the users through creation of cooperative usufructuary rights and vesting them in water users' cooperative societies. Implementation of this model of cooperative management of groundwater would require, among other things, enactment of a groundwater law for governing the use of groundwater, creation of groundwater management zones/districts/sub-districts, organising water users into some sort of formal associations and helping the water user bodies with technical information, funds and management advice.

14. The groundwater law should generally provide for the following, but the laws/rules should be region specific:
    (a) Delineation of water management zones/units/blocks.
    (b) Formation of water users and handing over the management to user associations or cooperatives.
    (c) Creation of well-defined usufructuary groups and individual rights in groundwater.
    (d) Framing of practicable rules and guidelines for regulating number, size and spacing of wells/tube wells.
    (e) Leasing by the state of usufructuary water rights to associations/cooperatives.
    (f) Allowing usufructuary rights to be sold among members of the associations/cooperatives.
    (g) Permitting associations or cooperatives to acquire private wells/tube wells for common/group use and install/construct new wells/tube wells.
    (h) Recharging of aquifers.
    (i) Fixing water charges taking into account the real resource cost of water.
15. In the case of surface water, there is a need for a similar legal and management structure with allocation of water rights to groups/

associations/cooperatives on the basis of present allocation of water and transfer of the management to these groups.

16. At the user's level, out of several options, only volumetric rationing of water, coupled with unit pricing can result in effective reduction in demand for water and groundwater draft.

17. Social engineering of the formation of water users associations becomes much easier under the public trust framework (as against the private property regime), mainly, due to five main factors: (a) there are no conflicts of ownerships; (b) state is not the dictating partner; (c) local-level flexibility is feasible in organising the association and managing the water resources; and (d) it is possible to clearly define the user charges.

18. For success, sustenance and viability of participatory or water users institutions, several factors are important, which should be recognised. The first precondition is that it should emerge from an earnestly-felt need of the stakeholders or water users. Such a need may arise out of the stress conditions imposed by the earlier system of management. The second set of factors is the intrainstitutional requirements that include definition of membership, role of members, mechanism set in order to resolve conflicts, qualities of leadership, financial stake of members, and the size of the institution. The third set includes, such extrainstitutional factors like financial institutions, marketing system, other local organisations, policy environment, availability of infrastructure and mechanism of interaction with the state. And the fourth set of factors relate to, transparency in financial matters and decisionmaking within the organisation.

19. Continuous watch on water balance (demand and supply) situation across agroclimatic and hydrogeological zones will be necessary to demarcate water grids, on the lines of power grids, to promote and plan management of water resources, including interbasin transfers of water.

20. A mere five per cent saving of water in agricultural sector can spare sufficient water to meet the demand of non-agricultural sectors. Substantial water is lost through evaporation in the unlined watercourses. The seepage is also not insignificant. Therefore, popularisation and encouragement of appropriate

water saving scientific methods and technologies should be inbuilt into transfer of technology packages in different agro-climatic regions and sub-regions.

21. National Agricultural Research System needs to emphasise development of water saving varieties/cultivars, agronomic practices, and mechanical devices suitable for different agro-climatic regions and sub-regions. In addition, research on economic efficiency on water use under alternate water property regimes also needs high priority.

15

Agricultural Scientists' Perceptions on National Water Policy

Introduction

Water is an essential resource for sustaining life and environment. In view of its limited availability and competing demands, it is imperative to utilise it with utmost efficiency. Its development and efficient use are necessary to meet the basic needs of biotic populations and maintain a congenial environment. At various stages in the history of Indian civilisation, knowledge has been building up in improving the development and judicious management of water resources. Our forefathers were conscious of the importance of this precious resource and the consequences arising from its mismanagement as is evident from:

> No grain is ever produced without water, but too much water tends to spoil the grain. An inundation is as injurious to growth as dearth of water.
>
> — *Narada Smriti XI,* 19

> Rishi Narada enquired from emperor Yudhistira whether the farmers were sturdy and prosperous and whether the dams had water for distribution in different parts of the kingdom.
>
> — *Kaushika Sutra* (3150 B.C.)

Highly prosperous early civilisations around Mesopotamian plains and the Yellow River, Indus and Nile river valleys are known to have perished when they failed to appropriately operate and judiciously manage large water bodies in irrigated agriculture. Equally glaring examples are available from our recent past in India when mismanagement of canal waters contributed to acute problems of land degradation. However, of late, the issues of environment, equity and

---

Policy Paper 1, 1995.

economic competitiveness which were not considered important a couple of decades ago, have appeared on the centre stage of land and water development programmes. Pressure groups on water- and land-related environmental issues throughout the world have created an awareness which was reflected in repeated international declarations culminating in the recent Rio-summit.

This awareness has led to formulation of national policies on land use, environment, forests and agriculture in India. Issues relevant to water, figure prominently in all these policies. However, there are differences in areas of focus giving conflicting signals which should be resolved for creating harmony in the implementation of these policies. On the other hand, under pressures of growing biotic populations, fresh water is subjected to several competing demands. In view of the above concerns, the Government of India adopted a National Water Policy in 1987. Although, the policy is an important instrument for the conservation and efficient and equitable use of water, yet critics feel that it has not emerged through extensive national debates among intellectuals, scientists, progressive consumer groups and policymakers.

The National Water Policy (see Annexure A-15.1) indentifies provision of water for drinking, Irrigation, flood control, hydropower, navigation, industrial and other uses in that order, as the primary objectives of water resources develoment. It proposes planned, integrated, multidisciplinary, scientific, and multiobjective development and management of water resources to meet the changing water needs. The policy also lays down that maintenance and enhancement of environmental quality, and social and economic growth with equity, be important considerations in meeting water resources development objectives.

The National Water Policy is essentially a statement of goals and intentions of the government. It has to be set into action through specific programmes, projects, and regulations. This is not easy because specification of objectives for individual projects is subjective and value based. Implementation of the projects involves multiple agencies and requires reliable predictions of responses of the water resource systems to natural inputs, human decisions and complex land-water interactions. These include physical, chemical and biological responses, and future economic, social, political and environmental impacts. Predictions of such responses are difficult because water resources systems are complex. Therefore, the implementation of the National Water Policy calls for a multidisciplinary scientific approach.

Agriculture, being the major user sector using over 80 per cent of available water resource potential, the National Academy of Agricultural Sciences is concerned about its use and consequences. A Round Table was held in August 1994, wherein scientists, technologists, sociologists, economists and policy makers discussed the issues relating to the National Water Policy in depth. This was followed by a national symposium during the Second Agricultural Science Congress in January 1995.

Water Resources (Supply)

The main source of sweet water on earth's surface is precipitation in its various forms. Precipitation is a limited, renewable, natural resource with variable distribution over both space and time. Within the agroecological regions it exhibits certain typical characteristics of distribution and amount received. The precipitation received on the land surface is partly infiltrated and converted into soil water, and used by vegetation. Part of infiltrated water evaporates and the rest seeps down to join the groundwater or appears as springs or streams through subterranean flows. The remaining part of the precipitation on the land surface takes the form of surface runoff or evaporation. Wherever the physiography permits, the surface runoff and groundwater are exploited for various uses.

Assessment of the water balance over space and time provides the basis for estimating the water resource potential. Based on this potential, water is allocated to various uses. In India, information on various aspects of precipitation are fairly well documented over long period of time by India Meteorological Department. Data on infiltration and evaporation losses which also affect the stream and river flows on the one hand and soil water and groundwater on the other are, in general, scanty. Thus, it is very difficult to assign any confidence limits on the assessment of potential and utilisable water resources, and reconcile the assessments projected by various expert panels based on updated scientific information. There is no institutional mechanism to bridge the growing gaps in information provided by various ministries on utilisation of the created irrigation potential. There is an urgent need to resolve the following issues in a given time-frame:

a. There has been a continuous progress in the development of methods in assessment of surface and groundwater resources. However, they need to be reinforced with improved scientific knowledge and precise field data on surface and sub-surface hydrology.

b. The concept of irrigation potential needs in-depth examination to precisely define all terms related to water so as to quantify its use in agriculture. It is necessary to express the irrigation potential in volumetric terms rather than on irrigated crop area basis, as the latter does not account for variations in climatic, soil and crop characteristics. The discrepancies between the data on irrigated area based on land records by the Ministry of Agriculture, and the records of potentials created and utilised by the Ministry of Water Resources need attention (Figure 18.1).

Figure 15.1

Irrigation Development in India: Gap in Irrigation Potential Created and Utilised

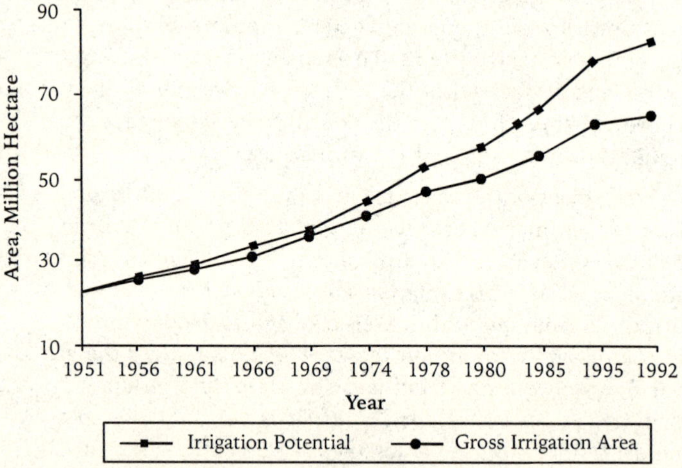

As part of the above, discrepancy arises from differences in perceptions and definitions for determining the potential use, an expert panel needs to be constituted to develop uniformity in approaches between the concerned government departments.

c. Sourcewise information on minor irrigation (ponds and other sources and groundwater separately) needs to be systematically collected and documented.

d. There is a need for coordination among ministries of Water Resources, Agriculture and Rural Development on potential developed under the integrated watershed programmes. A perspective plan with a time-frame for development of the ultimate potential when the country will be fully covered under this programme, needs to be evolved.

e. There is lack of national consensus on the role of water in continuing food security. There is growing divergence between the projections by the planners and actual water use by competing non-food crops. According to the Eighth Plan document, the area under irrigated foodgrains is projected as 62, 70, and 78 million ha at the end of Eighth, Ninth and Tenth Five Year Plans, respectively, which leaves a gap of around 10 to 15 million ha in actual irrigated area under foodgrains.

f. Besides river basin- and sub-basin-wise records, presentation of data at agroecological unit level is desirable for realistic developmental planning.

g. The draft National Water Information Bill approved by the National Water Board envisages access to the computerised databank to central and state government agencies. However, it does not provide access to research organisations. All the relevant data must be made accessible to various authorised scientific research organisations so that research inputs can be effectively incorporated in the planning and operation of water resource projects.

h. Tanks are as old as human civilisation. They are the earliest rain harvesting structures used by our ancestors for sustainable growth of human life. They were treated as community assets and maintained by the community both for domestic and irrigation needs. Neglect by the administration and the community during the past two centuries led to the decay of these valuable water-supply systems. In the recent past, however, there have been efforts to restore these tanks to their earlier status. Besides the role of governmental and non-governmental organisations, there is a need to strengthen the role or research and development components in this huge task, particularly with respect to tank design and management practices which can contribute to reduction in evaporation losses, prevention of silting and protection from pollution.

Water Resources (Demand)

Demands of water for domestic, agricultural, industrial, power generation, navigation and recreation are dynamic. Irrigation demand for agricultural sector also varies with time and space due to the differences in agroclimatic conditions, changing cropping patterns, market mechanisms and land uses. The scientific studies with respect to these are very few, and hence, hardly any confidence on the estimates could be afforded on the present and projected water uses.

In the backdrop of the above, the following issues need attention:

a. The quantitative projections of allocations to meet demands of various sectors are not yet specified since actual observations are very limited in most of the sectors. For realistic estimations/projections, database needs to be strengthened.

b. Irrigation demand with respect to agroecological zones considering current and anticipated cropping patterns is yet to be established.

c. There is an urgent need for developing a time-bound research agenda to work out the demands of various sectors in a dynamic frame and develop a basis for conflict resolution among competing demands by various sectors, and within each sector, on considerations of equity, ecology and economics.

Need for Rational Use of Water Resources

The major source of water, namely rainfall, is unevenly distributed giving rise to regions of high water resource availability and extremely dry regions. This is further compounded by its monsoonal character, with about 85 per cent of annual rainfall occurring over 4 to 5 months, with actual number of rainy days varying between 4 and 56, during a year. Water storage can help to even out temporal imbalances in water availability during monsoon season as well as across years. So far, only one-third of identified storage capacity (650 km^3) has been achieved. This backdrop calls for appropriate conservation of water resources in rivers, lakes, ponds, soil profile and aquifers, and its rational use during lean periods.

In view of the ever increasing demands for water, there is need for augmenting the resource through interbasin transfers, artificial recharge of aquifers, use of marginal quality water, conjunctive use of surface and groundwater, runoff harvesting in rainfed areas, watershed development, etc.

Agriculture being the major user of water in India, there is a need to enforce rational use of water in this sector. This would encompass development of appropriate cropping patterns, efficient water conveyance and application methods, and irrigation scheduling systems as also rational water pricing, and farmer education and training.

Water resources and energy management related issues represent other grey areas demanding attention of water scientists. These issues have an important role in efficient management of water resources, particularly the groundwater. Most of the states are yet to develop an energy plan along with alternate sources of energy for utilisation of untapped reservoirs of groundwater, particularly in the northeastern region. The National Water Policy has not dealt with this aspect appropriately. An interlinkage between the Energy Policy and the Water Policy of the country is necessary.

Experience gained during the last 45 years has underlined the importance of proactive action with reference to potential negative consequences of taking irrigation water to unirrigated areas. Learning from this experience, we should hereafter make an adequate provision for research and training in each major and medium irrigation project, so that sustainable farming and land-and water-use systems are developed even before water becomes available for irrigation. This will then enable farmers to take to sustainable and scientific water-use practices right from the day irrigation water becomes available and thereby, help to avoid problems like waterlogging and salinisation.

The National Water Policy stresses the need for intensification of research efforts and provides an indicative list of topics on which research efforts should be pursued. But the emphasis therein has been on physical, scientific and technological systems. The sociological, institutional, operational and economic aspects of management which are vital for the success of agriculture, have not been indicated. Therefore, the list needs to be supplemented with items like:

- development of methodologies for assessing irrigation performance,
- development of management information systems,
- improvement of water delivery and application systems,
- water quality protection,
- conjunctive use of surface and groundwater,
- disposal of drainage effluents,
- recycling of industrial and agricultural effluents,
- effective use of marginal quality water, and
- water-use efficiency enhancement through crop improvement and related agronomic practices.

In the past, in most states, the irrigation system used to be centrally managed by a single agency which was responsible for water resource development as also for primary, secondary and tertiary level management. The tasks of water allocation, distribution and monitoring also rested with it. This had to be changed to improve performance. In the seventies, mutidisciplinary authorities were constituted but were not implemented in the true spirit. They have taken the shape of one other department, rather than an integrated multidisciplinary team. Their restructuring in the desired direction is an immediate need.

A consultative group comprising scientists from agricultural and other institutions and departments dealing with development and use of water resources, both at national and state levels, should be constituted for scientific planning of water resource development, management, and utilisation and for strategic guidance to the multidisciplinary authorities.

Linkages between the Ministry of Water Resources, State Departments of Agriculture and Allied Sectors, State Agricultural Universities and ICAR institutions, about the canal water management, reservoir sedimentation, and conjunctive water use planning, execution and management should be strengthened. Institutional linkages need to be forged amongst the Ministry of Environment, Ground Water Board, ICAR institutes and other institutes dealing with the subject to identify researchable issues and undertake research programmes on pollution of groundwater arising from intensive use of agrochemicals in production agriculture.

Skyrocketing cost of development of water resources are making water resources projects, economically unsustainable (Figure 18.2). If market mechanism is allowed to play its role, the chances are that not only enough funds will be generated to develop the water resources, but also the quality of service would improve and the use of water will become efficient. A case in point is that of promoting the groundwater markets that have developed in states like Gujarat and Tamil Nadu.

Figure 15.2

Cost of Creation of Irrigation Potential through Major and Medium Projects

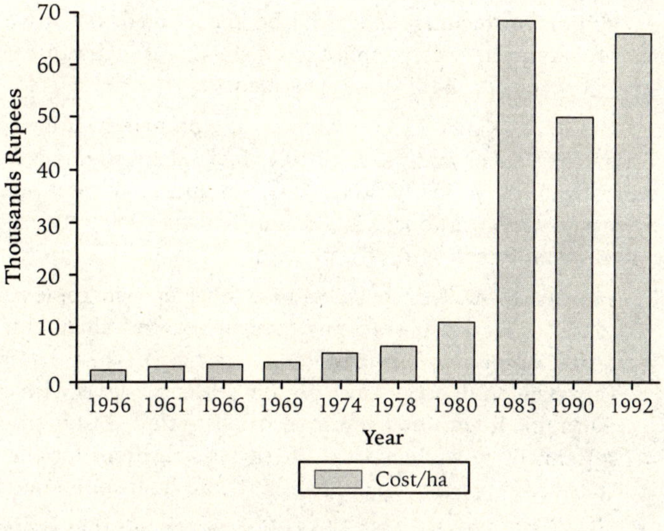

Current rates of canal water and electricity charges based on horsepower of the motor, irrespective of energy used are conducive to wasteful use of water. As enumerated in the National Water Policy, the pricing structure of water is overdue for revision so as to reflect the realistic cost of supply of water as well as to be a deterrent against wasteful use of water. The pricing shall be based on volumetric basis.

Rain Water Management Issues for Dryland Agriculture

Dryland areas in India extend to over 100 million ha of the cropped areas. Here the source of water for crop sustenance is rainwater. The rainfall in these areas varies from 400 mm to more than 1,500 mm per annum, which is received in one or two short rainy seasons. The annual evapotranspiration demand is about 1,500 mm in the high rainfall areas and above 2,000 mm in the low rainfall areas. Appropriate rainwater management is the key to sustainable agriculture in the dryland areas of India. Water use policy issues for different dryland agroecosystems are:

a. *Arid Subtropics:* In this ecozone, water use policy should aim at progressively increasing area under agroforestry and agropastroral land use that have an animal component, and decreasing the annual cropped area. Soil and water conservation by vegetative cover and dune stabilisation techniques should be encouraged. Incentives on promotion of efficient water application methods for high value crops, like arid horticulture, be liberally provided.

b. *Semi-Arid Tropics (SAT):* This zone comprises subtropical northern Indian plains and the peninsular Indian SAT. In this ecozone, the annual rainfall varies between 500 mm and 1,500 mm and the annual evapotranspiration is 2,000 mm or less. Water use policy suggested for different agroecosystems of SAT are:

i. *Subtropical SAT:* Key considerations in water management need to be, conservation of rain water *in situ* and channelling of runoff water into groundwater for recharge. There has to be emphasis on drainage during rainy season in areas with >750 mm rainfall and intensification of post-rainy season cropping systems by providing 1 to 2 life saving irrigations for ensuring stable crop production at increasing thresholds of productivity.

ii. *Dependable Tropical SAT:* The policy for sustainable water use should comprise the creation of a series of interlinked watersheds both in the Vertisol and Alfisol regions. This will permit recapturing the past heritage of Kakatiya and Chola dynasties who developed networks of interlinked small, medium and large tanks involving small groups of individuals, communities and society at large.

iii. *Undependable Tropcial SAT: The* soil and water conservation policy for these areas must emphasise the sustainability of post-rainy season cropping by conserving water in the soil profile.

c. *Bimondal Rainfall Areas:* Rejuvenation of the tank irrigation systems and introduction of limited irrigated-dryland agriculture are the relevant key policy issues for these regions.

d. *High Rainfall Areas of the NE India and Western Ghats:* The key policy issues here are related to land use. The area under agroforestry systems must be increased to sustain agricultural productivity. This region is exposed to excessive soil erosion. This calls for strengthening of research programmes for generation of integrated technology involving engineering and biological inputs for erosion control.

Environmental Concerns in Water Resources Development and Utilisation

The National Water Policy states that all the water resources projects should be planned and executed so as not to cause adverse impact on the environment and the ecology. However, in view of the continuing low irrigation efficiency in the case of major and medium projects, environmental problems like waterlogging, soil salinisation and impairment of water quality have manifested at many places. Only about one-third of the water diverted for irrigation effectively reaches the crop, and the remaining water either runs off or moves down to groundwater body to cause environmental problems. It is estimated that about 24 per cent of the areas irrigated by major and medium projects is threatened by waterlogging and associated problems. Such ill effects are direct consequences of inadequate planning combined with unscientific water management practices.

Efficient management of the developed water resource needs to be the central concern for the two-fold advantage of bringing about enhancement in water-use efficiency and simultaneously, reducing the environmental degradation. In fact, optimising the efficiency of the use of existing water resources should be the first charge of the government.

There has been a general lack of rigorous analysis of the likely environmental impacts associated with a particular project and the exploration of possible modifications or alternatives to shape an informed decision so that environmentally unacceptable and unexpected results could be avoided. What is needed now is that the exact nature of the environmental impacts should be monitored with specific reference to the prevalent cropping systems. This will help to obtain information to evolve corrective strategies and also to provide a better basis for future decisions to develop irrigated agriculture without the environmental damage.

The ecological characteristics have not often received due attention in planning the development of the regional water resources. Developmental patterns tend to be stereotype, imitating the success stories of some regions blindly. Consequently, the positive impacts on agricultural productivity and overall economic growth have not been achieved as targeted while the negative impacts on the environment are discernible. The major irrigation projects in the high rainfall and the black soil regions having accumulated salts in the soil profile, bear testimony of the resultant environmental degradation. In the case of new projects, assessment of the proposals should be made in the context of ecological features and probable ecological effects and realistic benefits. The total project cost should include realistic environmental protection costs. Scientific procedures are now available for such an assessment and it should be further strengthened by employing modern research tools and techniques.

In view of increasing demand of water from domestic and industrial sectors, curtailment of fresh water supplies to agriculture is likely and increasing use of marginal quality water for irrigation is inevitable. Research is needed to develop site-specific management strategies for sustainable agriculture under such situations. At the basin level, provision has to be made for removal of some salts out of the system. This should hold good even in the case of fresh water irrigation, since large quantities of salts are added annually which would eventually cause deterioration in soil and groundwater quality.

With increasing agricultural, domestic and industrial intensification, pollution of groundwater is likely to increase. Groundwater once contaminated is damaged for a long time to come. This should be an important concern since prevention of groundwater pollution is always considered to be a better strategy than adopting ameliorative measures after contamination. This aspect was ignored in the Clause 7.1 of the National Water Policy and needs attention.

In the industrial uses of water for disposing of heat and other wastes, even though the amount of water used is small, its polluting effect on water quality is relatively large. Industry-specific effluent standards have already been developed and wherever they do not exist, they need to be developed. What is needed is an effective regulatory effort for strict enforcement of these standards, since these effluents ultimately find their way into agricultural lands.

An extensive water quality monitoring system for surface and groundwater is required. This will help in identifying the kind, extent and the sources of pollution, and bring to focus the need for undertaking corrective steps. The current monitoring network is obviously inadequate. Moreover, it is always from the states. States will have to be encouraged to build up their secondary and tertiary networks as extensions to large-scale primary network of the Government of India.

The research and development effort in reconciling the conflicts arising from growing intensification of coastal irrigated agriculture and conservation of coastal ecosystem is extremely inadequate. There is an urgent need to establish a research network to take care of multidisciplinary diverse problems relating to the management of coastal ecosystems.

Environment management system with appropriate infrastructure needs to be set up within the water resource management framework and be made an integral part of it for strengthening environmental input in the water resources projects. This should include rigorous analysis of environmental features, improved prediction of ecological effects with possible modifications and options, and regular monitoring of the environmental parameters/impacts to evolve environmental protective strategies.

On-Farm Water Management

There is an immense scope for improving conservation, distribution and on-farm utilisation of irrigation water. The first step for efficient use of any resource is to conserve the same for optimising its availability at the site of use. The water available for irrigation must be allocated optimally to competing regions and crops. The other means for efficient on-farm water management include, optimum scheduling of irrigations based on suitable methods, conjunctive use of water from different sources, improved agronomic practices, etc. Under irrigated agriculture, provision of adequate drainage is crucial for efficient water management.

Apart from providing right amounts of water at the right time and right place, attaining of high water-use efficiency requires that crop yield be maximised with given amount of water. For this, it is essential to adopt improved soil and crop management practices to overcome the constraints to high yield. Tillage, mulching and fertilisation, and combinations thereof have been observed to substantially increase crop yield and, thus, increase water-use efficiencies by reducing evaporation and deep percolation components of soil water balance. Instances can be cited from several areas in respect of quite a few crops where judicious use of crop production inputs along with irrigation water has contributed to the miximisation of the crop yields. "Compared with around 1.4 mt of wheat production in erstwhile Punjab (Current Punjab plus Haryana) during 1950-51, it increased to 20.57 mt of wheat from the same region. This is more than three times of the total wheat production of 6.46 mt in the country during 1950-51. During the same period, the productivity of wheat increased from around 7 quintals to 38.6 quintals per ha in this region. The above inspiring success came from judicious integration of irrigation with a balanced package of wheat crop production inputs."

Issues Demanding Urgent Attention

a. As much as 50 per cent of the water released at the project head-works is lost in transmission to the canal outlets. Additional loss occurs in water courses which is directly proportional to their lengths and duration of water flow. Lining of water courses reduces these losses to bare minimum and has found favour with the farmers of command areas. Lining of main canals may also be considered if losses are excessive or there is a threat of waterlogging and soil salinisation. Where the groundwater can be exploited for irrigation, there is hardly any justification for lining as the farmer has better control over groundwater which adds to timeliness of water applications. Where groundwater is unfit for irrigation, seepage from canal network can play havoc with soil productivity.

b. Optimal allocation of available irrigation water among competing regions and crops is possible with the knowledge of water production functions of crops. At present, precise production functions are lacking and there is an urgent need to develop the same at least for important crops in different regions. Improper allocation of water undermines the equity aspect. For example, in Maharashtra, only 3 per cent of total cropped area occupied by sugarcane is claiming 76 per cent of irrigation water, while other crops are denied even a lifesaving irrigation.

c. Efficient use of irrigation water requires that water be applied to growing crops at appropriate times and in adequate amounts. Where sufficient water is available to fully meet the crop water requirements, irrigation must be applied before yield-or quality-reducing stress occurs. Scheduling irrigation with limited water is a big challenge to agricultural scientists and needs rigorous research. Available information on simulation and optimisation models can be used to develop optimal irrigation schedules to crops.

d. Modern irrigation techniques like sprinkler and drip should be promoted where water is scarce, and the topographic and soil conditions do not permit efficient irrigation by conventional methods. Promotion of such water-saving devices should be an objective of the National Water Policy.

e. Irrigability of command areas, needs to be ascertained and land should be developed, and farmers educated about irrigation management before letting water into the command area. Concept of irrigability of soil has to be considerably fine-tuned with respect to different crops. There is considerable scope for research on this account and should be included in the list of topics to be researched.

f. Consolidation of landholdings is essential to enthuse farmers to invest in irrigation system at the farm level.

g. A group of experts concerned with canal water distribution and efficient on-farm use of irrigation water together with economists and, sociologists and progressive farmers, be constituted in each canal command with the following mandate:

   i) To overcome the existing mismatch between crop water requirement and supply of water because of rigid canal schedules.

   ii) To work out the procedures to ensure uniform supply to

head-end and tail-end farmers by making compensation in time for the reduced tail-end discharge.

iii) To develop a multidisplinary performance assessment system for assessing the impact of irrigation projects from the points of view of economics environment and equity. Such a performance assessment system should be built into irrigation management practices. This will help to initiate timely mid-course corrections, when needed and keep the efficiency of the system high.

This group shall be an integral part of the CADAs.

Elimination of Constraints to Higher Yields

a. Conjunctive use provides a greater control on timeliness of irrigation and should be encouraged by making adequate energy available to farmers at a reasonable cost. In order to ensure sustained availability of groundwater, average annual withdrawals should not exceed average annual recharge. Where fresh water is in short supply, groundwater of marginal quality could be advantageously used in combination with good quality water or for alternate irrigations. Recommendations concerning saline and sodic water developed by Indian scientists from field experiments are now available and should be used.

b. Development of conjunctive use system requires knowledge regarding geology of groundwater basin and aquifers, hydrology of surface and groundwaters, existing surface and groundwater facilities and storage and transmission characteristics of the basins. Although, some efforts have been made in India to predict groundwater behaviour using simulation and optimisation models, these need to be strengthened.

c. Drainage has to be an integral part of the irrigation system, particularly when perennial irrigation is contemplated. This consideration does not find a mention in the National Water Policy. Hence, provisions in Clause 10 on irrigation will have to be adequately amplified to cover this critical aspect.

d. Excess water in the root zone causes aeration stress in crops other than rice. This situation develops on fine-textured, low permeability soils after a heavy rain/irrigation or in waterlogged soils. Surface and sub-surface drainage has been shown to improve crop production.

Development of Human Resource in Water Science and Technology

Water science and technology deals with the study of nature and occurrence of water, assessment of its quantity and quality, collection, distribution, management, use, recycling, pollution treatment and the associated economic aspects. Thus, water science and technology is central and basic to many aspects of water resources development and management, operation and is mutidisciplinary in nature.

Appropriately trained manpower is required in all the state and central departments to facilitate modernisation and management of irrigation systems and other works related to improved water management.

At present, a cadre of water management technologists to extend the known technology concerning optimal irrigation scheduling and agronomic practices for increasing water-use efficiency of crops is lacking. There is strong need to create such a cadre in the state departments.

There is a need to start postgraduate degree programmes in water science and technology to train manpower to cater to the research and development needs in this area for persons with bachelor's degree in agricultural engineering, civil engineering or agriculture. The universities have necessary capability for initiating such programmes. Modern developments in the areas of crop sciences, physical sciences, computer sciences, social sciences, management sciences, operations research etc., can be advantageously utilised in the teaching and training of professional staff in the areas of water resources management.

Short-term training programmes for in-service officers could appropriatly be utilised by the development departments along with proper financial and promotional incentives and institution of departmental fellowships. This will be a suitable mechanism to upscale job-related skills of the technical staff which can enhance their capabilities in tackling many a practical field level problems in water management.

Water and Land Management Institutions (WALMIs) have done a good job in training different levels of in-service staff of irrigation departments and this activity should continue. They should be linked closely with agricultural or engineering universities for a healthy mutual interaction and optimising the use of educational resources, particularly libraries and highly trained faculty.

Though the various agencies associated with water science and technology at the state, national and international levels have been making efforts to impart training and create public awareness, a systematic approach is yet to be followed. More emphasis should be placed on courses dealing with water issues in programmes for farmers, village-level workers, public as well as informal professional education.

The science and technology support to research in water sciences has suffered because of inadequacy of investment and trained manpower. Fortunately, the existing universities have the capability to generate the human resources through formal education and in-service training at all levels for functionaries involved in water management and the farmers. The facilities need to be appropriately strengthened for this purpose.

Of late, there has been a growing realisation for improving the quality of life of rainfed farmers. The current national and international initiatives *vis-à-vis* the magnitude of these problems are inadequate. A special effort needed for these areas to disseminate existing improved technologies and to further refine site-specific soil and water conservation technologies through research. Constitution of national/state water management cadre as suggested earlier will need to be appropriately strengthened and extended, to meet the demands in rainfed areas.

Water Law: Policy Direction

The National Water Policy is silent on this aspect. However, the following issues demand attention:

a. The doctrine of riparian rights, basis of water legislation in India needs to be reinterpreted in the light of perceived societal necessities of the time and more importantly, the need to be in tune with the constitutional values. Interpreting Article 21 of the Constitution of India expansively so as to include right to water as a fundamental right to life, the Supreme Court has given a new dimension to principles of water laws. Further, the Constitution in Article 39(b) (Directive Principles of State Policy) has made it incumbent on the state to ensure that the ownership and control of the material resources of the community are so distributed as to best subserve the common good. Water being a precious natural and common prosperity resource, water law must take into account the principles of access, justice and social equity.

b. Law needs to recognise the hydrological unit of water resources and provide for conjunctive use of surface and groundwater, resulting in integration of administration of surface water rights and groundwater rights. Groundwater should be presumed to be tributary to natural streams.

c. Law must also incorporate notions of accountability of water bureaucracy to ensure access to the availability of water and provide for expeditious remedies for violation of rights. Water-related disputes may be entrusted to water 'Adalats' for speedy resolution of disputes.

d. There is an urgent need to have legislation on groundwater. The legal regulations have, of necessity, to be region specific, namely groundwater abundant regions, scarcity regions and islands surrounded by oceans. The Government of India has already formulated a model groundwater bill but the states have not yet established the legislative measures in that direction. Proper climate needs to be created in the country for the acceptance of such legal measures.

e. Development and regulation of interstate river waters, mandates cooperation amongst basin states along with the existing procedures for compulsory adjudication of disputes by tribunals. Constitution may be amended to incorporate a provision on interstate agreements (compacts) in respect of interstate rivers.

f. The existing law on water quality needs to be effectively implemented for prevention of pollution of surface and groundwaters. Groundwater pollution being more serious and hazardous than surface water pollution would require different institutions for prevention and abatement of pollution.

Water Rights

There is need to institute an effective water rights regime which should incorporate the economic, technical and organisational requirements. From an economic point of view, water rights system should allow (i) definition of water rights on a private/individual basis and (ii) transferability/exchange of water rights among users. Defining rights on an individual basis establishes an upper limit for water withdrawal and use. Such limits would provide incentives for individuals to use water more efficiently. Similarly, transferability or renting out of water rights enables individuals to directly realise themselves the economic benefits emerging from selling the amount of water they saved by their conservation efforts.

Although, private rights system is necessary for ensuring efficiency, it is not sufficient for promoting equity and ecological security in water use. For this purpose, the rights system should be designed in such a way as to allow effective social control not only to set limits for collective withdrawal but also enforce and monitor water allocation and distribution across regions and individual users. In this respect, it is necessary to design the private rights system itself within the public trust framework. This framework provides for a hierarchical system of rights and duties applicable for the state, community, and individuals.

Under the public trust framework, the state has the duty and responsibility for establishing legal guidelines for instituting a water rights system, setting limits for overall water withdrawal in a regional context, providing technical information, and coordinating water allocation across regions/communities. The village-level community organisation (e,g. *panchayats*, user groups, etc.) have the duty for allocating individual water rights consistent with the legal guidelines established by the state-level legislation, enforcing the monitoring of water withdrawals by individual, creating a forum both for promoting water rights transfers as well as resolving allocational disputes among the users. The individual users have water rights that specify their water entitlements which they can either use on their own farms or sell to others. Notably, water rights of individuals within the public trust framework are not absolute but subject to the regulatory rights of the state and communities.

The water rights system operating within the public trust framework allows social control and decisionmaking at the stage of water rights distribution and decentralised private decisionmaking at the stage of actual water utilisation. It provides an institutional synthesis with greater potential for conflict resolution. Moreover, such a system also proves to be a rare policy instrument that could simultaneously address all the three major goals of sustainable development, i.e. ecological security, economic efficiency and social equity in the context of water resources.

While ecology and equity requirements can be taken care of at the stage of water rights allocation, where social control is exercised to set limits for collective withdrawals as well as to ensure more equitable distribution of water across regions and individuals, economic efficiency is achieved at the water utilisation stage by allowing transferability and water exchanges that ensures the water to flow into the most productive uses and regions. The transferability of water rights is an important requirement for promoting socially optimal water resources use and income distribution.

Research studies, especially on establishing water balance at the regional and village levels, identifying organisational forms, developing distribution criteria etc., are extremely inadequate and should be strengthened urgently. The economic and ecological compulsions for an institutional change will eventually overwhelm issues like political exigency, practical difficulties, etc., which seem to have delayed the institutional change so far. It is about time now to formulate a viable national policy for establishing a water rights system in India. The National Water Policy, limits itself to management of irrigation systems by farmers and does not go beyond that, into the aspects of water rights and the need for establishing and regulating them. The existing provisions of Clause 12 need to be substantially expanded.

Social Issues

The National Water Policy emphasises that:

> Efforts should be made to involve farmers progressively in various aspects of management of irrigation systems, particularly in water distribution and collection of water rates. Assistance of voluntary agencies should be enlisted in educating the farmers in efficient water management.

It is now well recognised that strengthening government assistance to farmer management systems is more cost-effective compared to assistance to government-managed systems. Farmers' participation may be encouraged in respect of watershed development, water harvesting, soil conservation, wasteland development, village tank operation, social forestry, mitigation and management of water-related natural hazards and irrigation water management. The recent amendment to the Constitution of India devolving powers to the people through the *panchayat raj* system should be advantageously used.

The following issues have not been adequately addressed in the past and need detailed examination for development of appropriate subsystems:

a. Type and elements of the modes of transfer of the management of water resources to the Water Users' Associations (WUAs) (e.g. decisions on cropping systems, water releases, collection of water rates, maintenance of infrastructure, restoration and management of village tanks, organising inputs etc.).
b. Changes needed in the roles of governmental agencies and farmers' organisations.
c. Legal policy, infrastructural and institutional issues.
d. Evaluation of the impacts of management transfer.
e. Adequate training of farmers' organisations.
f. Role of NGOs and other catalytic agencies.
g. Renovation of physical systems before affecting transfer.
h. Provision of adequate funds for major repairs and proper guidelines for use of financial resources and technical advice,
i. Role of WUAs in deciding cropping patterns.

ANNEXURE A-15.1

NATIONAL WATER POLICY

The Need for a National Water Policy

1.1 Water is a prime natural resource, a basic human need and a precious national asset. Planning and development of water resources need to be governed by national perspectives.

1.2 It has been estimated that out of the total precipitation of around 400 million hectare metres in the country, the surface water availability is about 178 million hectare metres. Out of this, about 50 per cent can be put to beneficial use because of topographical and other constraints. In addition, there is a groundwater potential of about 42 million hectare metres. The availability of water is highly uneven in both space and time. Precipitation is confined to only about three or four months in a year and varies from 10 cm in the western parts of Rajasthan to over 1000 cm at Cherrapunji in Meghalaya. Further, water does not respect state boundaries. Not merely rivers but even underground aquifers often cut across state boundaries. Water as a resource is one and indivisible—rainfall, river waters, surface ponds and lakes, and groundwater are all part of one system; water is also a part of a larger ecological system.

1.3 Floods and droughts affect vast areas of the country, transcending state boundaries. A third of the country is drought-prone. Floods affect an average area of around 9 million hectares per year. According to the National Commission on Floods, the area susceptible to floods is around 40 million hectares. The approach to the management of droughts and floods has to be coordinated and guided at the national level.

1.4 Even the planning and implementation of individual irrigation or multipurpose projects, though done at the state level, involve a number of aspects and issues such as environmental protection, rehabilitation of project-affected people and livestock, public health, consequences of water impoundment, dam safety, etc. On these matters common approaches and guidelines are necessary. Moreover, certain problems and weaknesses have affected a large number of projects all over the country. There have been substantial time and cost over runs on projects. In some irrigation commands, problems of waterlogging and soil salinity have emerged, leading to the degradation of good agricultural land. There are also complex problems of equity and social justice in regard to water distribution. The development and exploitation of the country's groundwater resources also give rise to questions of judicious and scientific resource management and conservation. All these questions need to be tackled on the basis of common policies and strategies.

1.5 The growth process and the expansion of economic activities inevitably lead to increasing demands for water for diverse purposes—domestic, industrial, agricultural, hydropower, navigation, recreation, etc. So far, the principal consumptive use of water has been for irrigation. While the irrigation potential is estimated to have increased from 19.5 million hectares at the time of Independence to about 68 million hectares at the end of the Sixth Plan, further development of a substantial order is necessary if the food and fibre needs of a growing population are to be met. The country's population which is over 750 million at present, is expected to reach a level of around 1,000 million by the turn of the century.

1.6 The production of foodgrains has increased from around 50 million tonnes in the fifties to about 150 million tonnes at present, but this will have to be raised to around 240 million tonnes by the year 2000. The drinking water needs of people and livestock have also to be met. In keeping with the objectives of the International Drinking Water Supply and Sanitation Decade Programme (1981-1991), adequate drinking water facilities have to be provided to the entire population in both urban and rural areas and sanitation facilities to 80 per cent of the urban population and 25 per cent of the rural population by the end of the decade. Domestic and industrial water needs have largely been concentrated in or near the principal cities, but the demand from rural society is expected to increase sharply as the development programmes improve economic conditions in the rural areas. The demand for water for hydro- and thermal- power generation and for other industrial use is also likely to increase substantially. As a result, water, which is already a scarce resource, will become even scarcer in future. This underscores the need for the utmost efficiency in water utilisation and a public awareness of the importance of its conservation.

1.7 Another important aspect is water quality. Improvements in existing strategies and the innovation of new techniques resting on a strong science and technology base will be needed to eliminate the pollution of surface and groundwater resources, to improve water quality and to step up the recycling and reuse of water. Science and technology, and training have also important roles to play in water resources development, in general.

1.8 Water is one of the most crucial elements in developmental planning. As the country prepares itself to enter the 21st century, efforts to develop, conserve, utilise and manage this important resource have to be guided by national perspectives. The need for a national water policy is thus, abundantly clear; water is a scarce and precious national resource to be planned, developed and conserved as such, and on an integrated and environmentally sound basis, keeping in view the needs of the states concerned.

Information System

2. The prime requisite for resource planning is a well-developed information system. A standardised national information system should be established with a network of databanks and databases, integrating and strengthening the existing central and state level agencies and improving the quality of data and the processing capabilities. There should be free exchange of data among the various agencies and duplication in data collection should be avoided. Apart from the data regarding water availability and actual water use, the system should also include comprehensive and reasonably reliable projections of future demands for water for diverse purposes.

Maximising Availability

3.1 The water resources available to the country should be brought within the category of utilisable resources to the maximum possible extent. The resources should be conserved and the availability augmented by measure for maximising retention and minimising losses.

3.2 Resource planning in the case of water has to be done for a hydrological unit such as a drainage basin as a whole, or for a sub-basin. All individual developmental projects and proposals should be formulated by the states and considered within the framework of such an overall plan for a basin or sub-basin, so that the best possible combination of options can be made.

3.3 Appropriate organisations should be established for planned development and management of a river basin as a whole. Special multidisciplinary units should be set up in each state to prepare comprehensive plans taking into account not only the needs of irrigation but also harmonising various other water uses, so that the available water resources are determined and put to optimum use having regard to subsisting agreements or awards of tribunals under the relevant laws.

3.4 Water should be made available to water short areas by transfer from other areas including transfers from one river basin to another, based on a national perspective, after taking into account the requirements of the areas/basins.

3.5 Recycling and reuse of water should be an integral part of water resource development.

Project Planning

4.1 Water resource development projects should, as far as possible, be planned and developed as multipurpose projects. Provision for drinking water should be a primary consideration. The projects should provide for irrigation, flood mitigation, hydroelectric power generation, navigation, pisciculture and recreation wherever possible.

4.2 The study of the impact of a project during construction and later on, human lives, settlements, occupations, economic and other aspects should be an essential component of project planning.

4.3 In the planning, implementation and operation of projects, the preservation of the quality of environment and the ecological balance should be a primary consideration. The adverse impact, if any, on the environment should be minimised and should be off-set by adequate compensatory measures.

4.4 There should be an integrated and multidisciplinary approach to the planning, formulation, clearance and implementation of projects, including catchment treatment and management, environmental and ecological aspects, the rehabilitation of affected people and command area development.

4.5 Special efforts should be made to investigate and formulate projects either in, or for the benefit of, areas inhabited by tribal or other specially disadvantaged groups such as Scheduled Castes and Scheduled Tribes. In other areas also, project planning should pay special attention to the needs of Scheduled Castes and Scheduled Tribes and other weaker sections of society.

4.6 The planning of projects in hilly areas should take into account the need to provide assured drinking water, possibilities of hydro-power development and the proper approach to irrigation in such areas, in the context of physical features and constraints such as steep slopes, rapid run-off and the incidence of soil erosion. The economic evaluation of projects in such areas should also take these factors into account.

4.7 Time and cost overruns and deficient realisation of benefits characterising most irrigation projects should be overcome by upgrading the quality of project preparation and management. The under-funding of projects should be obviated by an optimal allocation of resources having regard to the early completion of ongoing projects as well as the need to reduce regional imbalances.

Maintenance and Modernisation

5.1 Structures and systems created through massive investments should be properly maintained in good health. Appropriate annual provisions should be made for this purpose in the budgets.

5.2 There should be a regular monitoring of structures and systems and necessary rehabilitation and modernisation programmes should be undertaken.

Safety of Structures

6. There should be proper organisational arrangements at the national and state levels for ensuring the safety of storage dams and other water-related structures. The central guidelines on the subject should be kept under constant review and periodically updated and reformulated. There should be a system of continuous surveillance and regular visits by experts.

Groundwater Development

7.1 There should be a periodical reassessment on a scientific basis of the groundwater potential, taking into consideration the quality of the water available and economic viability.

7.2 Exploitation of groundwater resources should be so regulated as not to exceed the recharging possibilities, as also to ensure social equity. Groundwater recharge projects should be developed and implemented for augmenting the available supplies.

7.3 Integrated and coordinated development of surface water and groundwater and their conjunctive use, should be envisaged right from the project planning stage and should form an essential part of the project.

7.4 Overexploitation of groundwater should be avoided near the coast to prevent ingress of seawater into sweet water aquifers.

Water Allocation Priorities

8. In the planning and operation of systems, water allocation priorities should be broadly as follows:

- drinking water
- irrigation
- hydropower
- navigation, and
- industrial and other uses.

However, these priorities might be modified, if necessary, in particular regions with reference to area-specific considerations.

Drinking Water

9. Adequate drinking water facilities should be provided to the entire population both in urban and in rural areas by 1991. Irrigation and multipurpose projects should invariably include a drinking water component, wherever there is no alternative source of drinking water. Drinking water needs of human beings and animals should be the first charge on any available water.

Irrigation

10.1 Irrigation planning, either in an individual project or in a basin as a whole, should take into account the irrigability of land, cost-effective irrigation, options possible from all available sources of water and appropriate irrigation techniques. The irrigation intensity should be such as to extend the benefits of irrigation to as large a number of farm families as possible, keeping in view the need to maximise production.

10.2 There should be a close integration of water-use and land-use policies.

10.3 Water allocation in an irrigation system should be done with due regard to equity and social justice. Disparities in the availability of water between head-reach and tail-end farms and between large and small farms should be obviated by adoption of a rotational water distribution system and supply of water on a volumetric basis subject to certain ceilings.

10.4 Concerted efforts should be made to ensure that the irrigation potential created is fully utilised and the gap between the potential created and its utilisation is removed. For this purpose, the command area development approach should be adopted in all irrigation projects.

Water Rates

11. Water rates should be such as to convey the scarcity value of the resource to the users and to foster the motivation for economy in water use. They should be adequate to cover the annual maintenance and operation charges and a part of the fixed costs. Efforts should be made to reach this ideal over a period. While ensuring the assured and timely supplies of irrigation water. The water rates for surface water and groundwater should be rationalised with due regard to the interests of small and marginal farmers.

Participation of Farmers and Voluntary Agencies

12. Efforts should be made to involve farmers progressively in various aspects of management of irrigation systems, particularly in water distribution and collection of water rates. Assistance of voluntary agencies should be enlisted in educating the farmers in efficient water use and water management.

Water Quality

13. Both surface water and groundwater should be regularly monitored for quality. A phased programme should be undertaken for improvement in water quality.

Water Zoning

14. Economic development and activities including agricultural, industrial and urban development, should be planned with due regard to the constraints imposed by the configuration of water availability. There should be a water zoning of the country and the economic

activities should be guided and regulated in accordance with such zoning.

Conservation of Water

15. The efficiency of utilisation in all the diverse uses of water should be improved and an awareness of water as a scarce resource should be fostered. Conservation consciousness should be promoted through education, regulation, incentives and disincentives.

Flood Control and Management

16. There should be a master plan for flood control and management for each flood-prone basin. Sound watershed management through extensive soil conservation, catchment-area treatment, preservation of forests and increasing the forest area and the construction of check-dams should be promoted to reduce the intensity of floods. Adequate flood-cushion should be provided in water storage projects wherever feasible to facilitate better flood management. An extensive network for flood forecasting should be established for timely warning to the settlements in the flood plains, along with the regulation of settlements and economic activity in the flood plain zones, to minimise the loss of life and property on account of floods. While physical flood protection works like embankments and dykes will continue to be necessary, the emphasis should be on non-structural measures for the minimisation of losses such as flood forecasting and warning and flood plain zoning, so as to reduce the recurring expenditure on flood relief.

Land Erosion by Sea or River

17. The erosin of land, whether by the sea in coastal areas or by river water inland, should be minimised by suitable cost-effective measures. The states and union territories should also undertake all requisite steps to ensure that indiscriminate occupation and exploitation of coastal strips of land are discouraged and that the location of economic activities in areas adjacent to the sea, is regulated.

Drought Management

18.1 Drought-prone areas should be made less vulnerable to drought-associated problems through soil-moisture conservation measures, water harvesting practices, the minimisation of evaporation losses, the development of the groundwater potential and the transfer of surface water from surplus areas where feasible and appropriate. Pastures, forestry or other modes of development which are relatively less water-demanding should be encouraged. In planning water resource development projects, the needs of drought-prone areas should be given priority.

18.2 Relief works undertaken for providing employment to drought-stricken populations should preferably be for drought proofing.

Science and Technology

19. For effective and economical management of our water resources, the frontiers of knowledge need to be pushed forward in several directions by intensifying research efforts in various areas, including the following:

- hydrometeorology;
- assessment of water resources;
- snow and lake hydrology;
- groundwater hydrology and recharge;
- prevention of salinity ingress;
- water-harvesting;
- evaporation and seepage losses;
- economical designs for water resource projects
- crops and cropping systems;
- sedimentation of reservoirs;
- the safety and longevity of water-related structures;
- river morphology and hydraulics;
- soils and material research;
- better water management practices and improvements in operational technology;
- recycling and reuse;
- use of seawater resources.

Training

20. A perspective plan for standardised training should be an integral part of water resource development. It should cover training in information systems, sectoral planning, project planning and formulation, project management, operation of projects and their physical structures and systems, and the management of the water distribution systems. The training should extend to all the categories of personnel involved in these activities as also the farmers.

Conclusion

21. In view of the vital importance of water for human and animal life, for maintaining ecological balance and for economic and developmental activities of all kinds, and considering its increasing scarcity, the planning and management of this resource and its optimal, economical and equitable use has become a matter of utmost urgency. The success of the national water policy will depend entirely on the developmental and maintenance of a national consensus and commitments to its underlying principles and objectives.

16

Impact of Inter River Basin Linkages on Fisheries

Introduction

Human societies have always tried to expand the spatial extent of availability of water by the diversion of streams or rivers. The idea of drawing water from the rivers in eastern India, which have larger run-off, in comparison to certain places in the peninsular region, where the precipitation levels are much lower, can be seen as an extension of that practice. In the 19th century, Sir Arthur Cotton had thought of a plan to link rivers in southern India for inland navigation. Though partially implemented, the idea was later abandoned because inland navigation lost ground to the railways. In more recent times, the 'Ganga-Cauvery Link' proposal mooted by former Irrigation Minister, Dr. K.L. Rao was examined and found impracticable because of the very large financial and energy costs involved, and the 'Garland Canal' idea put forward by airline pilot Captain D. Dastur, was termed as technically unsound.

The recent revival of the idea of interlinking of 'surplus' basins with 'deficit' basins has been the result of work done by the National Water Development Agency (NWDA) and bears a conceptual continuity with Dr. Rao's proposal. The Himalayan rivers component and the peninsular rivers component constitute the two parts of the river-linking project. The Himalayan segment envisages 14 links, *viz.*, Manas-Sankosh-Tista-Ganga, Kosi-Ghagra, Gandak-Ganga, Ghagra-Yamuna, Sarda-Yamuna, Yamuna-Rajasthan, Rajasthan-Sabarmati, Chunar-Sone Barrage, Sone Dam-Southern tributaries of Ganga, Ganga-Damodar-Subernarekha, Subernarekha-Mahanadi, Kosi-Mechi, Farakka-Sunderbans and Jogi-Gopa-Tista-Farakka. The general idea is to transfer waters from 'surplus' eastern rivers to 'deficit' central, western and southern regions.

Policy Paper 28, July 2004.

The peninsular segment involves 16 links, *viz.*, Mahanadi (Mani Bhadra)-Godavari (Dowlaiswaram), Godavari (Inchampalli low dam)-Krishna (Nagarajunasagar tail pond/Pulichintla), Godavari (Inchampalli)-Krishna (Nagarjunasagar), Godavari (Polavaram)-Krishna (Vijayawada), Krishna (Almatti)-Pennar, Krishna (Srisailam)-Pennar, Krishna (Nagarjunasagar)-Pennar (Somasila) Pennar (Somasila)-Cauvery (grand Anicut), Cauvery (Kattali)-Vaigai-Gundar, Ken-Betwa, Parbati-Kalisindh-Chambal, Par-Tapi-Narmada, Damanganga-Pinjal, Bedti-Varda, Netravati-Hemavati and Pamba-Achankovil-Vaippar. The project is expected to provide irrigation facility to 25 m-ha land through surface water and additional 10 m-ha land through groundwater as also generate 34,000 MW of power.

As regards impact on fisheries, there could be positive as well as negative impacts. With creation of more water basins and canals, the resource for fisheries and aquaculture is expected to increase, while on the other hand, there could be mixing of fish species between the river basins, loss of certain amount of biodiversity as also entry of some invasive fish species.

Indian Fisheries

Fisheries is an important economic activity in the country, with over seven million people involved in the sector, majority of whom live in 3,937 coastal villages and fisher-hamlets, along major river basins and reservoirs in the country. The sector has also been one of the major contributors to foreign exchange earnings through exports, to the tune of Rs. 6,800 crores annually. India is the fourth largest fish producer in the world and second in inland fish production. The fisheries sector contributes 4.5 per cent to the agricultural GDP and 1.4 per cent to the total GDP. The country is endowed with about 8,118 km of coastline with some of the richest fishing grounds in the world and estimated potential of fish production from the seas is 3.9 million metric tonnes. The main inland fishery resources include about 1.20 million hectares of brackish water area, 2.38 million ha of fresh water ponds and tanks, 3 million ha of reservoirs, besides about 1,91,000 kms of rivers and canals. The potential fish production from inland waters is 4.5 million metric tonnes. The present total fish production of 6.12 million metric tonnes is expected to increase to about 8.4 million tonnes, with the annual growth rate of around 4.5 per cent. Both inland and marine fisheries directly or indirectly depend on the river systems. The different river systems, *viz.*, Ganga, Brahmaputra, Mahanadi, Godavari, Krishna, Cauvery, Narmada, Tapi and their tributaries have been studied over the years, with regard to hydrobiological conditions, biota, distribution and abundance of fish species, and so on. Major fish species along with the endemic and endangered ones have been listed for the river systems and the effects of impoundments documented. New approaches of regulatory measures, river ranching, protected habitats and sanctuaries are also being evaluated. The aspects of minimum flow and environmental flow in river systems are being debated in the context of aquatic biodiversity and fisheries management.

International Scenario

International experience in linking of rivers has shown both beneficial and harmful aspects. Diversion of water in upper reaches in the case of the Aral Sea in Central Asia (in former USSR) resulted in loss of over 60 per cent of its surface area and two-thirds of volume. The cause is attributed to a vast expansion of irrigation in the Central Asian Republics beginning in the 1950s, which greatly reduced inflows to the Sea. China has officially launched the world's largest water diversion project, which will divert water from the Yangtze River in the south to the country's dry north, including Beijing. The project, valued at US $ 59 billion, may cost twice as much as the ongoing Three Gorges Hydroelectric Project. The project will be the biggest of its kind in the world and the largest engineering programme in China. It consists of three canals running about 1,300 km through the country's eastern, middle and western parts.

The Spanish National Hydrological Plan (SNHP) has proposed a massive transfer of water from the Erbo river in the north of the country to the Valencia and Murcia rivers in the south, which suffers severe water shortage difficulties due to such things as intensive agriculture and tourism. There, however, have been reactions regarding the loss of Europe's most ecologically important wetlands and the proposal is being reviewed.

Impact on Fisheries

Positive Impacts

- *Canals and reservoirs*: The proposed interlinking of rivers would comprise more than 36 major dams and 30 canal links. In addition, there will be many more irrigation canals and barrages. These major reservoirs, canals and other water harvesting structures will add to the potential fishery resources of the country.

- *Rejuvenation of lakes and rivers*: The rivers and lakes in the water recipient zone will bring benefit with increased water perennially, congenial habitat and consequently, higher fish production.

Negative Impacts

- *Loss of habitat*: River interlinking might affect fish feeding and breeding habitats in the rivers and lakes in the water donor zones due to lowering of water volume and enhanced siltation load. The flood plains and wetlands connected with donor rivers would also be affected. River run-offs provide energy for a number of vital processes in downstream estuaries, delta and coastal areas. Reduced river discharge could result in loss of coastal habitats such as mangroves, coral reefs, sea grasses, estuarine and delta regions.
- *Water quality changes*: Significant changes in water quality of rivers, lakes, estuaries and coastal waters could occur due to changes in sediment load, nutrients and contaminant levels. The levels of toxicants and contaminants in donor rivers may go up owing to reduction in self-purifying functions subject to changes in flow regimes.
- *Loss of biodiversity*: Each river system has distinct groups of biota different from other water bodies. When environment is altered, they are affected, with particular threat to endangered and endemic species. The linkage of rivers could also lead to loss and homogenisation of genetic diversity of fishes.
- *Changes in land-ocean interactions*: River is a critical component of the delta-estuary-coastal sea ecosystem. Un-impounded rivers provide energy for a number of vital processes in downstream estuaries, delta and coastal areas, upon which healthy fisheries are dependent. The linkage of rivers could alter the timing and quantity of river discharge into the sea, which may alter the river-mediated land-ocean interactions and coastal fisheries. A recent study conducted by Central Marine Fisheries Research Institute, Kochi, on 'Impact of

Dams on River into Sea and Changes in the Coastal Waters' has shown that reduction in river discharge into the sea will adversely affect the water chemistry and productivity profile of coastal waters and the estuarine fisheries.

In view of the above concerns, NAAS organised a Round Table[1] on 'Impact of Inter-linking of river basins on Fisheries' to have a comprehensive assessment of the plan on interlinking of rivers, discuss the similar proposals in other countries, understand the aquatic biodiversity and fish distribution in Indian rivers and project the fisheries scenario in the event of the linking of rivers. The recommendations emerging from the deliberations are given below.

Recommendations

- *Data mining*: Large volume of raw data on water quality, productivity, aquatic ecology and fish biodiversity are available for river systems from various parts of the country. Most of these data are not properly analysed and interpreted. Hence, there is an urgent need for data mining and analysis to gather base line information on water quality, aquatic ecology, productivity and fish biodiversity for various river systems.
- *Modelling studies and scenario analysis*: Software-aided modelling studies are required to assess the impact of changes in river runoff on water chemistry, productivity, aquatic ecology, biodiversity and fish production. The impact of interlinking of river basins on fisheries will vary from one river system to another and from region to region. Hence, detailed scenario analysis and simulation studies have to be carried out for each of the river systems to assess the impact under varying flow conditions.
- *Scaled-down approach*: Before initiating large-scale river basin linking at national level, it will be appropriate to conduct

1. Round Table held during 21-22 May 2004, under the convenership of Dr. S. Ayyappan, Deputy Director General (Fisheries), ICAR. The Round Table, which was basically an interactive meeting of different stakeholders, was inaugurated by Sri S. Regunathan, Secretary, Department of Animal Husbandry and Dairying, Govt. of India; with Dr. S.Z. Qasim, Former Member, Planning Commission, chairing the Meet. With participants from a number of organisations, the discussions and the plenary session were chaired by Dr. V.L. Chopra, President, NAAS. Experts from various govt. departments like Joint Secretary, Department of Animal Husbandry; Member, Task Force on Interlinking of Rivers; Ministry of Water Resources; International Water Management Institute; World Fish Center; and ICAR participated in the discussions.

studies on linking adjacent river systems with 10 similar ecosystems. Such scaled-down approach will be very useful to assess and understand the environmental impacts of river linking on aquatic ecosystem.

- *Environmental flow*: Environmental flows may be broadly defined as the 'provision of water for freshwater-dependent ecosystems to maintain their integrity, productivity, services and benefits in cases when such ecosystems are subject to flow regulation and competition from multiple water users'. There is an urgent need to assess the minimum environmental flow in donor rivers. It is anticipated that the knowledge base created through this process will strengthen the level of understanding of methodologies available for use, their relative strengths and deficiencies, and their potential for application under various circumstances.
- *Loss-gain statement*: There is an urgent need to work out a loss-gain statement on the possible impact of interlinking of river basins on fisheries, using the available data.
- *River ranching*: River ranching of fishes is being carried out in various parts of the country by different agencies. It is necessary to evaluate the usefulness of river ranching for enhancing the fish production in our river systems. River ranching with fish seed in altered scenario needs to be addressed in a proper perspective.
- *River siltation and dredging*: The siltation pattern in donor and receipt rivers could change due to interlinking. Heavy siltation of rivers, canals and lakes is already an acute problem in the Himalayan region. There is urgent need to assess the usefulness of river bed dredging for improving the river ecology and fish production.
- *Impact on estuarine and coastal waters*: River run-offs provide energy for a number of vital processes in downstream estuaries, delta and coastal areas. These processes include transport of nutrients, organic matter and nutrient-rich silt, oxygen enrichment, entrainment of nutrients in bottom sediments, dilution and flushing of pollutants, etc. The most widely discussed and well-documented impacts of large-scale hydrological alterations (river damming and river diversion, for

example) on marine systems are, reductions in water and sediment discharge. Immediate attention should be given to study the impact of altered river flow due to river interlinking on biodiversity pattern of estuarine and coastal fauna.

- *Biodiversity studies*: It is necessary to study and identify the endangered species which could become extinct due to interlinking. Based on existing biodiversity lists of exotic and native species, it is necessary to identify the prevalent differences between the rivers likely to be linked and project the post-linking scenario.
- *Taxonomic studies*: Taxonomists and taxonomy related studies are required to assess and document the aquatic biodiversity patterns. Hence, there is an urgent need to give stress to taxonomic studies and developing human resources in the field of taxonomy.

17

Harnessing and Management of Water Resources for Enhancing Agricultural Production in the Eastern Region

Introduction

The eastern region comprising Bihar, West Bengal, Orissa, parts of eastern Uttar Pradesh and eastern Madhya Pradesh is bestowed with abundance of rainfall, river flows and fertile soils. The ultimate irrigation potential of this region is 52.73 Mha. which is about 47 per cent of that of the entire nation. Surface water is the predominant source of irrigation in this region. The irrigation potential of major and medium projects is 19.66 Mha, while that of minor irrigation projects is 33.07 Mha. including 12 Mha. from groundwater. Almost all the major irrigation projects are diversion type. In spite of the fact that the region is served by several rivers such as Ganges, Mahanadi, Sharada, Gandak, Kosi and Sone, irrigation has been uncertain and undependable.

Historically, the eastern region was the most prosperous agricultural tract of the country. During the triennium ending 1950-51, the yield of foodgrains in this region was 6.44 t/ha while that of the northern, western and the southern regions was 6.08, 3.90 and 5.54 t/ha, respectively. However, with the advent of the input intensive new agricultural technology following the Green Revolution, the region lagged behind the other parts of the country. Public investments in agriculture in this region for the period 1969-1985 were grossly inadequate to be compatible with the requirements of input intensive agriculture and as such this region has not benefited by the Green Revolution.

Currently, the average foodgrain yields of the eastern region are far lower than those of the other regions of the country. The average yield

---

Policy Paper 3, April 1998.

of rice in the region is 1350 kg/ha while the all India average yield is 1746 kg/ha. Similarly, yield of oilseeds is also much lower than the national average. However, with the foodgrain yields plateauing in Punjab, Haryana and western Uttar Pradesh the increase in foodgrain production to meet needs of the increasing population of the country, has to come from this region. Thus, there is an imperative need to improve the foodgrain production from this resource-rich region.

Because of its vast water resource potential, rich soils, and the gap between the potential and actual yields of crops, the region could become the future granary of the country. The National Water Policy document of the National Academy of Agricultural Sciences (NAAS) (1996), drew attention to the vast untapped potential of water in this region and inadequacy of the current programmes, projects and policies to harness this potential.

The NAAS considered it necessary to focus attention on water related issues of the eastern region and organised a seminar on 'Harnessing and Management of Water Resources for Enhancing Agricultural Production in Eastern Region'. The seminar had four themes: (i) Harnessing surface and groundwater resources for increasing agricultural production; (ii) Rainwater management for sustainable agricultural production; (iii) On-farm water management, issues and remedies; and (iv) Socioeconomic aspects and human resource development in water resource management.[1] The themewise recommendations of the seminar are presented below.

Recommendations

1. Harnessing Surface and Groundwater Resources for Increasing Agricultural Production

1.1 Groundwater Potential Remains Largely Untapped

1. The region is rich in groundwater especially in the plains. However, the utilisation of this potential is just 5.6 per cent in eastern Madhya Pradesh, 8.4 per cent in Orissa, 19.2 per cent in Bihar, 24.2 per cent in West Bengal and 35.8 per cent

1. Seminar organised in February 1997. It was jointly hosted by the NAAS, Orissa University of Agriculture and Technology, Bhubaneswar and Water Technology Centre for the Eastern Region, Bhubaneswar. Several government officials from six states namely, Andhra Pradesh, Bihar, Madhya Pradesh, Orissa, Uttar Pradesh, and West Bengal, besides those of central government, the ICAR institutes and agricultural universities participated in the seminar.

in eastern Uttar Pradesh. For bringing a dramatic change in the agricultural scenario of this region, it is imperative to utilise this resource judiciously. This calls for action on technological and legislative fronts.

2. The exploration, survey and estimation of groundwater potential both in terms of quantity and quality should be strengthened. Efficient well-construction technology for exploitation of low yield aquifers in hard rock areas should be evolved for drought mitigation. The technology package should include both engineering and biological components. Conjunctive use of surface and groundwater should be encouraged in irrigation project areas to optimise the water use and to alleviate the degradation of soil and water resources. Various technologies for groundwater recharge such as, use of dug-wells, ponds, water harvesting structures in drains and rivers should be studied for feasibility and identified for their implementation depending on availability of the finances. Enactment of groundwater legislation should be expedited to protect groundwater from overexploitation and pollution.

1.2 Village Pond Culture System Needs Revitalisation

3. The region has a traditional culture of having ponds both at individual level in the backyard, and community level as village ponds. The hydrology of village ponds, taking into account surface inflow-outflow, storage, seepage, and recharge etc., should be studied to assess the availability of water for developing appropriate strategies for agricultural production in the region. These strategies should be tested in typical case studies for validation and the experience utilised to develop policy of rejuvenation and multipurpose uses of ponds with special emphasis on crop production. Harvesting surplus monsoon runoff through artificial recharge by cost-effective techniques should be encouraged.

1.3 Strengthen Collaboration and Coordination of Research on Water Use

4. Presently, the water resources developed for irrigation are used for irrigation only, while ample scope exists for their

multipurpose use. However, proven technologies are not available for such use. The research on multipurpose use of irrigation water needs to be strengthened through collaborative research efforts among the various organisations, namely, the ICAR institutions, SAUs, state and central government agencies.

5. The extent of coastal saline and alkali soils should be estimated and appropriate amelioration measures, cropping/farming systems should be developed for such areas.
6. A mandatory mechanism should be envisaged to bring about coordination amongst research institutions, SAUs, and state government agencies for proper exchange of information to facilitate establishment of a National Data Bank, and to develop strategies of research as well as implementation of water resource development and watershed management programmes.

2. *Rainwater Management for Sustainable Agricultural Production*

2.1 High Rainfall Potential and Problems Need Studies

7. The high rainfall of the region, instead of being a blessing, becomes a curse because of the recurring floods, waterlogging and crop damage, every year. Therefore, proper rainwater management holds the key for effective exploitation of the rich bounty of the rainfall.
8. Scientific approach to rain water management that accounts for appropriate hydrologic analysis for identification of wet spells, dry spells, critical dry spells, assessment of runoff, design of farm ponds, seepage control with inexpensive lining materials, critical growth stages of crops, design of appropriate irrigation systems, suitable timing and assessment of benefits of supplementary irrigation to rainfed crops, etc., should be adopted for obtaining the best results from rainfall. Field evaluation studies ought to be carried out to test and evolve economically viable technologies of rainwater management.

2.2 Watershed Management-The Key to Efficient Water Resource Development and Use

9. The programmes of watershed management and soil conservation should be designed to classify and identify lands of various degrees of degradation and work out appropriate measures to rejuvenate them. Preference should be given to good marginal lands and more fragile ecosystems. Emphasis should be on harnessing indigenous techniques of soil and water conservation based on farmer's experience.

10. Water use planning should be done on micro watershed basis with government support and people's participation in blocks and districts on the pattern of land use planning for optimal and sustainable water resource development. Teams of leading scientists should be formed to explore and develop technology packages for better integrated water resource utilisation in the region.

2.3 Make Use of Experience of Water Management at Damodar Valley Corporation (DVC)

11. A number of small-scale erosion control and water harvesting structures have been constructed in DVC, Hazaribagh, which have not only reduced runoff and erosion but also assisted in large-scale conservation of water and increasing area under agriculture. The design and installation of such small-scale soil and water conservation structures in the DVC should be properly documented for exploring their use in various watersheds in the eastern region. The research organisations working in the eastern region should collaborate with the DVC to assess and evaluate these structures for their adaptability.

2.4 Waterlogging and Flooding Problems

12. One of the major reasons of waterlogging and flooding is impediment of water flow due to construction of railway and road embankments without adequate provision of cross-drainage. There is need for closer coordination amongst the departments of railways, roads, irrigation and flood control to provide adequate cross-drainage to ensure natural flow of water.

13. A programme of construction of farm ponds/micro-reservoirs along low order streams should be advocated to intercept runoff, reduce peak flow, enhance groundwater recharge and create irrigation potential for increasing agricultural production.

3. *On-farm Water Management: Issues and Remedies*

3.1 Increasing Water Use Efficiency

14. Efficiency of water use in canal command areas in eastern region is quite low. Field to field irrigation is prevalent and this not only reduces irrigation and nutrient efficiency, but also makes crop diversification difficult. Although, appropriate interventions with new water management technologies have been developed through several on-farm water management studies they are yet to be extended to command areas of distributaries and branch canals.

3.2 Improving Drainage

15. Drainage congestion as well as surface flooding in canal commands, pose serious threat to sustainable irrigated agriculture in eastern region. Appropriate surface drainage technology needs to be evolved integrating preventive and curative measures. Adequate research backup with appropriate cost benefit assessment is necessary for development of efficient drainage systems.The large permanent water bodies, such as *Chaurs, Manus, Taal*[2] etc., should be used for safe disposal of drainage water and as irrigation sources. These wetlands are a part of our environment and suitable technological packages should be developed keeping socio-economic conditions in view.

3.3 Ensuring Conjunctive Use of Water

16. Conjunctive use of surface and groundwater must get a high priority in developmental programmes of this region not only to augment water supply but also to lower water table and contain secondary soil salinisation. Adequate supply of power/diesel and funding support is absolutely essential for groundwater exploitation in the region.

---

2. *Chaurs, Manus, Taal* are the local names of large permanent water bodies.

3.4 Introduce Crop Diversification

17. For effective crop diversification and increase in cropping intensity low water requiring, high value crops are needed. More flexibility of irrigation system to meet water demands, in terms of time, amount and locational accessibility should be ensured. Appropriate marketing and processing facilities hold the key to the success of the diversified agriculture and measures should be taken by the government to meet this need.

3.5 Need for Collaboration Between Different Agencies

18. To address water management issues in an integrated way, a strong collaboration between the agencies dealing with irrigation system management and on-farm water management is required.

4. *Socioeconomic Aspects, and Human Resource Development in Water Resource Management*

4.1 Farmers' Irrigation Cooperatives Needed

19. The farmers' irrigation cooperative movement should be encouraged. The operation and maintenance of irrigation minors and subminors must be entrusted to the farmers' cooperatives. Concession in water rates can be given to these cooperatives for the services rendered by them. Suitable changes should be made in the Irrigation Act to revitalise the system. At present, the revenue officers such as *Tehsildars* are declared as Irrigation Officers. This practice should be discontinued as these officers have little exposure to the complexities of modern irrigation system. The District Agricultural Officer who has overall responsibility of agricultural production for a particular district and knows various aspects of agricultural developments including irrigation is technically competent to become Irrigation Officer so as to meet the requirement of the Irrigation Act and the rules made thereunder. Alternatively, the Irrigation Engineer responsible for water delivery can also be designated as Irrigation Officer.

4.2 Develop Water Management Centres at Different Levels

20. There should be Water Management Committees at district, block and *panchayat* levels involving District Agricultural Officer, Block Development Officer, *Tehsildar*, irrigation field functionaries, agricultural engineers, economists, and farmers' representatives. The committees will take decisions on, (a) timing of canal closure for annual repair and maintenance, (b) design of suitable cropping pattern for each block within the irrigation command, (c) delineating the waterlogged and high water table areas for rice and fish farming, (d) control of stray cattle menace specially during dry season, (f) avoiding mismatch between timing of water delivery and crop needs in canal command areas, (g) laying out field channels and drains and making suitable provisions for their maintenance at *gram panchayat* level.

4.3 Assessment of Improvement in Agricultural Productivity as a Measure of Irrigation Efficiency

21. The primary objective of irrigation system is to improve agricultural production to specified degrees. However, the performance evaluation of a system is done only in terms of area irrigated with no emphasis on quality of irrigation and its productivity. It is, therefore, necessary that besides the gross irrigated area, the quality of irrigation and productivity of the area is also recorded and used in evaluation of performance of irrigation systems.

4.4 Encourage Farmers' Association

22. The farmers' association could purchase pump sets and accessories and other essential farm equipments like, seed drill, bund former etc., the cost of which could be financed by banks under Preferential Group Loan Scheme, and cooperatives of *Panchayat* could render custom hire services for pumping water from open wells, tanks and rivers to the cropped areas of the farmers. Provision for custom hire service should be made for repair and maintenance of tube wells and pump sets at the *panchayat* level. For electric pump sets uninterrupted power supply should be assured for at least eight hours a day. If this is not possible, a rotational system of power distribution should be introduced.

4.5 Entrust Management Functions to *Panchayats*

23. The management and maintenance of tanks, *jheels* and lakes specially in West Bengal and Orissa should be vested with *Panchayats* through appropriate legislation. Fish culture in permanent water bodies like *Taals, Jheels* etc., will help in providing food and employment besides reducing breeding of mosquitoes and incidence of diseases.

4.6 Encourage Rice-Fish Farming

24. Rice-fish farming accommodates crop diversification and reduces the investment risks in rainfed lowland rice cultivation. The system also generates year round employment in the farm and ensures high productivity and profitability besides assuring conservation of ecosystem. This needs to be improved through integrated use of crop and fish culture technology.

4.7 Treatment of Industrial and Municipal Water

25. The conventional treatment of industrial and municipal waste water and its appropriate recycling can provide an additional source of nutrient rich irrigation water for improving productivity of vegetable and other crops. This approach should be extensively tested by the soil scientists and agronomists for various soil types and cropping systems and where the potential exists the policy and implications should be determined and steps taken to implement them.

4.8 Need for Creation of Specialised Cadre of Water Management

26. So far, there is no specialised cadre of water management in the country, hence the subject does not attract any attention. We suggest that the graduates and postgraduates in agriculture, agricultural engineering, and civil engineering should be given sufficient theoretical and practical training in this subject through specialised postgraduate programmes in water science and technology. The faculty of the institutions should be provided appropriate exposure and training in national and international institutions having such specialised expertise. Water resource management in agriculture is a

multidisciplinary subject matter area and hence, a multidisciplinary training should be provided to all personnel involved. Special courses need to be designed for transfer of technology to the farmers connected with water use and management. Training on watershed management should be an important and integral part of human resource development, particularly in the rainfed areas.

4.9 Strengthening Water Users' Agency through Training and Appropriate Legal Acts

27. In accordance with the spirit of the National Water Policy (1987) of Government of India, farmers should be made partners in management of distribution system and distribution of water. This can be realised by organising farmers into registered Water Users' Associations (WUA) at appropriate hydrologic units of about 300-600 ha, with appropriate transfer of certain operation and maintenance (OM) responsibilities to them. The responsibilities of WUAs can begin from minor downwards leaving the department to pay greater attention to management of head works and main/branch canals. Suitable legislative action should be taken to provide legal status to WUAs. The managerial aspect of personnel of irrigation department should be highlighted and they should be suitably trained.

18

Disaster Management in Agriculture

Introduction

Human vulnerability to disasters is an age-old phenomenon. Disasters play havoc with people's lives. Their aftermath results in death, destruction, misery and trauma. The unique geoclimatic conditions of our country make this region vulnerable to natural disasters. There are varied reasons for occurrence of disasters. Modernisation and industrialisation have disrupted the natural resource balance. This, alongwith alarming increase in population, depletion of resources, lack of knowledge and skills and rising economic disparities are considered to be mainly responsible for the vulnerability of society to these catastrophies. The situation, though more or less similar globally, is quite distressing in countries of South Asia, such as India. The long history of disasters, and the resulting sufferings, are reasons enough to ponder over their manageability.

One of the major repercussions that disasters have, is on agriculture. A substantial part of the population depends on agriculture for its livelihood. Agricultural activity is adversely affected by any unforeseen weather changes or variations in physical conditions. This gets accentuated in case of cyclones, floods and droughts resulting in disruption of people's livelihood and adding to the risk, damage and stress of disasters. Drought is a perennial feature, 16 per cent of India's total area is drought prone and approximately, 50 million people are annually affected by droughts. About 68 per cent of total sown area of the country is susceptible to drought at one time or other. Under this latest dispensation, another 'sub-humid' category, having an adverse water balance, has been added to arid and semi-arid category. The drought of 1987-88 was one of the worst in India. Drought situations had also affected about 11 states, in 1999-2000.

Policy Paper 27, July 2004.

Food losses due to pests, disease, wild animals, insects and weeds are considerable, about 35 per cent of world crop production is lost in spite of pesticide and control programmes. Insect outbreaks are generally the result of a combination of temperature, monoculture of crops, introduction of plants to new locations, weather pattern, and migration (ecological factors). In India, locust infestation-prone areas are—Rajasthan, Gujarat, Punjab, Haryana, Andhra Pradesh, Karnataka and Maharashtra.

Some lessons learnt from drought of 2000 in Rajasthan:

- There should be a specific policy for drought prone areas.
- Disaster Management Policy has to be area-specific and forecasting should be dependable and timely.
- Relief must be timely.
- Proper crop planning/land use planning should be done.
- Fodder banks should be created.
- Occupational diversification should be made available.

Disaster management has presently evoked a methodological approach that focuses on systematic observations, their analysis and dissemination of likely events to structure informed application of resources in time. The most critical goal of disaster management activity is to reach out to the people promptly in the remotest areas and reduce vulnerability to disasters. This calls for a multipronged strategy involving the government at various levels, international agencies, non-governmental organisations, community based organisations, and the panchayati raj institutions to put together resources and efforts in managing disasters.

The world over, any measure, be it preparedness, mitigation, relief or rehabilitation in the wake of disasters, is primarily regarded as the responsibility of the government. But despite the vast resources at its command and the power to take necessary steps, due to the range of activities involved, the government needs to involve other key stakeholders in this exercise. While long-term preventive and preparedness measures are taken up, the unprecedented nature of the disasters calls for nationwide response mechanisms with clear-cut assignment of roles and functions by various institutions at the central, state and district levels. This, alongwith other initiatives, forging efforts with international agencies, non-governmental organisations, panchayati raj institutions with emphasis on education and training, can go a long way in making a dent in disaster management in agriculture. To discuss issues of development in the area of disaster management in agriculture, National Academy of Agricultural Sciences (NAAS) organised a one day brain-storming session.[1] The role of the following was deliberated upon:

a) Government of India;

b) International agencies;

c) Non-governmental organisations;

d) Panchayati raj institutions and

e) Education and training in disaster management.

Given below are the major recommendations which emerged from deliberations at this meeting.

Recommendations

- The importance of disaster management in agriculture was recognised and its further strengthening through a comprehensive disaster management policy by the Government of India providing a renewed focus on anticipatory preparedness, prevention and rehabilitation was strongly recommended.
- State government needs to be more proactive in handling disaster related issues leading to effective management in agriculture.
- In the context of National Calamity Contingency Fund (NCCF), the existing process of preparing a memorandum of state government, sending a team from central government, etc., should be reviewed, and made quicker and more convincing.
- The cropped area throughout the country is vulnerable to one or the other pest disease, insects and weeds, therefore

1. NAAS organised a one day session on Disaster Management in Agriculture at New Delhi, on February 27, 2004 under the convenership of Dr. Panjab Singh, Director, School of Agriculture Sciences, Indira Gandhi National Open University (IGNOU), to critically examine all the pertinent issues comprehensively. The meeting was attended by concerned subject matter specialists, agriculture planners and administrators, social scientists, educationists, grassroot-level workers and representatives from international agencies and NGOs.

development and implementation of an effective and regular pest surveillance system was recommended, which would continuously monitor pest activity through an exclusive network.

- A national plan for pest control should be established for development of regulations, regarding imports and exports of plant material, pest forecasting and damage assessment components, adoption of international pest management strategies, extension services to engage farmers' support and provide pest control supplies, and a training component to update and discover new methods of control. In affected areas, the contingency plan prepared by the Department of Agriculture to face any natural calamity should be put into action to bring relief to the farmers.
- Natural calamities and disasters have transborder causative factors and impacts, demanding collective and international preparedness and action.
- It was felt that networking of international agencies would go a long way for channelling multifaceted assistance ranging from pre-disaster, response, recovery and rehabilitation. The forging partnership should ensure immediate, medium-term and long-term livelihood interventions.
- Effective interagency and intercommunity communication and educating the stakeholders, help in facing and mitigating the calamity effects. Hence, linkages for experience sharing, database transfers, action plans and strategies should be put in place. It would also make the initiative cost-effective and timely.
- It was strongly felt that in addition to the technical and managerial skills in relation to disaster and related activities importance should be given to disaster management education also to provide knowledge to the learners on disasters preparedness, mitigation and rehabilitation; create awareness about effective disaster response in various emergency situations; equip the learners with tools for meeting emergency medical requirements; incorporate gender sensitive, empathy-based disaster management approach; and inculcate new skills and sharpen the existing skills of government

officials, voluntary activists, professionals and elected representatives for effective management.

- Role of disaster management education and training have been emphasised in planning and implementation of disaster management strategies. The workshop recommended that the education should be designed to provide comprehensive knowledge on different types of hazards, disaster management techniques and impediments in the way of disaster reduction and should address to the community needs.
- The workshop recommended that the state agricultural universities should introduce, develop and launch a course on disaster management through multidisciplinary approach and NAAS to be the nodal hub for disaster management in agriculture with focus on research, education and training.
- To empower people at grassroots levels, educational programmes, such as awareness- and application- oriented certificate programmes should be designed. The potential of open and distance learning system should be used to reach the masses.
- Disaster management in agriculture curricula must be developed and certified to establish joint standards of practice across the nation. This is necessary to ensure uniformity in standards of humanitarian and mitigation practices in the region.
- There is need to strengthen research support system to evolve continuous process of analysis, action/reflection which will help to decipher the gaps in disaster management plans. Towards this end, the institutions/universities may be encouraged to undertake the multidisciplinary research projects.
- NGOs and community organisations/bodies play an important role in disaster preparedness and management. There should be well coordinated network between NGOs, community organisations/bodies and the government, and also among NGOs themselves. Better coordination would result in wider reach and more comprehensive approach for disaster management.

- For dealing with location-specific coping strategies based on time-tested traditional knowledge, blended with modern science and technology should be included in training and education of disaster management in agriculture.
- Academicians in the universities and research institutions, and NGOs, who are already involved and working in the field of disaster management should be involved in converting a disaster into a new opportunity with improved technology inputs and application, and they should be encouraged for their active participation specially in disaster management in the field of agriculture.
- Promotion and protection of livelihood in agriculture and evolution of alternative livelihood strategies, were emphasised to improve the livelihood standards in rural society.
- The impact of agricultural practices on environment paving way for ecological disaster was identified with respect to disaster management in agriculture.
- It was observed and recommended that the rainfall regimes in the preceding post-monsoon season and summer must be taken serious note of, while assessing the situation likely to arise in the coming monsoon months.
- The workshop recommended that special meetings, at least two meetings, of *Gram Sabha* and *Gram Panchayat* should be held to manage disaster. These meetings are necessary to plan preparedness activities before monsoon, at least.
- Capabilities of PRI functionaries should be enhanced through education and training programmes in the field of agriculture and disaster management, which will help to develop well-trained self-help groups and community based organisations.
- The workshop emphasised that the village development committees should be formed and motivated to monitor and evaluate the works performed by PRI before and after disaster.
- There is an urgent need to make *Gram Sabha* fully operational to ensure regular meetings; effective contributions in planning and implementation; and identify local needs for disaster management in agriculture.

- Basic infrastructure and transport facilities should be provided to the panchayati raj institutions to help the residents in evacuation and providing relief and implementing disaster management and development programmes effectively.
- The importance of database communication and information sharing was fundamental to the decision support system and its further strengthening through geospatial information system (GIS) was strongly recommended. GIS can be used as decision support system to identify and integrating, monitoring and predicting hydrological hazards, which are the major environment risks for Indian agriculture. It is emerging as pioneering technology of great significance and may serve as a powerful tool in the development planning and governance, together with disaster management education.

Section IV

Agrobiodiversity and Biosafety

19

Belowground Biodiversity in Relation to Cropping Systems

Introduction

Soil, rhetorically described as the 'final frontier', harbours a largely unknown microbial universe whose diversity in quantitative terms is indeed staggering. There are more than 10^{16} prokaryotes alone in soil. Recent estimates show that the soil microbial diversity would be at least three times greater than thought earlier.

The belowground biodiversity is not a passive entity; it is a continuum of the life in soil. There is an intimate link between the belowground biodiversity (BGBD) and the processes and sequences of the biodiverse life aboveground (AGBD). Although, laser-based technology and the tools of new biology have now made it possible to see through the largely opaque soil and to study the belowground communities *in situ*, it is well nigh impossible to make a direct estimate of the structural and functional diversity of the belowground communities, and to relate these with the biodiversity aboveground. Although the 'last frontier' has not been breached, the few installations made here and there have shown that BGBD as provider of ecosystem goods and services is vitally linked to ecosystem functions, and hence, the welfare of the Universe.

The agroecosystems consisting of the temporally variable and spatially heterogeneous cropping systems are candidates for BGBD search in relation to the contributions made by the belowground communities towards sustainability of land and soil productivity under the different cropping systems. The search has led to more questions than answers and the researchers are now concerned more than ever to make BGBD search relevant to the systems of agricultural production.

Policy Paper 36, May 2006.

Belowground Biodiversity—The Provider of Ecosystem Goods and Services

A provider of ecosystem goods and services, the belowground biota, as drivers and managers of the natural soil processes, are irrevocably linked to land and soil productivity, which forms the basis of sustainable crop production. The more important aspects of these processes include:

- Nutrient cycling.
- Regulation of soil organic matter dynamics.
- Soil carbon sequestration and its impact on greenhouse gas (GHG) emission.
- Modification of soil physical structure.
- Assistance to plant nutrient acquisition mineralisation, fixation and mobilisation of nutrients.
- Enhancement of plant health, and biotic and abiotic stress tolerance.

Shape and Size of the BGBD

The shape and size of the belowground biodiversity is as complex as that of the aboveground. The BGBD consists of the prokaryotes and eukaryotes: the *microflora* bacteria, fungi, algae etc., *micro-* and *mesofauna* protozoa, nematodes, arthropods, mites, etc. and the *macrofauna* insects, earthworms, small vertebrates, etc. These diverse biotic groups and the roots of plants that grow in the soil, interact with other species, and constitute the life in soil. Apparently, the extent of diversity in numbers of various biotic groups varies directly with their relative dimensions. For every species of vertebrates and earthworms in a soil, there may be hundreds of species of fungi and thousands of species of bacteria. Mathematical interpretation of the diversity abundance distribution relations of the micro-organisms in soil, has allowed prediction that there could be 10^7 prokaryotic taxa in a gram of pristine soil.

The genomic equivalent of prokaryotic diversity of a pasture or a forest soil may be anywhere between x 20-30 larger than that of an average arable soil, although the relative abundance of culturable microorganism populations in soils under such contrasting land utilisation systems may be nearly the same. Considering the small number of environmental DNA clones generally analysed under ecogenomic studies, one is possibly looking at the tip of the iceberg even with the state-of-the-art molecular techniques, such as DGGE, SSCP and tRFLP.

Rhizosphere—A Hot Spot of Belowground Biodiversity for the Benefit of Cropping Systems

Root exudates, sloughed off mucilage, dead cells, and other metabolites and by-products are responsible for changed composition and dynamics of the soil biota in the root-soil interface, the rhizosphere. The hallmark of the rhizosphere lies in the physical and chemical interactions between the roots and the biotic communities of the soil that govern many soil processes influencing the growth of plants. In view of such unique interactions, the 'rhizosphere' is considered as a 'hot spot' for the complementary relations between the aboveground plant life and the belowground biota.

Molecular tools of ecogenomics have allowed preparation of fairly accurate rhizosphere microbial fingerprints of several crops under varied systems of production, particularly with reference to the biogeochemical cycling and turnover of nutrients nitrogen, phosphorus, iron, sulphur etc., plant growth promotion by non-nutritional mechanisms, and suppression of deleterious microorganisms in the soil-root interface. Use of species-specific molecular probes coupled to gene sequencing, and now the microarrays, permits functional assessment of the rhizosphere biodiversity *in situ*.

Threats and Likely Causes of BGBD Loss in Ecosystems

Belowground biodiversity loss is a complex problem involving multiple causative factors. To ascertain the causes of the loss, it is imperative first to know what determines the biodiversity or the species richness of communities belowground. In spite of the advances made in BGBD search, it is yet difficult to disentangle the roles of the soil, environment and aboveground biota in determining the BGBD. However, the evidence, like that of the neutral North American deserts having a higher bacterial diversity than the biodiversity rich South American acidic rain forests, suggests that the life sustaining (or threatening) abiotic soil environmental factors (for example, pH) may be more important than the aboveground biotic diversity in determining the BGBD. Examples of the kind now available with other belowground biotic communities embolden this argument.

Available evidence suggests that soil biotic communities are more resilient to anthropogenic perturbations than the aboveground diversity, but disturbances in aboveground biota (such as that are caused by land clearing, resettlement, monoculture of crops and trees, use of agrochemicals, intensive tillage and such other practices which negatively affect aboveground biodiversity) may have large impact on the BGBD, and might negatively affect the self-regulatory capacity of the ecosystems. It seems that the high-energy driven intensive agricultural production systems have introduced a kind of vicious cycle, whereby leading the loss in BGBD. As a consequence, the soil biological function, are provided more and more with energy intensive external inputs, bringing further peril for the BGBD, and hence, for the sustainability of the production systems.

Belowground Biodiversity—Methodology Base for Study

While the traditional methods of sample collection and direct estimation of the faunal component have remained largely valid, the techniques of assessing microbial communities have now undergone a tremendous change. Assessment and identification of the diversity can now be accelerated by using automated methods (e.g. Biolog), and broad scale measures of diversity, e.g., phospholipid fatty acid (PLFA) profiling, fatty acid methyl ester (FAME) profiling, DNA hybridisation and reassociation kinetics, etc. These techniques have largely overcome the limitations of the cultivation-based techniques.

Molecular techniques of analysis of the community or species diversity at genetic level based on polymorphism of indicator genes extracted from environmental samples have emerged as alternatives that allow a higher resolution and avoid many of the limitations of the traditional methods. The indicator genes may be functional genes, e.g. genes encoding proteins that perform particular metabolic functions relevant to the ecosystem.

In view of the above background, NAAS organised a brainstorming session[1] on "Belowground Biodiversity with Special Reference to

1. Session held on March 10, 2006, Barkatullah University, Bhopal (MP) with Dr. B.N. Johri, Emeritus Scientist and Editor NAAS as the Convener. Each theme was introduced by a lead speaker followed by panelists' interventions and open-house discussion of the issues raised.

Cropping Systems" and deliberated at length on five major themes: (i) Sustaining Biological Basis of Agriculture, (ii) Development of Tools of BGBD Research, (iii) Climate-Land-Soil and BGBD, (iv) Belowground and Aboveground Biodiversity Interface and Linkages, and (v) Priorities and Researchable Issues.

A short plenary session summarised the main points and recommendations.

BGBD—Concepts and Questions

The thematic deliberations addressed the following concepts and questions:

- 'Microbial diversity in soil is beyond practical calculation'. Whatever we talk about, diversity is based on our 'observations' made from seemingly, the observed facts. Considering the vastness of the microbial diversity and the variable entity of the soil, it is incomprehensible that we can reach to the bottom of the microbial diversity in soil by empirical observations alone. Mathematics greatly helps us in making observations and drawing conclusions from the observed facts. We possibly need simple mathematical theories to guide us in exploring the BGBD and making predictions.
- 'Soil biodiversity is an abstract aggregated property of species in the context of communities or ecosystems'. Can differences in community or ecosystem function be assigned to the community diversity or species richness of communities belowground? Functioning of terrestrial ecosystems, plant biodiversity, productivity, variability and stability seems to depend heavily on community diversity of soil biota. Functional diversity rather than taxonomic diversity (community structure) or species richness *per se* is the major determinant of ecosystem functioning. It may be thus, more important to understand the linkages between the actions of a key species or the functional groups and ecological functions of different ecosystems than to search for the diversity index or the species richness and try to relate the same to ecosystem or community functioning.
- What determines the belowground biodiversity and species richness of communities? How important are soil abiotic

factors *versus* aboveground biotic diversity and interactions in determining the resulting belowground community? The tenets of species-area relationship, as applied to microorganisms, prokaryotes and eukaroytes, reveal that community diversity and species richness increase with the area owing to the spatial complexity of habitat structure even within a constant biotic frame aboveground. Environment affects species, and the outcome of their interactions, but not the diversity or ecosystem processes directly. The commonly observed BGBD loss in intensively managed arable soils compared to the native biodiversity of the ecosystems may be a reflection of the intensity and duration of exposure to the changed environment (e.g. pH) under the systems and practices of intensive arable soil and crop management.

- The ecological tenets of structure-function relationship of ecosystems as empirically applied to soil microorganisms would show that (i) maintenance of a diverse and functional microbial community i.e., functional biodiversity is essential for soil sustainability, and (ii) loss of biodiversity inherently makes the biological systems less resilient to environmental stresses, such as those introduced by intensive soil and crop management practices. This sets the agenda for BGBD search in relation to sustainable productivity management of land and soils for the crops. The key issue here is, whether it is possible to maintain a tight fit BGBD at threshold level within the ambit of sustainable production system.
- Can the positive relationships between soil (biological) quality, soil functions, and soil productivity as clearly discernible in the undisturbed natural ecosystems be extended to agricultural production systems and exploited for sustainable production management of crops? The biologically driven alternative agricultural production systems and practices have provided empirical information about the changes in soil biological processes and the quality traits. One important question to answer in this context is whether the observed changes and benefits are in integration or in segregation of the different land use systems?
- Can we delineate keystone BGBD markers across the functionality of different production systems? The state-of-the-

art of the BGBD search in relation to the status of soil health and productive functions of different land use systems relies heavily on soil biological quality parameters (soil organic matter (SOM), microbial biomass, soil enzymes, etc.). For linking BGBD functions with the functionality and stability of different production systems and generating principles and practices of their sustainable management, we need broad-based biotic indicators. Against the current state of our knowledge, it is difficult to envisage that the available indicators will be exclusive and cut across the different production systems and their functions. Instead, system and function-specific biotic indicators, which satisfy the stringent requirements of functionality, reliability and reproducibility will help in advancing the cause of BGBD search for cropping system management. Could new molecular tools help us to reach such indicators?

- The nexus between system productivity and soil organic matter (SOM), especially the more labile part of it, through aggradations of fertility and physical structure underwrites the SOM-BGBD sustainability linkage. Organic matter management being central to soil sustainability management, can a working model be developed based on SOM or any relevant part of it including its role in buffering the soil reaction, to provide for the indicators of SOM-BGBD sustainability linkage?
- Carbon sequestered in soil is three times more compared to that in aboveground biomass. In view of the mycorrhizal fungi functioning as a main carbon sink in soil with direct links to the atmosphere, can the arbuscular and ectomycorrhizal fungi as distributed in the ecological zones be considered as indicators of carbon cycling and the interconnected ecosystem processes *vis-à-vis* plant production and GHG management?
- Nematodes, especially the bacteria feeding ones, are good indicators of soil disturbance. Can nematodes be developed as indicators of soil stress in disturbed production systems? Could ants be used as keystone species in cropping systems? Can the role of collembolans as ecosystem engineers at micro-level be exploited to relate to ecosystem health?
- Can any indicator ratio or quotient, such as that of the

proportion of r and k strategists or that of termites *versus* earthworms be used to monitor production system sustainability?

Recommendations

- Considering the enormous diversity and heterogeneity of the microbial forms and spatial variability of habitat structure even within the similar production systems, research effort is needed to develop mathematical models for assessment of the BGBD-cropping system interaction. The microbial ecologists, crop production specialists, molecular biologists and statisticians need to collaborate in this endeavour.
- In view of the alarming scenario of global warming *vis-à-vis* GHG emission, BGBD contributions towards carbon sequestration in soil should receive active consideration. The impact of GHG atmospheric warming nexus on BGBD ecosystem functionality links needs precise understanding. The agricultural production systems, such as wetland rice should receive adequate attention from the point of view of reducing the risk of GHG emissions.
- In view of the ongoing degradation of the biological base of soil productivity in the prime agroecosystems (such as the rice-wheat system of the IGP), efforts should be directed to precisely assess, evaluate and monitor the BGBD structure and functioning in select agroecosystems. Inventorisation, identification, characterisation and conservation of the soil microorganisms using molecular and traditional tools should be a priority objective. Agro-ecosystems such as dryland agriculture, hill agriculture, coastal zone agriculture and mixed farming should be the immediate targets.
- Intensive agriculture such as that practised in rice-rice or rice-wheat cropping systems over a large area of the country, utilises heavy doses of chemical fertilisers and pesticides, whereas these crops show low belowgroun: aboveground biomass ratio on account of considerable ground—and surface water utilisation. This has brought about considerable changes in the belowground habitat, and introduced stresses and strains that are likely to result in at least short-term disturbances in microbial community structure and

functioning in the soil and the rhizosphere. Forms that degrade xenobiotics are likely to be abundant under such situations. These production systems offer scope for studying the extent of soil degradation that allow microbial functionality at an acceptable level under the broad principles of BGBD conservation and management in intensive cropping systems. Hence, long-term restoration studies of the degraded systems shall help arrive at quantifiable indices of soil microbial diversity-functionality relationships of the agroecosystems.

- Attempts should be made to work out the threshold level of the functional microorganism groups and their phenological dynamics in the rhizosphere in relation to the turnover and cycling of nutrients for better management of sustainability of the production systems. The holistic approach should take care of the metabiosis in the rhizosphere that determines the successional colonisation of niche by individuals and population groups.
- Plant breeding has generally tended to ignore the functional contributions of microorganisms in soil and rhizosphere (e.g. AMF and Rhizobium symbioses), and the plant genetic traits like enhanced association with and response to the beneficial microbial community groups in soil. Crop-genotype-specific differences in supporting soil biological processes can open a new area of plant breeding and biotechnology for exploiting soil biology for better production management of crops and cropping systems in a sustainable manner. The conventional and modern tools of plant breeding can be taken advantage of in breeding crop cultivars with enhanced response to the specific belowground communities. Such varieties shall be of greater advantage in case of alternative, low-input systems of production.
- The root exudates microorganisms interaction in the rhizosphere can be a potential area of soil biological management for sustainability. The genetic control of root exudation remains largely unknown, and if known, shall offer scope for modification by genetic engineering to exploit rhizosphere build up of the beneficial microbiota and to facilitate suppression of the deleterious indigenous ones. Also,

there is scope to exploit allelochemicals in this context to suppress unwanted weeds, pathogens and herbivores. Bioinformatics can be a useful tool to advance the concept.

- Fine roots, < 30 um are considered, in a broad sense, a part of the soil biota and functionally interact with the microbial community groups in many diverse ways. The interactions between fine root architecture and soil microorganisms in relation to dynamics and turnover of nutrients, herbivory, escape and tolerance to deleterious microorganisms, etc., should be studied employing modern tools, like radiotracer techniques to generate data for genetic manipulation and modelling.
- The major representative crops and cropping systems should be utilised to generate data on interventions directed towards the re-establishment of soil biology using biomarkers, such as the AMF, nematodes, earthworms, etc., so that the interventions with organics and biologicals can be based on sound scientific knowledge. Data generation is particularly important where new crop types inbreds, hybrids and transgenics are involved.
- Considering the steady increase in the exploitation of transgenic crops, it is important to take up, on priority, assessment of the ecological impact of their residues and rhizodeposits on BGBD structure and function; as for example, the overall impact of transgenes directed to suppress the insects or the nematodes, or the impact of crop genotypes modified in the pathway of recognition and defence to fungal pathogens or AMF colonisation. This should also take care of the horizontal gene transfer from the transgenics to the microorganisms. This information, apart from putting transgenics use on sound ecological understanding, shall help in understanding the microbial gene expressions in soil.
- The IPM practices provide a platform for studying the relations of the ecologically-driven IPM interventions with other soil processes as also the signal transduction mechanisms against pathogens and pests. This may be applicable in a reverse way on the likely impact of biofertilisers and PGPRs (AMF, Azospirillum, Rhizobium, fluorescent Pseudomonas, etc.) on signal transductions for defence against the pathogens. Such

information is likely to go a long way to design appropriate plant disease and pest management practices based on sound ecological principles.

- A more coherent and directed effort is necessary to assess the indigenous diversity, especially of the growth promoting soil microorganisms. This will help in extending the advantages of the natural bioresource base (that the BGBD is) to intensive agriculture and maintenance of soil health.
- To meet the goals stated above, it is essential (a) to put in place a research initiative on long-term ecological research (LTER) for varied ecosystems to investigate the BGBD cropping or agroecosystem functional linkages, and to utilise the information for more balanced and productive cropping systems, and (b) to upgrade soil microbiology, including molecular biology research tools and methods. These efforts will not be successful without the well-trained and dedicated manpower in the fields of molecular and morphometric taxonomy, biosystematics, molecular ecology, and bioinformatics—BGBD-cropping system interaction research. This will require considerable investment for laboratory upgradation, introduction of teaching and training programmes, and adequate research support to the institutes and universities. Creation of Centres of Excellence around outstanding microbial ecologists will be a desirable step in this direction.
- The recently established National Biodiversity Authority for implementing the Biodiversity Act (2002), should set up an expert group to develop guidelines for the conservation, and sustainable and equitable use of soil *microflora* and *mesofauna*.

20

Transgenic Crops and Biosafety Issues Related to their Commercialisation in India

Improvement of agricultural production and productivity as well as the future versatility of agricultural production are dependent on the rational utilisation of technologies. We stand at the convergence of an incredible array of new technologies, such as recombinant DNA technology, information technology and high throughput genomics, to enhance our understanding of the structure and function of the genomes and to apply this information for plant and animal improvement. Products arising from modern biotechnology such as genetically modified (GM) or transgenic crops are providing new opportunities to achieve sustainable productivity gains in agriculture.

The transgenic crops were grown worldwide in about 68 million hectares in 2003. The engineered traits include insect pest resistance, herbicide tolerance and virus resistance. The first, and as yet the only, GM crop permitted for commercial cultivation in India is the 'Bt-cotton', conferring resistance to a lepidopteran insect pest, the bollworm. Transgenic crop acreage in India is currently about 100,000 hectares. Following its commercial introduction in 2002, the Bt-cotton has found overwhelming support from farmers to the extent that almost all the domestic and multinational seed companies in the country are transferring the Bt transgene (and its variants) into their agronomically superior hybrids. In addition, transgenic research and development in a number of other crop plants for various agronomic and quality traits is being pursued by several public sector institutions, in India.

The commercialisation of transgenic crops has sparked off intensive debates worldwide regarding biosafety and the impact of this powerful technology on agriculture, human health and environment.

Policy Paper 29, December 2004.

The biosafety concerns of transgenics owe their origin to the fact that tools of recombinant DNA technology have clubbed all biological systems into a 'single gene pool' providing access to even genes from completely unrelated or sexually incompatible organisms. Therefore, the transgenic technology, while providing unlimited scope for crop improvement, also imposes tremendous responsibility on the scientific community and the regulatory authorities towards ensuring biosafety.

A number of biosafety issues of transgenic crops are being debated with highly polarised views. It is imperative that rational science-based decisions are taken to resolve the contentious issues and evolve clear-cut national policies so that the benefits of this technology are effectively harnessed for achieving sustainable growth in agriculture without unacceptable risks either to human health and/or the environment.

The National Academy of Agricultural Sciences (NAAS), in collaboration with the XV International Genetics Congress Trust, initiated a national debate for addressing the issues related to research, biosafety testing and commercialisation of transgenic crops.[1]

The following issues were critically discussed during the Session:

- Relevance of transgenic crops in Indian agriculture.
- Need for prioritisation of crops/traits for transgenic research and development.
- Policy related to use of selectable markers in transgenic research.
- Policy related to biosafety testing for transgenes and transgenic events.
- Meeting the institutional requirements for biosafety testing.
- Approaches to analyse the impact of environmental concerns related to transgenic crops.
- Pre- and post-release evaluation and monitoring of transgenic crops.

---

1. Views of a large number of stakeholders were obtained electronically, and these views aided in the preparation of a 'base paper' for discussion. A brainstorming session was organised under the Convenership of Dr. R.P. Sharma, at NAAS, New Delhi, on December 1, 2004, on 'Transgenic Crops and Biosafety Issues Related to their Commercialisation in India'. The Session was attended by representatives from various public and private sector institutions, besides policy makers.

- Public-private partnership in transgenic research and development.
- Managing intellectual property rights (IPR) issues related to transgenics.
- Roadmap for effectively bringing the transgenic technology from laboratory to the farmers' fields.

Recommendations

The major recommendations that emerged are given below:

- India should promote genomics and transgenic research and development keeping in view the short-term and long-term requirements of the nation, besides maintaining the global competitiveness in agriculture. The vast plant, animal and microbial biodiversity should be mined for genes of importance to sustain advances in crop and animal productivity. The transgenic approach must be judiciously integrated into the crop breeding programmes based on specific needs, particularly in cases where conventional breeding is not feasible or effective.
- For effective utilisation of the technological, financial and human resources in the public sector, it may be useful to identify priorities for transgenic research and development based on objective criteria. Besides technology development for public good, the public sector institutions must play a vital role with regard to knowledge generation so as to fill the critical gaps in relation to transgenic development. Technology generation and capitalising on the available technologies should go hand-in-hand.
- In terms of the area under transgenic crops worldwide, herbicide tolerance as a trait, ranks first. Although, herbicide tolerant transgenics offer specific advantages, in the Indian context it is not, at present, a priority trait, since, it is preferable to adopt technologies that are labour-diversifying rather than those that could be labour-displacing. Also, while setting priorities for transgenic research and development, it is important to consider the social dimensions, including employment impact, besides environmental impact.
- The coming years will witness the use of food and non-food

transgenic plants as vehicles for mass production of novel vaccines, antibodies and other therapeutic proteins. During production and commercialisation of the pharma-transgenic crops it should be ensured that they do not contaminate the food chain. Appropriate guidelines should be formulated for production and handling of the pharma-transgenic crops. The developments in medical or pharmaceutical biotechnology utilising crop plants as vehicles should be monitored and the linkages with agriculture, analysed.

- Although, there is no scientific evidence of the adverse environmental effects of selectable markers (like antibiotic resistance genes) that are widely used during transgenic development, research efforts on developing and utilising 'clean gene' technologies or marker-free transgenics should be strengthened. Also, the information database regarding biosafety of the selectable markers should be upgraded to facilitate objective decisions by the regulatory authorities. Certain selectable markers like kanamycin and hygromycin resistance genes, that are proven to be safe, and other antibiotics that have outlived their clinical utility may be utilised till better alternatives are available.
- A transgene that has already undergone extensive biosafety tests should not be treated as new, even if it is a new transgenic event. Where adequate evidence is available that the recurrent parent genetic background of a notified/ registered genotype is nearly restored (through field data and/ or molecular data), only the agronomic performance and the level and stability of transgene expression may be analysed by two year trial data (one year of simultaneous large-scale Genetic Engineering Approval Committee (GEAC) and ICAR trial, and another year of ICAR trial). Even in case of a structurally altered transgene with no significant modifications in protein conformation, the toxicity and allerginicity tests need not be carried out, provided the predicted antigenic epitope remains the same and the level of expression of the transgene is within the defined limits.
- The toxicity and allerginicity tests for the transgenics should be prescribed on a case-by-case basis, based on the transgene, the crop and the economic product(s). Institutions with

relevant expertise and facilities should be identified, upgraded, accredited and networked for enhancing the efficiency of biosafety evaluation of the transgenics. The scientific data coming from reputed national/international institutions must be duly considered for testing toxicity and allerginicty.

- The major environmental concerns arising from the possible release of transgenics should be evaluated on a case-by-case basis depending upon the gene, the crop, the trait and the target geographical location(s). The priorities and parameters of environmental impact assessment of the transgenics should be identified so that empirical data are generated on specific aspects (like gene flow and transgene invasiveness) through well-designed experiments proactively initiated by the scientific organisations. It is also important to strengthen the institutional mechanisms for analyses of the environmental/ ecological effects of the transgenics.
- The Monitoring and Evaluation Committee (MEC) should be made more effective and relevant to the needs. The MEC may be a decentralised state-level committee constituted by the Review Committee on Genetic Manipulation (RCGM), with the active involvement of SAUs. The committee must maintain rigorous standards across the country. Clear guidelines should be formulated regarding the parameters and methodology of monitoring and evaluation.
- A major bottleneck for the transgenic R&D in India, is the non-availability of suitable genes, promoters and vectors, besides efficient transformation methods (for some crops). The expertise available in the country for these important technological components must be strengthened and effectively utilised in a directed fashion. A 'National Repository' may be established to maintain relevant genes/ promoters/vectors etc., and to facilitate access to the same through appropriate Material Transfer Agreement (MTAs)/ MoUs for transgenic research and/or commercialisation.
- Researchers/institutions involved in transgenic R&D must consider the implications of the IPRs related to the transgenics, particularly in the context of commercialisation. It may be useful to establish a single window advisory service in

India, for providing advice on IPRs pertaining to a transgenic project before its initiation.

- To effectively translate the potential transgenic events into commercially viable products, the following approaches may be considered:
  a. In the public sector institutions, transgenic development and deployment should be undertaken through a cohesive research programme in two phases. The funding agencies must ensure that the project is in a network mode with relevant expertise and intra- and inter-institutional linkages before approval. Funding for Phase I should cover all aspects related to technology development. After the review committee certifies the success of Phase I, funds for Phase II (testing and commercialisation) should be automatically released.
  b. Public sector may demonstrate a 'robust' transgenic line at the greenhouse level, and find an industry partner, wherever appropriate, for effectively taking it through commercialisation, including regulatory trials, value capture and sharing mechanisms.
  c. Public and private sector may join for developing a transgenic product concept and together seek resources for undertaking the project with a clean value-sharing arrangement.
- There is need for launching a genetic literacy movement in schools and colleges on the rapid developments taking place in the area of molecular genetics and genomics, so that there is a better understanding in the country of the opportunities and risks associated with recombinant DNA technology. This will help to promote the safe and responsible use of the tools associated with the new genetics in the country in the fields of food and agriculture, medicine, industry and the environment.

21

Biosafety of Transgenic Rice

Rice, wheat and maize, members of the grass family, are the major grains that sustain humanity. Of these three, rice is the most important since it is the staple food for over half of the world's population. Most rice farmers and consumers are in Asia, where approximately 70 per cent of the world's 1.3 billion poor people live. Of the 840 million people suffering from chronic hunger, more than 50 per cent live in areas dependent on rice for food, income and employment.

While the world population is continuing to grow, rice growth rate has decelerated from 2.3 per cent during the 1980s to 1.1 per cent during the 1990s. A major worrying aspect is the productivity decline witnessed in the favourable irrigated rice growing regions from where major part of rice production is derived. Yield stagnation in many Asian countries, limited possibilities of expanding arable land due to rapid industrialisation and urbanisation, depleting water resources and elevated wages of farm labour, are the main constraints for increasing rice production. In addition to these constraints, decline in native soil productivity, increase in salinity, alkalinity and waterlogging are further adding to productivity decline. The canopy structure of the new rices encourages pests and pathogens, which further inflict serious damage. Recognising the pending crisis in the state of rice production, the FAO, on November 3, 2003 declared, 2004 as the International Year of Rice.

Developing new rice varieties with higher yield potential and innate genetic capacity to stabilise production are vital to bridging the yield gap between favourable irrigated and the unfavourable rainfed environments. It is also a major challenge. The pace and extent of achieving this objective is constrained by limited availability of

---

Policy Paper 24, December 2003.

suitable germplasm. Therefore, conventional breeding is handicapped in gathering the necessary ingredients for meeting the dimensions of the task. Newer approaches, tools and technologies are needed to overcome these constraints, which conventional crop improvement programmes are finding difficult to overcome. The newer technologies will also have to be seed embedded if they are to meet the demand of the vast majority of resource-poor farmers, holding less than a hectare of land, so that more rice can be produced from lesser land, using lesser water, labour and chemicals.

During the last two decades or so, spectacular developments in modern biology have ushered in the era of biotechnology, bioinformatics and genomics. The power of these technologies enables us to discover genes of importance for human welfare from anywhere in the plant or animal kingdom. Genetic engineering enables us to move genes of interest between organisms, including those that are nor crossable. Transgenics, carrying novel traits for disease and pest resistance, are already proving to be a boon for even the resource-poor farmers in China. While these new tools and technologies are poised to lend a major impetus to war on poverty, hunger and disease, indeed, there are concerns in respect of any potential harm to human health and environment. To address all such concerns, genetically modified crops (GM crops) are going through an intensive testing for toxicity and allergenicity that is unprecedented in the history of agriculture. Even though, individual products of genetic engineering may pose more risk, less risk, or the same risk as anyone or another of the products of conventional breeding, deployment of GM rice in major rice growing areas of Asia, merits a comprehensive evaluation in view of the fact that these regions happen to be the centres of diversity for rice. The merit and demerit of not using GM technology in developing new crop cultivars need to be weighed against any possible risks associated with using it.

To discuss issues of development and deployment of GM rice in rice growing areas of Asia, Africa and Latin America, NAAS convened a meeting at Chennai. The meeting was attended by concerned subject matter specialists including, rice breeders, molecular biologists, entomologists, economists, nutritionists, social scientists, NGOs and a farmer representative.[1]

---

1. NAAS convened a meeting in Chennai, (Oct. 27 to Oct. 30, 2002), in collaboration with Syngenta International to critically examine all the pertinent issues comprehensively.

The areas of scientific discussions were:

- The biology and ecology of rice crop and its wild, weedy relatives and land races.
- Transgenic rice for commercialisation.
- Potential impacts of transgenic rice on human- and animal-health.
- Developing a risk assessment model and methodology for transgenic rice.
- Societal, economic and ethical issues affecting transgenic rice

Given below are the major recommendations, which emerged from deliberations at this meeting.

Recommendations

1. The importance of conventional plant breeding was recognised and its further strengthening through incorporation of new scientific and technological tools, including biotechnology was strongly recommended.
2. There are real benefits to be had, by all sections of the society, from the application of biotechnology, in particular the recombinant-DNA (rDNA) technology. A broad-based consultative process must be conducted to rank priorities for transgenic rice development suited to national needs on an ongoing basis.
3. Developing rice varieties having greater tolerance to drought, submergence and salinity, better resistance to pests and disease, and rich in micronutrients, appear to be achievable goals through rDNA engineering and should be pursued actively.
4. A comprehensive science-based risk assessment system must be practised for genetically engineered crops—including rice—that takes into account issues relating to centres of diversity and unique social and economic aspects.
5. Currently, genetically enhanced transgenic rice conferring high level of resistance to stem borers and leaf folders are in advanced stage of development in some countries. Transgenes encoding δ-endotoxins of *Bacillus thuringiensis* (Bt) are the commonly used insecticidal genes in view of their good record

of specificity and efficacy against target pests, and its non-toxicity to humans, other mammals and birds. The workshop recommended that to enhance its efficacy, wherever possible such transgenes must be put under the control of target tissue- or stage-specific regulatory elements rather than the constitutive ones.

6. Importance of resistance management through integrated pest management was stressed in the workshop, as also the imperative to extend the usefulness of Bt and such other transgenes. Resistance management strategies for the predominantly tiny rice farms of Asia need to be fine-tuned in keeping with the local practices. The aspect of refugia, in particular, would merit special attention. The workshop recommended that studies be undertaken to assess efficacy of refuges and refugia provided by non-transgenic rice crops. Given the scenario of greater vulnerability to resistance breakdown in Asian rice farming systems, deployment of two-genes (having dissimilar mode of action) transgenics may be required to prolong resistance.
7. The potential of gene flow in rice exists and its occurrence has been scientifically established. This phenomenon must be addressed for transgenic rice through appropriate regulatory oversight, on a case-by-case basis. Given the importance of rice as a staple crop, regulatory agencies must develop and put in place a comprehensive monitoring mechanism to assess long-term environmental/ecological impacts from the release of transgenic rice.
8. The regulatory process relating to genetically engineered crops needs to be made transparent, participatory (involving all stakeholders), effective and efficient. Appropriate regulatory oversight based on sound scientific information should be an essential requirement for the approval and release of crops derived from rDNA technology.
9. International experience garnered across different species must be used to make appropriate case-by-case study of biosafety assessment and regulatory oversight while taking into account current difficulties experienced internationally in interpreting and applying concepts, such as the precautionary principle and substantial equivalence.

10. No credible scientific evidence is available indicating horizontal transfer of antibiotic-marker genes under natural conditions, and chances of such transfer are far greater from other microbes in the environment. The workshop recommended use of alternative systems of selection where feasible. The workshop also recommended use of 'clean DNA' (minus the vector background) technology for transformation.
11. The technological application of micronutrient-enhanced rice must be reviewed through the regulatory process, keeping in view the social, political and cultural implications.
12. Prior to commercialising herbicide resistant rice, issues relating to the potential development of resistance in the genetically modified crop, besides the ecological, social and cultural factors from such an introduction to the society, must be addressed.
13. The workshop recommended that transgenic rice producing pharmaceuticals and drugs must be assessed carefully for their impact on food and feed chain. It was strongly felt that rice crop must be avoided to produce such compounds as far as possible. However, if there was critical need to develop such transgenic rice, a distinct set of biocontainment protocols must be developed under rigorous oversight.
14. No scientifically valid environmental or ecological impact issue was identified with respect to releasing transgenic rice varieties in centre(s) of origin and centres of diversity.
15. Basic research relating to transgenic rice should be encouraged even as effective regulatory systems are put in place.

22

Conservation and Management of Genetic Resources of Livestock

Introduction

The diversity of livestock genetic resources is very wide, both in variety and variability in terms of species, breeds, populations and unique genotypes. This diversity has been recognised as a vital resource for sustenance of mankind. Judicious utilisation and enhancement of the quality of these resources is important to ensure their sustainability to meet future demands.

The Indian subcontinent is amongst the 12 mega biodiversity resource centres in the world where domestication of flora and fauna took place in the antiquity. The spectrum of bio-diversity is exceedingly vast and varied in this subcontinent. In domesticated livestock and birds, a large number of breeds/types of cattle, buffaloes, sheep, goats, pigs, horses, camels, mithuns, yaks, dogs, cats, poultry, ducks, geese, turkeys, guineafowls and pheasants have evolved over time through natural selection and some human effort.

From the Vedic to the post-independence period, improvement in livestock has been made to meet changing demands. The primary thrust in cattle has been on draught. However, with the introduction of mechanisation in agricultural operations and rural transport, the relative importance of draught has decreased considerably in large farms and may further decrease in future. It is likely that this situation will extend to small farm holdings too. Consequently, there will have to be a major shift in terms of utilities of bullocks, which have been, till lately the backbone of traditional system of agriculture and rural transport. This shift will have to be taken into account while developing policies for improvement and conservation of cattle.

Policy Paper 14, December 2001.

The NAAS organised a workshop[1] on 'Conservation and Management of Livestock Genetic Resources (LGR)' at GB Pant University of Agriculture and Technology, Pantnagar, in association with the Department of Animal Husbandry and Dairying (DAHD).

The workshop considered and debated issues relating to current livestock production systems, the present status of livestock genetic resources, strategies for conservation, technology platform and the improvement framework for future. The workshop discussed ways and means of making the livestock genetic resources as sustainable inputs into the economic and social development process such that they become wealth and employment generating enterprises of the future.

Current Status and Future Demand

Animal production contributes about 40 per cent of the total value to agriculture globally. Two billion people depend, at least in part, directly upon livestock for their livelihood. The 4,000-5,000 breeds/populations provide a broad variety of meats, milk and eggs, together with a broad range of other goods and services, such as draught, fibre, hides, skin, manure, etc.

With appropriate management, livestock enable sustainable use of marginal lands for food production, transforming otherwise unusable fibrous material into high quality protein. When poorly managed, particularly in fragile production environments, livestock species become one of the causes of desertification which is not due to the livestock per-se but to human inability to manage the livestock and vegetation resources in an integrated manner. It is therefore, necessary that sustainable use and further development of these resources is well planned and managed. Human population growth, progressive urbanisation and larger purchasing power of people will increase demand for livestock products. Delgardo *et al*. project a growth in total consumption of meat and milk at 2.9 and 4.3 per cent per year, respectively, for India. In absolute terms this converts to 170 million mt of milk and 40 mt of meat by 2020 AD.

Considering future trends in demand and supply of livestock products and services, attention for genetic improvement will be

---

1. More than 210 academics, activists, researchers, non-government organisation representatives, government officials, and representatives of industry from India, together with a number of scientists from other countries participated in the workshop.

required both for indigenous and exotic breeds. Although, there is a need to utilise between-breed genetic differences through crossbreeding for higher yields, greater emphasis is required on improvement of adapted indigenous breeds/types because of valuable adaptive traits they have developed over long periods of time through natural selection.

Continued human population pressure, economic advancement leading to growing demands for foods of animal origin, and environmental factors affecting lands will require sustainable intensification of livestock production to meet increasing human demand. This can succeed only when the livestock genetic resources are adapted to the prevailing production environment. Where changes to the production circumstances occur steadily over time, the indigenous adapted livestock genetic resources possess better biological capacity to respond to the new conditions. Where major production system changes are proposed to be introduced within short periods, superior genetic resources from elsewhere in the country or from similar production environments of other countries will be required. For example, with much improved genetic evaluation and deployment, genetic material developed under high-input, low-stress, short life-cycle production environments of developed countries has made valuable contribution to expanding the higher input, 'softer' production environments of India.

Two main considerations will guide future activities: (1) the conservation of livestock genetic resources and maintenance of diversity and (2) substantial increases in rate and efficiency of livestock production. The current and future challenge is how to effectively generate increased output per unit of input from livestock sector while preventing major loss in genetic diversity and environmental degradation.

Production Systems

The locally adapted indigenous breeds commonly show low absolute levels of production. However, productivity itself is often remarkably high, when proper account is taken of the production environment, the level of input and the length of the production life-cycle. Indigenous breeds survive, produce and reproduce in environments under which they have been evolved. The medium and low-input production environments common in India, face several environmental stress factors.

Different production systems are in existence in different regions of the country. No macro- or micro-level studies of input-output functions have been undertaken in respect of these production systems, nor has any study been conducted on output of livestock products. These production systems are characterised by specific socioeconomic structure of the society including caste, religion and social status. Many are driven by various ecological considerations. It is important that these systems are studied at both macro- and micro-level, and impact assessment of different livestock production systems is made on livestock and their products. Constraint analysis including strengths, weaknesses, opportunities and threats (SWOT) of different production systems is necessary for developing strategies for conservation and management of livestock genetic resources.

The migratory system of livestock raising for sheep, goats and cattle is an age-old tradition. This system involves movement across states and countries. How this system has influenced structure and function of currently available sheep and goat breeds should be studied and documented, so that new innovations consistent with social change can be introduced which cause minimal hardship to stakeholders but lead to larger benefits.

Livestock resources in India are undergoing unprecedented changes due to introduction of new technologies. Issues arising from these changes need to be identified and examined, so that effective solutions can be found to inevitable problems.

Technology Assessment

Major changes in livestock production have occurred during the past few decades due to introduction of several new technologies, e.g. control of diseases like rinderpest, artificial insemination, embryo transfer, and associated technologies (genomics, cloning and transgenesis). These speed up reproduction, shorten the gene recovery process and enable more efficient genetic improvement.

Despite the present low efficiency and high cost of cloning, this technology will have a major impact on the conservation of livestock genetic resources. Once practical sampling protocols are developed for important farm animal species and they become financially viable, the entire improvement and conservation platform will change. It is, therefore, necessary that research in newer biotechnologies is intensified, so that traditional genetic improvement methods like progeny testing and sire/dam selection schemes are bypassed in future. This is particularly relevant to India and other developing countries, where large data collection and its management is expensive and unreliable.

It will be necessary to identify key policy issues and options which are likely to affect the access, development and utilisation of such new biotechnologies in relation to the management of livestock genetic resources. These policy issues include access to knowledge and information generated by new biotechnologies, intellectual property management, access and benefit sharing, regulatory issues, international collaboration in the conservation of LGR and public perceptions and ethical issues.

At present, no framework is available within a single international instrument regarding policies on access, utilisation and conservation associated with the application of biotechnology to animal genetic resources. Although, several international instruments provide pieces of policy guidance, they need to be reviewed for crafting consolidated guidelines which are cohesive and comprehensive.

Strategy for Conservation of Livestock Biodiversity

A number of methods have been used for conservation of livestock genetic resources. These include *in situ* conservation of the breeds/ populations, cryopreservation of semen, ova, embryos, skin, blood, DNA fragments, etc. These methods are relevant when the breed is rare or near extinction. In India, the situation is not so acute as to call for large-scale *ex situ* conservation efforts. What is, however, necessary is technology evaluation and perfection at selected institutions which can be used whenever and wherever required.

It is recommended that research institutions of ICAR, agricultural universities and other research laboratories initiate programmes to study and identify valuable adaptive traits at all levels (phenotypic, genotypic, DNA/RNA levels) and locate structural genes/QRTs responsible for these traits. Special emphasis should be laid on: resistance to various diseases, resistance to harmful endo- and ecto-parasites, tolerance to large fluctuations in quantity and quality of feed, tolerance to non-availability of adequate quantity and quality of drinking water, tolerance to extreme temperature, humidity and other adverse climatic factors, adaptation to low capacity management conditions, ability to survive, regularly reproduce and produce for long periods of time.

The viability of a livestock genetic resource programme is enhanced when it focuses on traits that increase the economic value of the breed to the communities involved. The assumption is that the model of economic value provides for all inputs and outputs from the herd or flock, over time. A number of approaches have been used to stop or reduce the decline of livestock genetic resources, and these models can be mutually supportive for short- and long-term insurance. 'Wise use' forms a highly desirable form of conservation. The maintenance of a breed in its environment also satisfies the requirements of Article 8 of the Convention of Biological Diversity, which gives first priority to *in situ* conservation. It is suggested that 'wise use' should form the basis for framing conservation policies.

Cryo-conservation as a means of maintaining breeds, should also be considered where specific animal genetic resources are at risk of loss because farmers have not been profiting from their use under the preventing production circumstances *in situ.* One of the most useful aspects of cryo-conservation is its supportive role in genetic upgradation of breeds.

Realising that no clear-cut guidelines are available within the present system of governance and management of resources in India, the strategy has to combine genetic improvement and conservation of resource.

It is necessary that identification, characterisation, evaluation and documentation of the livestock genetic resources are completed in the next five years. A complete description of each breed should be generated on the basis of different profiles including habitat distribution, physical conformation, adaptation, production, reproduction and socio-economic aspects.

Information on demographic distribution of breeds, their population trends and utilisation patterns should also be generated quickly. National and state livestock census should be conducted on breeds and information on ecologies in which they perform.

A complete database should be developed on populations of different breeds within each livestock species of the country. The database should also identify factors threatening the extinction of breeds.

The National Bureau of Animal Genetic Resources (NBAGR) has

a significant role to play as a nodal agency for different aspects of breed characterisation and conservation. NBAGR should develop a time-bound action plan for breeds to be surveyed and characterised and determining conservation needs and strategies.

Differences in number of breeds of a species listed in literature published by different authors from India as well as from other countries, including FAO, has created confusion about the exact number of accredited breeds within each species. A panel of expert animal geneticists should be established in the Department of Animal Husbandry and Dairying of Government of India to list all descript breeds of different livestock species and to evaluate the entire data before granting accreditation. An estimate of value of each breed in its own native environment needs to be worked out so that proper assessment of its sustainable management can be planned.

The basic strategy will have to be, conservation through sustainable improvement and management. This will involve selection for important economic traits. A district-level improvement plan, with village as a unit will have to be devised. A village-level committee should be established which functions as Breeders' Association Unit, and is responsible for bull selection. This association/society/unit should try to maintain: (i) listing of all animals of each farmer, (ii) birth and death register, (iii) health cover register, (iv) breeding register and (v) monthly milk record register. All adult males not used in breeding should be castrated and for each castration the farmer should receive a reasonable compensation. A district-level monitoring committee, which will provide technical guidance and will involve district animal husbandry officer and all veterinarians in the district should be established. This will meet the WTO requirement for import and export of livestock and its products.

There may be situations where crossbreeding with an exotic breed needs to be resorted to. Under these conditions, it is essential to upgrade the production system. A well-defined breeding plan should be developed to avoid problems of future degeneration. Every effort should be made to ensure that the livestock farmers move from subsistence farming to a financially viable livestock enterprise. They should get access to credit on low interest rate from financial institutions and arrangements should be made by the Breeders' Association to provide services and goods as required and a sustainable market for the product.

If a breed is identified as rare or in danger of extinction, the farmers who maintain the animals of such a breed should get compensation at the rate of profit earned through crossbreds. The village group/association/society should also arrange to take up marketing of animal products.

The management and conservation of domestic animal diversity is very costly. For developing people's participation in the conservation strategies, Government of India must provide incentives to farmers.

Human Resource Development

Scientific and technical manpower must be adequately trained for correct and critical identification of each breed, its description based on phenotypic and genetic characteristics including the production environment. It is regrettable that there is no action plan for capacity building and human resource development in this sector. The Department of Animal Husbandry and Dairying, in conjunction with Indian Council of Agricultural Research, must come up with an R&D and HRD plan for this sector.

Utilisation of Genetic Resources

The national policy for management and improvement of livestock genetic resources should have clear aims and objectives and these should be based on analysis of livestock population, breeding structure, disease problems, availability of feed resources, changes in social structure, sustainability of production system in terms of market demand of products and prices. Sound genetic resource utilisation policy, relevant to different farm animal species under different input systems need to be prepared by Department of Animal Husbandry and Dairying, Government of India in consultation with State Animal Husbandry Departments, Indian Council of Agricultural Research, State Agricultural Universities and other stakeholders.

Animal Breeding Policy

The breeding policy, which must include mechanism of livestock conservation, needs a special debate with wider participation of stakeholders. Similarly, Animal Biodiversity Bill, before finalisation, should be given wide publicity to generate debate to make it more effective in the context of conservation of animal genetic resources. A policy framework needs to be established under which conservation and vertical genetic improvement are linked in a manner that leads to reduction in animal numbers.

Legal Issues

Legal measures relating to conservation of livestock genetic resources are non-existent in India. Appropriate legal instruments need to be worked out to cover all aspects of breed utilisation, conservation, trade and marketing. It will be necessary to have a legal framework to regulate mandates and operations of livestock farms of government or government supported institutional herds. For example, *goshalas* and *gosadhans* should be declared as national gene banks and be involved in breed conservation and improvement programmes.

In the WTO structure and IPR regime, access to genetic resources and their transfer is becoming restrictive. Restrictions on export of genetic resources out of India should be considered in a manner that does not hinder reasonable exchange. Similarly, it should be ensured that access to new technology is not denied to us and our institutional structure is strengthened to face a denial, if and when resorted to. For this, efforts should be made to build new institutions in the area of animal biotechnology devoted to cell cloning, embryo fusion, nuclear fusion, gene grafting, gene transfer between species and within species and futuristic sciences. A technical cell needs to be established in DAHD on IPR and WTO issues where competence is available for examining these issues and suggesting appropriate responses and legal solutions.

Convention of Biological Diversity (CBD)

For appropriate alignment to CBD process we should take the following measures:

- Assess the relationship of IPRs to access and benefit-sharing provisions, including development of guidelines or best practices for achieving equitable benefit-sharing from use of genetic resources. In particular, we should pay attention to mechanisms, such as certificates of origin, evidence of prior consent for accessing genetic resources, evidence of prior approval of indigenous and local communities for access to traditional knowledge, and disclosure of this evidence in patent applications.

- Evaluation of the impacts of international processes relating to IPRs, including TRIPs, on the objectives of Article 8 (j) of the CBD.
- Development of a protocol on the protection of indigenous and local community knowledge and resource rights.
- Providing inputs into the ongoing WIPO processes on 'new beneficiaries' which are assessing issues relating to protection of traditional knowledge and development of a code of conduct, or a protocol, on access and benefit-sharing, especially in relation to the resources and knowledge of indigenous and local communications, and of developing countries.

Traditional knowledge

Emphasis should be given to identify, collect, collate and comprehend the traditional knowledge available for the maintenance of various grazing and wasteland ecologies and the animal genetic resources on sustainable basis. The diversity in animal genetic resources (AnGR) has been maintained in India for a very long time (3,000 years) on the basis of traditional wisdom of our farmers. This traditional knowledge should be documented and researched. Traditional practices should be blended with modern methods to allow maintenance of livestock on natural feed/fodder resources in hill/desert/plain ecosystems.

Database Management

For breed characterisation work, methodology, sample size and data analysis package suitable for field data must be generated and standardised. The NBAGR should prepare a breed directory and an animal watch list for regular monitoring of livestock genetic resources by Department of Animal Husbandry and Dairying, Government of India. The data bank facilities are very good at the NBAGR, but feedback of data from resource agencies is very poor. It should be made mandatory by funding organisations, while sanctioning projects in animal husbandry, to send one set of data to the national animal data repository at NBAGR.

The NBAGR is undertaking the breed characterisation and

conservation programme in collaboration with other ICAR institutions/state agricultural universities, under a network programme. Such linkages need to be developed with state AHDs, NGOs and other stakeholders also, to expand this effort.

Species Specific Recommendations

General

There is large genetic diversity in livestock as reflected in important domesticated species and a large number of known and lesser known breeds/strains. It is imperative that extensive surveys be undertaken in the home tracts of the known breeds for their description and evaluation and identifying the need and approach to their conservation and improvement. Where the numbers are extremely small, immediate efforts should be made to conserve those breeds, preferably *in situ*.

Species-wise recommendations for breeds, which require priority attention for conservation, are as follows:

Cattle

Cattle breeds such as Red Sindhi and Sahiwal which have their breeding tracts in Pakistan and Tharparkar, for which we share the breeding tract with Pakistan, are available with a few institutional herds and private breeders. They should be further improved and conserved. Similar attention should be given to Gir, Kankrej and Ongole breeds of Gujarat and Andhra Pradesh. These could be utilised in grading up of non-descript cattle under harsh environments.

Lesser known breeds of cattle such as Punganur, Red Kandhari, Vechur, Bhagnari, Deoani, Lohani, Bengal, Chitagang Red, Nepalese Hill, Kachcha Siri, Tarai, Lulu, Sinhala, Umblacherry and Gangateri need to be studied as genetic resource and steps taken for their conservation and improvement.

A number of new breeds such as Frieswal, Karan Swiss, Karan Fries, Sunandini are in various stages of development from a crossbred base. Their further improvement and conservation is necessary.

Buffalo

Murrah and Nili Ravi are two most important breeds of buffaloes.

They need to be improved further. Other important buffalo breeds, *viz.* Surti, Jaffrabadi, Mehsana, Bhadawari, Nagpuri, Pandarpuri, need to be studied and improved through selection.

Lesser known breeds such as, Kaziranga, Toda, Marathwadi, Sambalpuri, Kalahandi and Paralakhemandi would require extensive survey for their description and evaluation followed by improvement and their conservation if required.

Sheep

Of the 42 breeds of sheep, all the breeds of Jammu & Kashmir and other indigenous breeds like Pugal, Nilgiri and Garole need immediate steps for preservation.

Goat

Of the 20 breeds of goats, Jamnapari, Barbari, Beetal and Surti are threatened by extinction and would need steps for conservation. These breeds have played an important role in genetic improvement of goats. Other breeds of goats like Black Bengal and Osmanabadi also need to be studied and improved.

Camel

In addition to four important breeds of camel *viz.* Bikaneri, Jaiselmeri, Kuchchi and Mewati, there are lesser known breeds such as Marwadi, Mewadi, Sindhi, Shekawati. There is a need for proper description, evaluation, conservation and improvement of these breeds.

Horse

Little efforts for description and evaluation of indigenous breeds of horses and donkeys have been made. Breeds like Marwadi, Kathiawadi, Zanskari and Spiti are some important breeds which need immediate attention for conservation and improvement.

Donkey

Nothing is known of genetic variation in donkeys in spite of their large variation in phenotypic characters. There is a need for conducting surveys for description and evaluation of different types of donkeys available in the country.

Pig

There is a large variation in pigs as reflected in size, colour, performance, etc. There is a need for proper description and evaluation of these types and steps need to be taken for their conservation and improvement.

Yak

Little is known about the genetic resources of yaks. There are differences in size and reproduction performance of yaks located in different regions. These differences need to be studied and utilised in conservation.

Mithun

Little is known about the genetic resources of mithun with respect to their physical conformation and performance and pattern of domestication. There are differences in size of mithuns found in different parts of North-Eastern states. Studies for description and evaluation of these types should be undertaken.

Poultry

A number of species of poultry *viz.*, chicken, duck, guineafowl and quail, make important contribution to food and income. There is large genetic variation in these species, which are required to be characterised and utilised for improvement and their conservation.

Pet Animals

Description and evaluation of indigenous and exotic breeds of pet animals (dogs, cats and birds) need to be undertaken and their improvement requires immediate attention, as these species are becoming very important as pet animals for household and other useful purposes for policing, defence and other duties.

23

Seaweed Cultivation and Utilisation

Introduction

The earliest record of use of seaweeds dates back to 2700 BC in the compilation on 'Chinese Herbs' by Emperor Shen Nung. Reports show that seaweeds have been a part of the Japanese diet since 300 BC. Seaweeds are mainly eaten in the Oriental countries like Japan, China, Korea and more recently, in USA and Europe. The Republic of Korea has the highest per capita consumption of seaweeds in the world. After human food consumption, the next most valuable commercial use of seaweeds is as raw material for extraction of phycocolloids (agar, alginate and carrageenan), which are used in several industries. With 20,000 species of seaweeds in the world, India possesses 434 species of red seaweeds, 194 species of brown seaweeds and 216 species of green seaweeds. Traditionally, seaweeds have been collected from natural stocks or wild populations. However, these resources were being depleted by over-harvesting and hence, the need for their cultivation. Today seaweed cultivation techniques are standardised, perfected and made economically favourable. Besides, industry prefers a greater stability through sustained supply of quantity and quality of raw materials. In order to prevent overexploitation of natural seaweed habitats and to meet the needs of industry in an uninterrupted manner, nearly all brown seaweeds, 63 per cent of red seaweeds and 68 per cent of green seaweeds are being cultivated. Top five cultivated seaweeds in the world are *Laminaria, Porphyra, Undaria, Eucheuma* and *Gracilari.* These together account for 5.97 million metric tonnes of seaweed production. Top 10 countries producing seaweeds are China, Korea, Japan, Philippines, Indonesia, Chile, Taiwan, Vietnam, Russia and Italy. The current phycocolloids (seaweed gels) industry stands at over US$ 6.2 billion. The world production of commercial seaweeds has grown by 119 per

---

Policy Paper 22, December 2003.

cent since 1984 and presently, 221 species of seaweeds are utilised commercially including 145 species for food and 110 species for phycocolloid production.

Seaweeds and Their Uses

Macroscopic marine algae, popularly known as seaweeds, form one of the important living resources of the ocean. Agar, carrageenan and alginate are popular examples of seaweeds—these have been used as food for human beings, feed for animals, fertilisers for plants and source of various chemicals. In the recent past, seaweeds have also been gaining momentum as new experimental systems for biological research and integrated aquaculture systems. Seaweed products are used in our daily lives in one or the other way. For example, some seaweed polysaccharides are employed in the manufacture of toothpastes, soaps, shampoos, cosmetics, milk, ice creams, meat, processed food, air fresheners and a host of other items. In several oriental countries like Japan, China, Korea, etc., seaweeds are a staple part of the diet. Some typical examples of seaweed applications are narrated below.

Agar is widely used in paper manufacturing, culture media, packaging material, photography, leather industry, plywood manufacturing, preservation of foodstuffs, dairy industry, cosmetics industry and pharmaceutical industry. Carrageenan is employed in food industry. Its value in the manufacture of sausages, corned beef, meat balls, ham, preparations of poultry and fish, chocolates, dessert gels, ice creams, juice concentrates, marmalade, sardine sauces is well known. It is also used in the manufacturing of non-food items like beer, air freshners, textiles, toothpastes, hair shampoos, sanitary napkins, tissues, culture media, fungicides, etc. The applications of alginate find place in frozen foods, pastry fillings, syrups, bakery icings, dry mixes, meringues, frozen desserts, instant puddings, cooked puddings, chiffons, pie and pastry fillings, dessert gels, fabricated foods, salad dressings, meat and flavour sauces.

Indian Scenario

Although, India has a coastline of more than 8,000 kms and harbours about 844 species of seaweeds, commercial cultivation is yet to take place in India. R&D efforts over the years have resulted in valuable information regarding biodiversity, ecological conditions suitable for farming, species that could be incorporated in the cultivation, etc.

India is rich in algal biodiversity, has large stretches of suitable areas for cultivation and has communities of traditional fisherfolk. However, Indian seaweed industry suffers from absence of commercial cultivation practices, lack of infrastructure for commercial cultivation and absence of policy support. Since, seaweed is not an important aspect of Indian diet, its cultivation remains a low priority area. Contrary to that, seaweed cultivation presents several opportunities such as carbon sequestration through seaweed farming, provision of nursery grounds for fish and shellfish, medium for pollution abatement and diversified uses as animal feed and fertilisers. Multiple and conflicting uses of coastal areas, grazing by fish and illegal exploitation of seaweeds are some of the threats in the area.

Distribution of seaweed species in India are, Gujarat 202; Maharashtra 152; Goa 75; Karnataka 39; Kerala 20; Lakshadweep 89; Tamil Nadu 302; Andhra Pradesh 78; Orissa 1; West Bengal 6 and Andaman & Nicobar Islands 34. India presently harvests only about 22,000 tonnes of macro-algae annually compared to a potential harvest of 870,000 tonnes, a mere 2.5 per cent. Commercial cultivation of macro-algae has barely begun and is facing continuous regulatory hurdles. Processing of macro-algae is limited to lower grades of agar-agar and alginate and is modest in quantity. Manufacturers of agar-agar are working at less than 50 per cent of their capacity and there is no manufacturer of carrageenans. Instead of being a major global producer and exporter, India remains an importer of macro-algal products. The principal cause for this gap between the potential and the actual results achieved with respect to commercial cultivation and processing of macro-algae is the lack of clearly enunciated policy on cultivation and utilisation of seaweeds.

The National Academy of Agricultural Sciences organised a Round Table on 'Seaweed Cultivation and Utilisation' to discuss various aspects of seaweed cultivation and utilisation in India pertaining to their resources and diversity, biomass estimation, sustainable exploitation, commercial cultivation and processing, herbaria and algal cryopreservation, introduction of exotics and problems faced by the industries. About 50 participants representing research institutes, public and private sector undertakings took part in the deliberations. Recommendations emerging out of the discussions are given below.

Recommendations

General

Commercial cultivation and processing of macro marine algae should be a national priority and taken up as a mission mode project. A new nodal cell for promotion of commercial cultivation, processing and marketing of macro-algae should be set up in the Ministry of Agriculture, similar to Bureau of Aquatic and Fisheries Resources, in Philippines and Indonesia. It will be responsible for coordinating with other concerned ministries and departments, such as Environment and Forests, Ocean Development, Biotechnology, Industry, Commerce, Coast Guard, Port Authorities, Cooperatives, Customs and Excise as well as the state governments. It will resolve issues of overlapping jurisdictions and certain anomalous regulations. The Cell would also formulate detailed policies for the promotion of commercial cultivation, processing and marketing of macro-algae.

Commercial Cultivation and Processing

- The nodal agency shall give clearance for cultivation of specific species of macro-algae as ecologically safe and maintain a list of such species. All indigenous species of macro-algae are considered as ecologically safe for mass cultivation. *Kappaphycus alvarezii* which was introduced to Indian coastal waters more than 10 years ago and has since been domesticated, is considered ecologically safe.
- Recently, natural incidence of *Kappaphycus alvarezii* has also been reported from Andaman islands. Ecological studies have been undertaken regarding the cultivation of the species and no adverse effects to the ecosystem by the species have been reported. Large-scale cultivation of *Kappaphycus alvarezii* can be undertaken in Andaman islands.
- Clearance for use of specific coastal waters shall be obtained from state coastal zone management authorities. In giving the clearances, SCZMAs shall consider the multipurpose utilisation of the coastal areas and the prevalent regulations.
- As long as the method of cultivation of macro-algae does not envisage construction of permanent structures and the cultivation is done below the low-tide mark, CRZ regulations should not be applicable to seaweed farming.

- As the cultivated macro-algae has to be globally competitive in its pricing, the levy for the use of coastal waters shall be similar to that charged by other major global producers of macro-algae, such as Philippines, Indonesia and China.
- Cultivation of macro-algae and the wet and dried macro-algae so produced shall be treated as agricultural cultivation and agricultural produce for the purposes of fiscal levies such as sales tax, income tax, excise, octroi, etc.
- Integrated cultivation of shrimps and seaweeds should be encouraged in aquaculture as seaweeds act as scrubbers in reducing nutrient load and cleaning the environment.
- Seaweed cultivation should be encouraged and undertaken all over the Indian coasts including Chilika Lake, Palk Bay, Andaman islands as well as Lakshadweep islands as seaweed cultivation is ecologically safe and does not damage the marine ecosystems.
- Agar-agar processors have depended solely on naturally occurring beds of *Gracilaria* for supply of raw materials, that have been overexploited. Similarly, processors of alginates depend on naturally occurring sources for their raw materials and no carrageenan industry exists in India. To ensure dependable supply of raw material, high priority should be accorded to cultivation of agarophytes, alginophytes and carrageenophytes.
- Proper scientific methods should be devised and used for harvesting seaweeds so that sustainable utilisation can be done without damaging the ecosystem.
- The anamoly in classification of agar under central excise tariff and customs tariff needs to be corrected.
- In view of the problems encountered in dealing with wastes, proper waste treatment methods should be used by the industries.

R&D Priorities

- Creation of a biodiversity database on seaweeds.
- Taxonomical studies, both classical and molecular, for cataloguing as well as improvement.

- Improvement of strains through biotechnological interventions (tissue culture, genetic transformation by protoplast fusion and hybrid strain production, embryogenesis and regeneration, gene mapping and molecular mapping).
- Standardisation of techniques for large-scale tank cultivation as well as open water cultivation.
- *In situ* conservation of germplasm and establishment of herbaria.
- Periodic resource evaluation and biomass estimation on a national basis.
- Diversified uses of seaweeds as feed, fodder, feed additives, fertilisers, biocides and antimicrobials.
- Refinement of processes of preparation of agar, alginate and quality control.
- Ecological and EIA studies pertaining to introduction of exotic species.

At times it is desirable to import exotic species/strains to improve quality/yield of indigenous strains. Exotic species/strains may also be introduced to widen the range of macro-algae that are cultivated and for the nation to derive economic benefits from the introduction. A specific example is that of high valued species of *Eucheuma* and countries where commercial eucheumatoids have been introduced for cultivation or experimental purpose and where commercial quantities are currently being produced for the carrageenan industry include Fiji, Philippines, USA (Hawaii, California, Florida), Kiribati (Christmas and Tarawa Islands), Tuvalu, Samoa, Malaysia, French Antilles, Tonga, Japan, Indonesia, Federal States of Micronesia, French Polynesia, Guam, China, Maldives, Solomon Islands, Tanzania, India, Cuba, Vietnam, Brazil, Venezuela, Kenya and Madagascar. However, introduction of exotics should not be done indiscriminately as it carries the risk of pathogen invasion, parasites, epiphytes and undesirable changes in the ecosystem. ICAR has considerable experience with introduction of terrestrial exotics and could lay down a policy for the purpose, including time and duration of testing and certification procedures for the introduction of exotic species of marine macro-algae. Before the species is introduced into open waters, three years of R&D work is recommended in the laboratory and field, to prove the species to be ecologically safe.

24

Conservation, Management and Use of Agrobiodiversity

The National Academy of Agricultural Sciences (NAAS), in collaboration with the National Bureau of Plant Genetic Resources (NBPGR) organised a workshop on 'National Concern for Conservation, Management, and Use of Agricultural Biodiversity', commonly referred to as 'agrobiodiversity'.[1]

It was the first systematic, country-level effort to deliberate on issues related to the recognition of 'agrobiodiversity' as an entity distinct from overall biodiversity and the priority areas needing urgent attention. The theme was considered important, in view of the recent global developments which have far-reaching consequences on policies, legislative measures and the action plans of developing countries, rich in genetic resources.

The workshop was arranged in four technical sessions, namely, (i) sustainable and equitable use of agrobiodiversity, (ii) assessment of diversity and infrastructural needs, (iii) eco-development concerns in natural *versus* agro-farming systems, and (iv) entitlement and mode of compensation to beneficiaries of benefits derived from PGR use/ IPR. Perception about each theme was introduced through invited presentations and was followed by brainstorming, open-house discussions and consensus building. The salient recommendations emerging from this exercise are presented here for wide circulation among the stakeholders and for follow-up action.

---

1. The two-day workshop held at the Central Potato Research Institute, Shimla, 15-16 October 1997, was attended by over 125 scientists and experts, members of governmental and non-governmental organisations, representatives of agricultural universities and a broad range of other stakeholders.

Policy Paper 4, April 1998.

The Backdrop

Several international developments relating to genetic resources that influence national policy development, include the Convention on Biological Diversity (CBD), the Conference of Parties (COP), the Subsidiary Body for Scientific, Technical and Technological Matters (SBSTTA), the FAO Commission on Genetic Resources for Food and Agriculture (CGRFA), the revision of International Undertaking on Plant Genetic Resources, the Global Plan of Action (GPA), the Trade Related Intellectual Property Rights (TRIPS) in the General Agreement on Tariffs and Trade (GATT/WTO) and the Sanitary and Phytosanitary Agreements (SPS) in the World Trade Organization (WTO). These developments have affected the intellectual property regime, the access to biological resources, the equitable sharing of benefits accrued from use of genetic resources and several other related issues. In India, measures taken to cope up with the emerging situation, include drafting of a Plant Variety Protection (PVP) and Farmers' Rights Act, a Biological Diversity Act and the formulation of a National Action Plan for implementation.

The CBD's Conference of Parties (COP-III) decision[2] in November 1996, to grant specific recognition to agrobiodiversity requires urgent follow-up by all countries. In India, serious concern has been expressed about the deteriorating condition of the nation's agrobiodiversity and a variety of suggestions have been advanced for policy refinement and action. The need to analyse relevant factors and refining workable solutions in the Indian context has been stressed. The workshop carried these ideas forward to develop appropriate policies and evolve mechanisms for implementing action plan on conservation, management and use of our agrobiodiversity.

---

2. ....in order to facilitate an integrated approach to agricultural biodiversity (Resolution 3/95 FAO Conference), the CBD welcomed the interest of the Commission...

   Further reference was made to decision 111/11, whereby the COP encourages Parties to develop national strategies, programmes and policies for conservation and sustainable use of agricultural biological diversity, according to 14 action-oriented goals...

   In this regard, it was noted that there is a need for appropriate mechanisms at the national level to ensure that agricultural biodiversity activities and plans will be effectively integrated into the national biodiversity strategies, plans or programmes that Parties have requested to prepare, and report on to COP-IV, as a priority activity in line with Article 6 of the Convention.

   (*Source*: FAO Document, CGRFA-7/97/Rep).

Agrobiodiversity

Agrobiodiversity occupies a unique place within biodiversity. It recognises that agriculture evolved from bioprospecting, selection and development of a few species from plant and animal kingdoms, to meet human needs of food, fibre and fuel. All biotic factors related to agriculture, such as, plants, animals, fish, reptiles, insects, birds and microbes are components of agrobiodiversity. The conservation, management and sustainable use of these organisms (and their wild progenitors/relatives) require specific attention.

The CBD recognised the global need to conserve and manage the biological diversity existing on this planet. Subsequently, in the COP decisions 111/11 and 111/15, the CBD gave specific recognition to agrobiodiversity in conformity with Resolution 3/95 of the FAO Conference (q.v.). It is obvious that humankind is concerned *inter alia* with phenomena like global warming and environmental pollution, which the CBD intends to handle through effective management of biological resources. It is also recognised that biological resources related to food and agriculture need to be managed with added care in future so as to meet the demands of increasing global population.

The COP-III observed that various policies, legislative measures and action plans on biological diversity will have a bias towards environmental/forestry/total bioresources related issues at the cost of agrobiodiversity.[3] It hence recommended that the latter must receive adequate attention by the member nations in their developmental plans. Global attention was also drawn to the need of ensuring partnership among all stakeholders for initiating conservation activities and ensuring access to these resources both for equity and benefit sharing.

India: A Rich Culture of Agrobiodiversity

The Indian gene centre is among the 12 megadiversity regions of the world. About 25 crop species were domesticated here. It is known

---

3. It was noted that the CBDs' and GEFs' focal points tend to be located in the ministries responsible for environment while the focal points for programmes and plans in the agricultural sectors tend to be located in the ministries of agriculture. In accordance with decision 111/11 of the COP and the recommendations of the GPA, it was noted that there is a need for liaison between the different focal points and government departments with a view to developing effective integrated approaches for the conservation of agricultural biodiversity.

(*Source*: FAO Document, CGRFA-7/97/Rep)

to have more than 18,000 species of higher plants including, 160 major and minor crop species and 325 of their wild relatives. Around 1,500 wild edible plant species are widely exploited by native tribes. These include, 145 species of roots and tubers, 521 of leafy vegetables/greens, 101 of buds and flowers, 647 of fruits and 118 of seeds and nuts. In addition, nearly 9,500 plant species of ethnobotanical uses have been reported from the country, of which around 7,500 are for ethnomedicinal purposes and 3,900 are multipurpose/edible species.

The traditional farming systems of India are relatively stable and in equilibrium. The species complexes in traditional farming systems exemplify coexistence of plants and human tribes, draught animals, friendly birds, beneficial insects, pollinators, earthworms, soil micro-organisms and biocontrol agents. Modern farming systems, which evolved in response to the growing needs of the human society to ensure food and nutritional security have progressively replaced traditional agriculture. More than half of the cultivated area under major crops is now covered by improved varieties and farming practices. Biotic diversity is maintained in modern agricultural systems primarily through cultivation of 'mosaic of improved varieties'. It is important that diversity is assured while attaining high production levels and profitability.

A combination of *ex situ* and *in situ* conservation approaches is required for agrobiodiversity conservation strategies in the Indian gene centre. The efforts of the Indian Council of Agricultural Research (ICAR) are exemplary, yet there is a long way to go. Fortunately, the country is well endowed with local resources and expertise and is open to international collaboration and exchange. Good progress has been made in past years in institution building and human resource development (HRD) on PGR. The largest gene bank in the world for *ex situ* conservation has been set up at NBPGR, New Delhi. It will be prudent to recognise this facility as our national heritage with needed support for its sustainability. Efforts need to be continued to strengthen and consolidate similar initiatives in the public sector.

Human tribes, particularly women, have a long tradition of preserving plant species and the agroecosystems. There is a need to preserve the traditional practices and learn from the available local wisdom. Some of the vital issues of contemporary diaiogue on genetic resources are not new to the Indian society but these need to be revisited. Benefit sharing and the entitlement of beneficiaries are among the important issues that need better definition and implementation mechanisms.

Resolution: The importance of agrobiodiversity must be recognised as distinct national issue of prime concern within the broader area of biodiversity.

Recommendations

National Action Plan

1. High priority should be given to developing a sound and workable National Action Plan on Agrobiodiversity.
2. Agrobiodiversity and the available indigenous knowledge should be documented urgently through a well-organised approach. Both the formal and informal knowledge available with the farming communities deserves this documentation.
3. A comprehensive mission mode programme should be implemented for all areas related to agrobiodiversity. The pace of collection and conservation of agrobiodiversity should be accelerated. Efforts on bioprospecting and effective utilisation of the collections should be intensified. It should be ensured that the collections are secure and safe. What is conserved must be protected as a national heritage.
4. High priority must be accorded to various researchable issues relating to agrobiodiversity, which could be pursued through active involvement of the National Agricultural Research System (NARS) and other stakeholders, such as the NGOs and the farming communities.

Agrobiodiversity Conservation

5. A base collection of 1.60 lakh germ plasm samples of various crop species and their relatives is presently conserved in the National Gene Bank at NBPGR. This facility holds invaluable national resource and hence, must attract the concessions available to buildings declared as national heritage.
6. A sample of all plant genetic resources available with various holder organisations/institutes/universities/communities in the country must be made available to the National Gene Bank for safe conservation for posterity.

7. Modern technologies, such as *in vitro* and cryopreservation, are needed for conservation of non-orthodox seed species, vegetatively propagated crop plants, medicinal and aromatic plants and other high value crops. The required support should be provided for research to develop the needed technologies. Responsibilities should be assigned to designated institutions for standardising protocols for specific plant species.
8. *In situ* on-farm conservation should be promoted. For an effective *in situ* on-farm conservation of traditional cultivars/ landraces, specific areas, practices, systems and species should be identified. A system at the national level is required to be evolved to assess farmers' views on and expectations from *in situ* on-farm conservation of genetic diversity. Suitable modes and mechanisms for providing needed incentives to farmers should be evolved so as to ensure safe and effective conservation of genetic heritage through on-farm practices.
9. Ethnic communities, particularly women, have played an important role in the conservation of traditional varieties, especially in fragile agro-ecosystems. The role of women and communities must be recognised and rewarded while implementing on-farm conservation strategies.
10. Pilot scale experiments should be undertaken to develop suitable conservation model(s) for *in situ* on-farm conservation. Indicator species should be identified for assessing the health of varying agroecosystems.
11. Conservation of available breeds/strains of animals, fish and microbes needs urgent attention. While recognising the efforts being made by the respective ICAR Bureau of animal and fish genetic resources, these programmes should be intensified in a decentralised approach involving all the concerned institutions and communities. Also, attention is needed to conserve agriculturally important micro-organisms. The initiative to set up a separate Bureau for their conservation is a welcome development.

Agrobiodiversity Management

12. Simple, effective and practicable mechanisms for prospecting agrobiodiversity and monitoring should be evolved. Selected

amateur groups, including the school and college students, should be enlisted for this purpose.

13. Genetic variability of native, under-utilised species of food crops, fruits, medicinal, aromatic and other economic plants should be documented on priority. It should be supplemented through need-based introduction of useful species. Selected, hitherto unexploited, species having future potential should be researched on and adopted.
14. There is an urgent need to adopt appropriate quarantine measures in the national interest. We must revisit the present National Plant Quarantine Policy, particularly in the context of bioengineered materials/genetically modified organisms (GMO).
15. Characterisation, evaluation and documentation of PGR should receive a high priority. Relevant improved tools and technologies, such as biotechnology, should be deployed in future.
16. The national information network and database on germplasm should be strengthened.

Germplasm Registration

17. Registration of plant germplasm should be encouraged and availed of by all concerned. Registration should be based on properly characterised, documented and researched PGR information.
18. Mechanisms should be evolved for the legal protection of landraces/traditional varieties while recognising the ownership of the farming communities, private or public breeders. Registration of these cultivars should be done at the level of *gram sabhas*, *panchayats* and blocks linked to the national registration authority, such as, the NBPGR.
19. Systems for registration of other agriculturally related bio-resources, such as, breeds/strains of animal/fish/microbes, should be similarly developed and implemented on priority.

PGR Awareness/Literacy/HRD

20. Considering the relevance of agrobiodiversity in the emerging global scenario, there is a need for creating awareness and

understanding about it among the public and Indian masses. Literacy campaign for conservation and sustainable management of agrobiodiversity needs to be initiated at the grass roots level, starting right with the school and *gram sabha/ panchayat* levels.

21. Suitable curricula for students and orientation courses for the teachers/trainers need to be developed on priority. To begin with, the ICAR, through its own set-up and state agricultural universities, should take a lead. The University Grants Commission (UGC) and various Central and State Education Boards can expand this programme further.

22. There is a need for literacy on PGR policy issues such as, plant variety protection, breeders' rights, farmers' rights, *sui generis* system, etc. Recommendations on policy and management issues on agrobiodiversity should be widely circulated. Literature on PGR-related happenings and who's who is not accessible to most people. In order to create greater awareness about agrobiodiversity conservation and management issues in the global context and also to evolve consensus at the national level, the draft text for biodiversity legislation should be widely circulated along with selected literature on CBD, TRIPs, UPOV-1978, FAO Undertaking on PGR, Leipzig Conference, Global Plan of Action, etc.

23. Emphasis should be laid on human resource development to build required expertise in basic PGR management aspects, namely, germplasm identification, collection, characterisation, evaluation, documentation and conservation. Simultaneously, re-orientation of technology generation is warranted. HRD should be further oriented towards the needed expertise, technology and awareness for germplasm regeneration and on-farm conservation.

Access and Benefit Sharing

24. A well-understood procedure for accessing the genetic resource materials owned by farming communities/individuals and a fair and equitable sharing of profits arising from their use should be established. The percentage of profit going to communities can vary depending on margin of profit. For example, 5 per cent in case of crop plants and 10 per cent for

medicinal and aromatic plants or plants of industrial value, etc. The revenue generated through benefit sharing should be transferred to a national gene fund and should be exclusively available for research and development relating to public good and/or for the community development activities.

25. Needed sharing of conserved materials among community/national seed banks would be desirable in the national interest. A fair and equitable mechanism should, therefore, be developed jointly by all stakeholders in order to safeguard the interests of all concerned.

National PGR Policy

26. A National Policy Advisory Committee with wide representation should be instituted to act as an advisory body to the central government on matters concerning agrobiodiversity conservation.
27. A national legislation on agrobiodiversity/genetic resources should be developed so that effective instruments are made available for conservation, management and use. In the proposed draft legislation on biological diversity, specific mention of agrobiodiversity distinct from the biological diversity, be made and appropriate provisions provided to establish a separate national authority for dealing with issues related to scope, access and conservation of agrobiodiversity. The Department of Agricultural Research and Education should be given the nodal role for these aspects at the national level in the best scientific interest and for required technical coordination.
28. A National Gene Fund must be established, keeping in view the dimensions of these activities and the urgency of the matter. It may include allocations from consolidated fund of India; royalty on finished products of proprietary nature, percentage of profits as emanating from equitable sharing of benefits accrued from the use of PGR and tax-free donations from the users of these genetic resources.
29. Most of the developing world looks towards India for suitable models for agrobiodiversity conservation, management and use. The country is well poised for required interactions with the developed world, being equipped with the required

institutional support and the human resource needed for generating improved technologies. Hence, our *sui generis* system for protecting agrobiodiversity should be both innovative and practical in dealing with all scientific, political and legal issues.

Section V

Agricultural Research and Education

25

Agricultural Policy

Redesigning R&D to Achieve Its Objectives

Background

The National Agricultural Policy visualises technological upgradation of Indian agriculture as a core element of the agricultural development strategy. The goals of sustainable agriculture, food and nutrition security, risk management as well as growth and agricultural trade depend critically on improved R&D processes in the sector. Specifically, it endorses the concept of regionalisation of agricultural research based on identified agro-eco regions, (location-specificity), use of frontier sciences, participatory and proprietary approaches in R&D, strengthening research-extension linkages, and a "well-organised, efficient and result-oriented agricultural research and education system to introduce technological change in Indian agriculture." It calls for a redefinition of the partnership between central and state governments, assigning an upstream role for the centre. It also charts a pathway of incentives, support systems, investment imperatives and policy to "actualise the vast untapped potential of Indian agriculture."

The NMS recognised that the Indian agricultural innovation system needs to prepare itself for the imminent changes in agricultural R&D and education systems necessitated by this challenge. Unlike other public support systems, the public R&D sector has had the tradition of being more open to reform. Several internal and external reviews have been undertaken over the last decade or two. In fact, for the central system—the ICAR, such scrutiny is mandatory at the level of each institute/centre at quinquinnial intervals. Yet, in view of the rapid and far-reaching changes in the

---

Policy Paper 18, December 2002.

agricultural scenario, emerging pressures on public systems across the board, and indications of slackening tempo in research and transfer of technology, the Academy felt the need for an interactive dialogue for further reforms in the agricultural R&D system in the country in the changed scenario. The contours of Indian agriculture are changing and, the national system must also articulate new paradigms.

The NAAS, therefore, organised a workshop[1] to explore these implications. This policy brief brings out some of the critical findings and recommendations of this dialogue for larger public debate and policy action.

Elements of Change in Agricultural R&D

There is a felt need for insights into the relationship between agricultural science and policy in the Indian context. While the recent National Agricultural Policy document of the Government of India provides an immediate focus, there are larger questions concerning the capacity of R&D to identify and respond to the crucial and durable elements of agricultural policy on the one hand, and to contribute effectively to policy formulation itself on the other. The nature, extent and direction of changes in agricultural R&D, education, and extension demand detailed deliberations among the stakeholders—the community of professionals, farmers, NGOs, policy makers; both from private and public sectors.

The objectives of the workshop were:

- To articulate and understand the context for change in agricultural R&D, education and extension in India, with a view to identifying some key issues that demand change in the institutional setting.
- To deliberate the challenges in and processes required for organising and managing R&D for sustainable agriculture.
- To examine the changing relationships in and new demands on technology development and diffusion components.
- To analyse the demand for change in two areas crucial to innovation for sustainable agriculture, i.e., agricultural education and natural resources research.

---

1. A two-day workshop under the Convenership of Dr. I.P. Abrol to explore these implications involving senior thinkers, leaders and others stakeholders in the national agricultural innovation system at New Delhi on 10-11 April 2002.

The sessions of the workshop were organised accordingly into four technical sessions, followed by a concluding session where the key strategies for change in R&D and the processes to guide this change were discussed. The following paragraphs discuss the key issues from the presentations and uninhibited discussion during each of these technical sessions, the highlights of the workshop's findings, the policy recommendations and strategic action points.

The Context for Change

- The past decade has witnessed significant changes in growth rates and trends in agricultural production/productivity, resource use in irrigated and rainfed agriculture, and gives us evidences of natural resource degradation. Agricultural R&D continues to appease itself with claims of past success, often limited to varietal release, while there is evidence of stagnation or even deceleration in TFP growth, and of declining productivity of disciplinary commodity-based knowledge in the face of deteriorating R&D environment and mounting agroecological problems. Other changes include increasing private sector presence and profit motives in agricultural R&D, an erosion of public sector commitment to basic and poverty oriented research, and the potential and uncertainties of emerging biotechnology and information technology regimes. The demand for sustainable agriculture and poverty alleviation, the need for introspection and evaluation in R&D, are other challenges demanding critical changes in R&D paradigms.
- Two other variables need to be factored in as we consider response of public R&D systems. The first emanates from globalisation of the economy—an accepted paradigm change in the NAP, and the other is growing pressure on public financial resources. The former brings in issues like global and regional comparative advantage and efficiency of production, while the latter underscores prioritisation and accountability concerns. The crucial importance of bridging the knowledge divide, which has grown in recent years was also highlighted in the workshop. The rhetoric of achieving growth, efficiency, equity and sustainability enunciated in the NAP will depend critically on adequate response of all public systems, including R&D, to these changes.

- Participants in the workshop emphasised the need for major paradigm shifts in agricultural R&D, education and extension systems and necessary structural and organisational changes therein. Factors like commodity/disciplinary orientation, centralised organisation and management, bureaucratic overload, inadequate accountability and incentive systems, non-existent internal and external linkages, etc. have been identified in earlier reviews of the agricultural knowledge system. These are generic elements of the context—several of which have been flagged since long. Workshop participants felt that goals and objectives are dynamic and evolving, but their functional rigidities need to be overcome to impart efficiency and accountability.
- Analytically the first step is to identify the main issues in the changing landscape of agricultural innovation and development that we need to tackle. What are the main changes that affect agricultural R&D, extension, education, and agricultural development in general? Once we have identified these changes, can we ask further questions about how each component and actor in the system has to change? These questions may have to do with the nature and extent of change, possible period or sequencing of change, or partners in change, new norms or rules (such as prioritisation, accountability, incentive and reward structure, natural resource accounting or poverty impacts).
- There are important milestones of success in green revolution technology. This success, however, must not perpetuate a 'business as usual' approach to the generation and utilisation of knowledge and technologies in the agriculture sector. The innovation system must now look for ways forward to the next stages of excellence in science and success towards agro-ecological and socioeconomic goals, towards an evergreen revolution in agriculture.

Organisation and Management of Research for Sustainable Agriculture

- The NAP is committed to build a well-organised, efficient and result oriented agricultural research and education system and more effective transfer of technology processes. There is a need for an analytical framework that can guide the transition of traditional and somewhat fatigued research, education and extension hierarchies to these goals. The participants were aware that there are rigidities, bureaucratic bottlenecks, and vested interests which undermine the ability of the system as it exists to affect needed changes.

- The changes in organisation and management of agricultural innovation that are needed to meet the policy goals of sustainable agriculture must begin with an understanding of the previous record of organisational changes. These changes were made to meet the challenges posed by technology generation and dissemination needs of the green revolution, during a phase of massive public sector expansion. Reorganisation of the ICAR (1966 and 1974), the establishment of the SAUs (since the 1960s), new programmes such as the AICRPs, etc. were some of the major organisational changes to achieve certain desired objectives, stated explicitly in policies such as those for national food security, development of rainfed agriculture, etc. The need to restructure the organisation of R&D has again become pressing in the national context of agricultural trade liberalisation, increasing agricultural imports, mounting green critiques of conventional agricultural production practices and technologies, demand for sustainable agriculture using less of fossil fuels, and the emergence of biotechnology and information and communication technology regimes. What makes the current situation difficult is stringency of public resources including trained human resource. Innovative ideas are now needed to cope with this diverse and complex agenda. These are agenda-driven changes, and past record on this has been fairly impressive.
- Organisation of the R&D system needs a relook from the structural point of view as well. For the research system, decentralisation and debureaucratisation have been suggested time and again; but financial/administrative decentralisation and flexibility is not enough. The organisation and management of R&D now needs to be tailored to criteria and parameters of sustainability and comparative advantage that are location/region specific. If result oriented is to be read as income growth, poverty reduction or resource conserving

agriculture, there must be organisational changes that increase both fundamental knowledge partnerships and understanding of rural livelihoods and ecosystems. Organisational change now necessitates getting out of the insular mode and working with stakeholder groups in rural areas as well as developing customer-contractor relationships with input industry, or processing industry. This calls for structural changes in organisation, focusing on state/zonal research units, redefining the roles of central/state institutions, and developing processes and mechanisms for participatory activities, and above all re-instilling the lost virtue of accountability.

Technology Development, Diffusion and Linkages

- The NAP enunciates that "application of science and technology in agriculture will be promoted through a regular system of interface between S&T institutions and the users." The changes in the organisation and methods of extension, as well as the research-extension partnerships and linkages that enable the NARES to achieve the policy goals need detailed discussions and participatory analysis at different levels of the agricultural innovation system. There are several effective partnerships and coalitions in technology generation, development and adoption in Indian agriculture. In order to develop appropriate interface it is important, especially for public organisations mandated exclusively to generate technologies or to disseminate the technologies generated, to learn lessons from these successful partnerships in knowledge generation and utilisation.

  The linear model of technology generation, diffusion and adoption in spatially and functionally differentiated organisations, must give way to a non-linear model of continuous participatory learning within the larger agricultural innovation system including private sector actors. The latter, the innovation systems approach, based on interactive learning and building partnerships with the relevant actors/agencies in the agricultural innovation system demands institutional and organisational changes promoting fusion of research and extension, greater control of stakeholders, multi-faceted structures (like ATMA) at local levels, etc.

- Finally, institutional learning to enable changes in agricultural R&D depends on the existence of adequate mechanisms in place for constant feedback and in-built processes to affect necessary changes. It is amazing that the public research/ extension systems have made little efforts in the past to learn from small but successful experiments undertaken by private sector or voluntary organisations. There is a need to strengthen social science research in agricultural innovation systems to enhance the social and ecological learning capacity of R&D organisations.

Addressing Sustainability Goals

- The term 'sustainability' figures prominently in the NAP document and pronouncements relating to agricultural research, extension and educational organisations. The central question in this regard is, "how have we modified, if at all, any or all these organisations to address our sustainability goals?" The NAP document states that sustainable agriculture is that which uses the country's natural resources—land, water and genetic endowment, in a manner that is: technically sound, economically viable, environmentally non-degrading, and socially acceptable. The immediate response of all concerned actors in the agricultural innovation system is to explore how the NARES can address these goals. At the outset, social scientists and natural science professionals need to explore whether 'economic viability' and 'environment friendly' criteria are inherently contradictory and, if so, how can these be reconciled. The NARES needs to identify and develop the sustainability criteria and parameters to assess *(ex ante)*, the different ecoregional approaches in R&D and their *(ex post)* impact.
- It is evident that there are several institutional and organisational problems within public agricultural R&D that impede the larger goal of agricultural education, which is needed to produce the dynamic manpower required to address sustainable agriculture. The critical role of agricultural education in achieving transition to sustainable agriculture as well as in maintaining the dynamic nature and academic excellence of the training in agricultural sciences needs emphasis. The system should look for ways to improve the

resources available for agricultural education, effectively reduce professional inbreeding and inefficient infrastructure in agricultural universities, and bring in more of basic sciences and social sciences that can provide a holistic agricultural education.

- A major component that can help identify and achieve the goals of sustainability in the agricultural system is the natural resources research sub-system to ensure sustainable use and conservation of natural resources. The natural resource management (NRM) perspective towards sustainable agriculture can draw from the successes of the green revolution with attendant widespread resource degradation and from rainfed areas where the poor quality and degradation of resources have posed a major constraint in enhancing productivity and improving livelihoods. Misguided policies that contradict scientific knowledge and farmers' wisdom about resource use and conservation have been the bane of Indian agriculture. There is a need for alternative policies and organisational formats that can help natural resources research in addressing resource management problems in different agro-ecosystems, using different technological options and partnerships with various actors in the innovation system. A strong social science component within the system is essential to address such concerns.

Issues Highlighted at the Workshop

- The NAP recognises the crucial importance of science in agriculture and the need to revamp public R&D system. Based on this premise, the workshop highlighted the following issues: Changes in agricultural R&D are imminent. The successes of the past, especially the green revolution, have been important, but the research system should now look for new ways forward to achieve new quantitative and qualitative milestones. This will need strong commitment on the part of the state in terms of support as well as flexibility.
- The agricultural innovation system of the country needs an internal thinking mechanism, and a more professional approach much like the TAC of the CG system to identify, steer and evaluate arenas for agricultural knowledge and

technology generation and use. The traditional departmental approach of the Ministry of Agriculture is no longer tenable. Both central and state systems need to develop this mechanism.

- There was dissatisfaction with the productivity and efficacy of conventional disciplinary and commodity-based research. There is a need for a shift to trans-/multi-disciplinary, issue based research, with regionally differentiated strategies for each agro-ecological region and farming system. The R&D system needs to be reorganised in this mode. The present insular, uni-directional extension system needs to be replaced by a broad-based and decentralised agricultural innovation system.
- There was explicit demand to reduce the administrative overload and inefficiency in the ICAR and the SAUs, move towards more decentralised and flexible decisionmaking in research new norms of accountability to replace GoI regulations, and restore the thrust on quality to ensure excellence and globally competitive science.
- A framework of agricultural science oriented to the goal of sustainability is necessary to complement the production/ productivity goal, if knowledge generation is to serve the cause of agricultural, ecological and socioeconomic sustainability. This would entail institutional and organisational changes in agricultural R&D.
- Overall agreement on the need to enhance basic sciences social science and interdisciplinary course contents to improve agricultural education with relevant professional training that suits the demands of agro-ecosystems, industry, and regional socioeconomic contexts.
- The goals, expectations, and mandates of R&D institutions may be determined, but their performance will be determined by the O&M structures at central (ICAR) and state (SAU) levels. In the context of the new challenges, these need to change in significant ways. As part of public systems, these institutions have been straightjacketed. There is need to unshackle these rigidities.

- The participants expressed serious concern about three issues—relevance, quality, and accountability in public R&D structures including extension systems at central as well as state levels. These need serious and professional attention. Past record on bureaucratic effort has not been effective, if not counter productive.

Policy Recommendations

The policy regimes of the past, which had put in place a huge public R&D system, have changed. These systems themselves have started showing signs of attrition. This has far reaching implications for the future of Indian agriculture. Unfortunately, there is little indication of awareness or response on the part of the apex bodies in some cases ill-conceived and hasty response either at central or state levels. The deliberations of the workshop suggest following policy initiatives and action points to address this:

- The present departmental mode of organisation and management public systems in agricultural research and extension need to be moulded in an innovation system framework. This implies demolishing dichotomies like research/extension, plan/non-plan, centre/state, public/private, agricultural/rural, and so on. More effective mechanisms must be identified and developed to internalise stakeholder involvement in decisionmaking and improve partnerships across the board. There is need to shift from an hierarchical and linear technology generation-diffusion model to a non-linear and holistic learning mode.
- This will necessitate creating and nurturing a broad-base 'think tank' at the ICAR and SAU levels. The model of the Technical Advisory Committee (TAG) of the CGIAR with its own secretariat and loop in the funding process may be adopted with suitable modifications. At the level of institutes also, there is need to expand and back up bodies like the Research Advisory Committee (RAG), Governing Councils, etc. to play this role effectively.
- Bureaucratic and financial norms of functioning and accountability processes have proven inadequate for meeting scientific goals. These need to be replaced by more autonomous., flexible and performance-centred processes and

management climate. The NATP illustrates changes and initiatives in this direction. These need to replace (not supplement) the existing system.

- A progressive and dynamic R&D system to address agro-ecologically relevant strategies demands decentralisation of ideas, approaches and power in the conduct of research. The agro-ecological paradigm also implies a shift from commodity-based organisation of applied agricultural research and development.
- Since the productivity of disciplinary/subject matter research has reached a plateau, a system oriented, inter-disciplinary, issue-based approach is now necessary. This will necessitate reorganisation of research at functional levels (institutes and research stations).
- Basic and strategic research is the precursor of productive applied research, a fact which is being lost sight of in the quest for relevance and visible impact. Policy makers and planners need to take note of this disturbing trend. Central research entities (in ICAR) need to push their research upstream and funding agencies should accord priority to supporting basic and strategic discipline-based research also.
- Funding for public agricultural research extension and education continues to be stressed, more so at the state level, with conflicting policy signals. The government makes a commitment to raise public investment in R&D on the one hand, it enjoins ICAR and SAUs to mobilise their own resources on the other. This imparts uncertainly and inefficiency. It is high time that the issue of the level of investments, roles of public, private and international players, are discussed and sorted out in a consultative mode. Past efforts have been partial, *ad hoc* and ineffective.
- There is need for enhanced social science research capabilities both for policy analysis and to develop R&D strategies for specific agroecological regions and farming systems. There is need for greater interaction among natural and social sciences. This will help the R&D system to build bridges with policy making on the one hand and the endures (farms) on the other hand in interactive ways.

The Academy is deeply concerned about the erosion of scientific capabilities and performance in agriculture and rural science and technology, about growing public and policy apathy, and the far-reaching adverse consequences which may ensue. The workshop reiterated the urgency of redesigning the agricultural innovation system in a new mode. We hope that this policy brief will lead to greater debate and, finally, action to address these concerns.

26

Prioritisation of Agricultural Research

Background

The organisational structure and functional role of India's agricultural research system has evolved over a period of time to match the emerging demands. However, it is now being increasingly felt that there is need to bring about changes that will enhance our ability and capacity to deal with emerging challenges in a more efficient and effective manner. It is obvious that optimal use of available human, infrastructure and financial resources will become increasingly important. Amongst other steps, better priority setting and sharper targeting of research will be needed to make the system efficient. It is in this context that the National Academy of Agricultural Sciences sponsored a workshop on 'Prioritisation in Agricultural Research'.[1]

The workshop had the following objectives:

- Sensitise key research managers on the need to prioritise research as an integral part of research management process.
- Identify key issues that should be resolved to initiate and take the process further at different levels within the system.
- Suggest steps that will assist in internalising priority-setting mechanisms within the system at different levels.

Rationale

Priority setting is the process of defining a research portfolio that is consistent with country's agricultural policy, the research organisations' mission and the research programme's objectives. In

---

1. The workshop was organised by the Centre for Advancement of Sustainable Agriculture (CASA) on September 29-30, 2000 at Surajkund in Haryana and was attended by 35 senior research managers from within the Indian Council of Agricultural Research (ICAR)-State Agricultural Universities (SAU) system.

Policy Paper 15, December 2001.

the past, India's agricultural research institutions have been arriving at their research portfolios informally and without an explicit activity aimed at arriving at a set of prioritised research agenda.

Over time, the demands on India's agricultural research and education system have been fast changing. The system is now faced with the task of not only increasing the productivity of selected crops but also to address issues such as increasing disparity in agricultural growth, maintaining and enhancing the quality of natural resource base and reducing environmental degradation, poverty alleviation, emerging trade regimes, etc. These developments are rendering our research agenda highly complex. The need to be competitive worldwide would call for the research systems to pay greater attention to quality, cost-effectiveness and sustainability issues. The number and nature of stakeholders, which now includes farmers' organisations, private sector sponsored research, NGOs, etc. in addition to the farmers has also changed. These developments call for rationalisation of allocation of current and future resources for enhanced research efficiency. Institutionalising a systematic analysis of agricultural research priorities and integrating it with an effective monitoring and evaluation system, therefore, holds a key to making the system efficient and effective.

The thoughts and recommendations emerging from discussions on component issues are summarised below.

Operationalising Research Priority Setting

- In defining research priorities, it is necessary to involve both policy makers and the community of scientists working in the field and laboratories. Priority setting exercise has to be carried out at different levels. Three levels are usually distinguished: the national level, the programme level and the level of individual research institutions. At the national level, the relative importance of major research areas/sectors is determined. In the programme levels exercise, decisions are made on relative importance of different programmes within a sector and programmes that might cut across different sectors. At the level of research institutions, decisions are needed on the relative place of different research programmes/ themes in the portfolio and formulation of projects.

- In India, agriculture is a state subject and it is necessary that prioritisation exercise is done simultaneously at the level of each state within the overall national policy and priorities. As at national level, this exercise has to be done at three levels: sectoral, programme, and theme/project.
- Research prioritisation and associated resource allocation at the national level should be the responsibility of Director General of the Indian Council of Agricultural Research, the team of specialists representing different sectors and the planning unit consisting of professionals with specialisation in economic and statistical analysis and in computer support services. In recent years an array of methodologies and models of research priority setting have been developed—which take into consideration factors like efficiency, equity and sustainability. Additional criteria will have to be added considering the fact that 60 per cent of the people of India are dependent on agriculture for employment and sustenance. The exercise must also take into consideration the major opportunities that exist for India to increase its agricultural production through productivity enhancement to emerge as a major exporter of agricultural commodities and value added products and to remain food secure in the coming decades as population continues to multiply.
- In this exercise the leadership of ICAR will be guided by policy of the Government of India as defined by the Planning Commission and by the inputs of the Governing Body of the Council and members of the ICAR Society which reflect the views of the farming community, the states and other stakeholders.
- Following the broad allocation of resources for different sectors (e.g. crop science, horticultural science, animal science, fisheries science, natural resource management science and socioeconomic research) the team of senior management scientists led by the Director General should translate these broad national priorities into research programmes with clear objectives and goals to be achieved. In defining and refining research programmes, ICAR leadership should seek inputs from best of the available expertise by organising expert consultations and by drawing upon deliberations of important

national and international conferences and above all, on the basis of the economic analysis carried out by the planning unit. Decision to arrive at research programmes must also reflect interconnectedness of different sectoral problems and a recognition of the need of promoting involvement of multidisciplinary teams of scientists drawn from its network of research institutes. In the present set up of ICAR where sectoral research is organised division wise, there is a particular need to organise research in a programme mode. This is necessary to address sustainability concerns and to deal with scientific problems in a holistic manner.

- Following broad allocation of resources for different programmes, scientists in different research institutes should be called upon to translate these programmes into priority research themes, research projects and experiments to fulfil the mission and mandate of respective institutions. In formulating research themes and projects strong inputs are required from the grassroots organisation including farmers' groups, NGOs, industry, etc. The scientists will be required to formulate research projects taking into consideration the existing state of scientific knowledge and technologies, the progress already made and the feedback received from the farmers.
- To achieve programme objectives, it may frequently be necessary to involve more than one research institution (a research institute or an agricultural university under the ICAR systems or within the sister research systems e.g. Council of Scientific and Industrial Research or Indian Council of Medical Research etc.). The coordination of activities within research programmes must be entrusted to a senior scientist (as far as possible, Director of Institute) who will have overall responsibility of executing the programme in its entirety. It is clear that:
    - It is the responsibility of the programme convener (or the leader) to ensure complementation and coordination of programme components/themes) within different organisations.
    - It is the responsibility of the Director of the Institute to ensure complementation and coordination in the research

projects of different disciplinary/departmental units within the Institute.

- Prioritisation exercises, similar to those stated above, must also be taken up at the state level where the Vice-Chancellor and his senior colleagues undertake to define the major research priority areas and programmes and allocated resources to them. Scientists at the regional research stations and headquarters together will define research themes and formulate research projects and experiments, etc.
- In addition to the publicly funded research setup, there are a large number of professional societies and academies, which work towards promoting the advancement of science in their respective scientific disciplines or solving regional problems. These societies can play an effective role of providing inputs to decision makers for deciding directions and selecting priority research areas. There is need to sensitise these professional bodies to increasingly assume a role in defining and prioritising research agenda. In this task National Academy of Agricultural Sciences can take a lead role.
- Many organisations the world over have evolved a system whereby scientists source research funds through a competitive grant system. In this system, scientists in research institutes and universities are called upon to formulate projects in response to the priorities defined by professional peers. Only those projects, which respond to these priorities and have the right technical content, receive funding support. At present neither ICAR nor the state agricultural universities operate such a system. Recently a Competitive Grant Programme (CGP) has been initiated under the NATP project of ICAR. However, the grant programme is not specific to any identified research priorities and is somewhat open ended. It is recommended that CGP be initiated in relation to identified priorities on a few major themes of importance.

Manpower Needs

- Institutionalising priority setting in any system requires professional expertise. At present there is extreme paucity of adequately trained manpower in the areas of systems analysis, project monitoring and evaluation and requirement specific social sciences—areas that are important for undertaking prioritisation tasks. Priority setting and targeting of research require a wider use of social and economic criteria in deciding allocation of resources. There is, therefore, need of early review of the present status and future needs of social scientists as a whole (which includes personnel with specialised knowledge in areas of economics, sociology, statistics, management science, etc.) and take the follow-up action.

- Social scientists have a large role in defining opportunities in relation to gap analysis at the macro-level and undertake constraint analysis at the micro-level. In the past, weak social science component within the system has resulted in weak scientist-farming community interface. This has greatly reduced the efficiency and effectiveness of the technology generation and adoption process. There is need to take careful view of the system-wide manpower requirements in the area of social science both in short-term and over longer term periods.
- There is immediate need to organise short duration management oriented training courses and awareness workshops aimed at bringing about change in the mindset of research managers so that they appreciate the need and urgency of shifting towards greater systems perspective and participatory approaches for the developing and executing research agenda. The National Academy of Agricultural Research Management (NAARM) has the overall mandate in this regard and should take upon the task preferably in collaboration with expertise drawn from ISNAR or a similar reputed international centre. The target of these training programmes should be research managers at the ICAR headquarters and its institutes and the SAU system including the zonal research stations. These efforts will need to be persistent and result oriented.

In the past few years the National Centre for Agricultural Economics and Policy Research (NCAP) has developed expertise for tackling methodological issues of research priority setting. This should be further strengthened through appropriate collaborative links internationally. NCAP should also consider coordinating a research network involving groups engaged in priority setting within the ICAR Institute—SAU systems.

Database Requirements

- Availability of reliable, accurate and easily accessible database is a basic requirement for objective, transparent and participatory approach to formulation of research priorities at different levels. At present our system lacks such information base and poses serious impediment in making progress towards greater objectivity in research planning. It is recommended that ICAR considers initiating a system-wide programme on 'database creation and management' cutting across all the subject matter divisions at the ICAR headquarters, institutes and the SAUs. Indian Agricultural Statistics Research Institute (IASRI) should take a lead in developing, implementing and monitoring progress through a consultative mechanism. Towards this, IASRI should initially hold a series of workshops involving potential partners to define the basis and methodology for database generation and management, role of partners and an operational framework. IASRI's role should be both in providing research leadership (methodological issues) and coordination of otherwise decentralised activities. Director of IASRI should be designated programme leader/convener. He should be assisted by a programme advisory committee consisting of professionals drawn from a range of specialisation whose inputs will be critical to programme success.
- A fundamental requirement for rational decisionmaking is the availability of adequate and reliable information. Amongst others, availability of agroecological data is a primary need for developing a comprehensive framework for priority setting and evaluation of research at regional and national levels. This will call for agroecological zonation and integration of spatial biophysical and socioeconomic factors into regional/national framework for assisting priority setting. The objective is to assess the likely economic benefits of research programmes targeted to specific geographic areas, commodities and social groups to provide policy makers with an assessment of

patterns of benefits that will accrue under alternate targeting options. These approaches will also require developments of databases and evaluation models.

- It is recommended that small units should be established in each ICAR institute and in the SAUs (including at each of the zonal research stations) as nodal point for networking.

Methodological Issues

- In recent years considerable progress has been made in developing and employing methodologies for formal priority setting of research based on social, economic and other parameters. The approaches used included benefit-cost analysis, congruence, scoring methods, etc. Efforts are now needed to improve priority setting methodologies by systematically integrating the social and economic parameters and by involving non-social scientists in establishing links between socioeconomic data, technical information about costs, the probability of research success and the overall objectives of research programmes.
- It is important to realise that some commonly adopted approaches based purely on economic considerations have serious limitations. In particular, there are difficulties in prioritising natural resource management related research because the usual economic models are inadequate to capture the real value of research in this area.

  This is also true for socioeconomic research. Research to evolve appropriate methodologies for prioritising research in these areas, therefore, needs to be strengthened. In the absence of such methodology, informed judgement would be made at the policy making level.

Conclusion

The new and changing environment for agricultural research calls for urgent reforms in the way research is managed. Institutionalising priority setting mechanisms is a part of this bigger agenda which encompasses several issues. While the overall reform process has to be necessarily gradual, priority setting should take precedence and must be taken up in the first instance. This calls for commitment on the part of research manager at every level. To begin with, a series of internal discussions can help in orientation of managers by bringing about clarity on issues such as the different levels at which priority setting exercise should be undertaken, who should be involved in this exercise, etc. This should be followed by assigning roles and responsibility at different levels to achieve the objectives.

27

Scientists' Views on Good Governance of an Agricultural Research Organisation

Background

The National Policy on Agriculture (NPA) sets a goal of 4 per cent growth in agriculture, which would be environmentally, economically and socially sustainable. Retrospectively, agricultural growth rates have seldom touched beyond 3.5 per cent. Beginning with 1990s, the growth rates exhibited declining trends; widening further the divide between expected and attainable. Prospects of building agricultural production through area expansion have almost come to a standstill. What is thus required, is to increase productivity to refurbish national goal of 4 per cent growth in agriculture. While attempting that, agricultural science and technology must not lose sight of issues related to sustainability i.e., rise in productivity in harmony with protection of environment, building of natural resource base and favourable economics; globalisation—requiring farming to be more precise, dynamic and competitive—quality and price wise and in frontier subjects of the 21st century i.e. information technology and biotechnology. Thus, at the threshold of the new millennium, Indian agriculture confronts an entirely different set of constraints and challenges than in the past.

Agricultural education, research and development do not get funding commensurate to their importance. Against a justifiable rate of investment equivalent to 1 per cent of agricultural GDP, allocations have seldom exceeded 0.3 per cent. The situation seems less likely to improve due to waning interest and enthusiasm to invest in agricultural science and technology (S&T). Attainment of over self-sufficiency in food is a cause for creeping in of a sense of complacency in the minds of planners, policy gurus and fund administrators. Quite

---

Policy Paper 17, December 2002.

contrary to this perception, if futuristic agriculture has to attain and sustain a growth rate of 4 per cent, and remain environmentally benign, economically favourable and socially relevant, it will require far more infusion of science and funding commitment.

While National Agricultural Research System (NARS) has a genuine case for greater financial support, its agenda ought to address general concerns of funding agencies on returns to investments and perceptions of various peers and stakeholders on relevance and utility of S&T results. More than ever, NARS have to become increasingly responsive, participatory and transparent in internal decisionmaking and governance. Alignment of their findings to external scrutiny from the point of measurable contribution and high degree of social responsibility will be necessary. Public research systems must, therefore, find ways to continually improve performance and accountability by becoming more sensitive to farmers' livelihood security, productivity surge, quality of natural resources and climatic integrity and country's international obligations in the face of WTA and liberalisation of economy. In this pursuit, research organisations will be required to manage their programmes efficiently and effectively on the one hand and rewrite their research objectives on the other. It will, thus, be necessary to change the ways research organisations are governed and structured currently.

Seized with the problem that the traditional methods and means of steering public research organisations were not adequate to cope with the existing and emerging demands of agricultural S&T. The ICAR and the NAAS sponsored a workshop to discuss and capture scientists' perceptions on the subject.[1]

Governance and Good Governance Concept

Governance—Definitions and Concepts

- Governance is defined as the way a system or organisation is guided and steered.
- From NARS perspective, governance is related to guiding, actuating and steering its research, education and extension functions towards growth and improved performance.

---

1. A one-day workshop under the convenership of Dr. J.C. Katyal, on 28 September, 2002, "Good Governance: A Framework for Improving work Culture, Performance and Accountability in the National Agricultural Research System (NARS)."

- Growth and improved performance depend upon an enabling environment to perform (internal processes i.e., rules and procedures) and generation of S&T goods having relevance and utility to all stakeholders (NARS-stakeholders' interface i.e., external processes).
- Governance is the point where external (peers and stakeholders) and internal (rules and procedures) environments meet. It is through governance that NARS capture the concerns of diverse stakeholders—funding agencies/peers, clients, staff, farming and civil society in general, which are translated into programmes, projects and activities. These in turn are executed within the bounds of rules and procedures.

Good Governance

- Good or effective governance means strengthening the rule of law and a legal framework for organisational growth and maintaining development and delivery of quality output by building democratic, responsible and responsive system.
- Applied to NARS, good governance signifies decentralised working, authority to modify/formulate rules and procedures; ability to exercise them in an unbiased manner: and manage the environs and personnel assigned to research laboratories and fields and address the needs and perceptions of various clients and stakeholders. Remaining efficient in input use and relevant in application of its output are other important indicators of a good governed NARS.
- Good governance extends beyond management; while the latter refers largely to processes within the system or organisation, the former encompasses growing interactions between the organisation and its stakeholders (see Box 27.1).

Box 27.1

Good governance certainly rises above the routine application of internal administrative and financial rules and procedures in managing the affairs of organisations:

- It relates more to a method of management that ensures conforming to basic rules of society, "both those embodied in law and in ethical custom" (Cadbury, 1998).
- It is fundamentally a state of the mind. It cascades from a set of core values that remain relatively unchanged with time (Narayana Murthy, 2000).
- It has eight characteristics—consensus oriented, follows the rule of law, efficient and effective, accountable, transparent, responsive and equitable and inclusive (UNDP, 1997 and UN-ESCAP, undated).
- It is the principles and not just rules that must guide the processes.
- Good governance is not about establishing command and control structures for management but about empowering and enabling each individual to perform and deliver. It is about the creation of a framework of organisation's vision, structures, systems and people that embede the set of core values into its processes and which allow growth in synergetic interactions with all its stakeholders. Hence, good governance is goal and problem solving oriented, fair, just and transparent along the hierarchical toposequence.

The Analytical Framework of Good Governance

- Good governance framework is built around three dimensions—internal mechanisms, performance and accountability.
- The internal mechanisms relate to how decisionmaking roles and responsibilities are defined, accepted and applied to establish overall work culture. Legitimacy of the organisation (accepted role in the national hierarchy, external rules and procedures, control and powers delegated to it to function autonomously), extent of decentralisation of its governance structures (governing body, institute level committees and task forces up to the group or individual levels, internally imposed rules), and how the linkages with its external environment (stakeholders) are established describe broadly the canvas of internal mechanisms.

- Performance dimension—a key external variable of good governance—relates to the use of resource inputs with product outputs. Performance defines a system's viability and visibility. Non-performing assets—falling output corresponding to constant inputs or stagnating output against rising input, sooner or later, are destined to be disinvested and phased out. Besides efficiency with which inputs are used and acceptability of and profitability from the output produced are measured, performance management should ideally focus on indicators that assess long term value creation like human resource development, client service, product standards, quality of internal processes and organisational learning.
- Accountability—an external dimension of good governance, is defined as responsibility for performing those tasks or achieving those results for which the individual or the organisation has been delegated the necessary authority. Without grant of necessary authority, individuals and organisations are less liable to be held accountable for use of resources, production of quality output and the processes of fairness, responsiveness and honesty by which the output is produced and its impact and relevance are sustained.
- Accountability relates performance to relevance and social responsibility. By defining roles and responsibility and operational rules and procedures, internal processes, in turn, influence performance and accountability.
- Good governance requires that external and internal rules be supplemented by organisational and individual value systems and ethical codes.

Governance: The ICAR Case Study

Organisation and Growth of ICAR

- Indian Council of Agricultural Research was born as the Imperial Council of Agricultural Research on 16 July 1929 as a Society under the Societies Act of 1860.
- Society re-christened as the present-day Indian Council of Agricultural Research (ICAR, also referred to as the Council) in January 1966.

- ICAR is the apex national organisation for guiding, promoting, conducting and coordinating education, research and development in agriculture including animal husbandry and fisheries.
- Besides its headquarters at New Delhi, ICAR has a vast network of institutes and deemed to be universities (87 and 4) respectively, all India coordinated research projects (82) and krishi vigyan kendras (261). Thirty-four state agricultural universities, three central universities and one central agricultural university work closely with the ICAR.
- ICAR employs about 5,000 scientists, 9,000 technicians, 5,000 finance and administrative functionaries and 11,000 support staff.
- ICAR functions in a multi-layered hierarchy. It is linked to the government through Department of Agriculture and Education (DARE). Over the years, DARE has overwhelmed the functioning of ICAR. It asserts more as a command and control structure and less as its original facilitating role of linking ICAR with government.
- Agriculture Minister is the President of the ICAR Society. He guides the policy direction and is the final authority in the matters of appointments, deputations abroad, and disciplinary/ vigilance cases. With time, he has been saddled with making decisions and/or giving approvals/sanctions on cases which otherwise are well within the perview of administrative hierarchy.
- Director General, ICAR and Secretary, DARE is its Chief Executive Officer. He is responsible for and has necessary authority to manage the affairs of ICAR system.
- In technical functions, eight deputy directors general, each in a distinct subject area, support DG, ICAR. Depending upon the subject speciality, a DDG guides and facilitates functions of a number of directors of the institutes, project coordinators of AICRIPs and zonal coordinators of KVKs.
- In the matters of administration and finance, one Additional Secretary each (drawn from the cadre of Indian Administrative Service and till early 2001 the post was equivalent to Joint Secretary rank) assists him. Additional Secretary who looks

after personnel and administration matters, functions as Secretary ICAR. Whereas, Additional Secretary who is concerned with financial aspects, acts as Financial Advisor ICAR/DARE.

- In matters of administration and finance, DDGs depend solely on the decisions/opinions of the Additional Secretary, DARE/ Secretary, ICAR and FA, DARE. In this trichotomous set up while the DDGs are burdened with many responsibilities, matching authority remains either denied or ill defined. On many occasions, decisions of administration and finance follow a course, different than that which DDGs perceive and pursue. With the result optimum working environment suffers, mistrust grows and delays and indecisions become of common occurrence.
- The ICAR system thus serves a variety of stakeholders namely farmers, scientists, students, central and state government departments, funding agencies, etc. More than the size and a multi-layered hierarchy of the ICAR system, it is its accountability to a wide variety of stakeholders and responsibility to millions of clients that makes ICAR a complex organisation.

Box 27.2

Performance Record of ICAR

- Despite complexity of structure and functions, performance of ICAR has been commended and acclaimed both nationally and internationally.
- It has succeeded in building a scientific base for Indian agriculture by contributing to increases in productivity of wheat (4-fold), rice (3-fold) and sorghum and pearl millet (2-fold). Likewise, productivity of other food items like fruits and vegetables, milk, meat and poultry have multiplied several folds.
- Over the last 25 years, foodgrain production increased by more than 100 million tonnes and the total factor productivity grew at 1 per cent-plus per year. In this achievement, contribution of research varied between 30 per cent and 50 per cent with an internal rate of return ranging between 40 per cent and 60 per cent.

contd...

...contd.

- The past successes have led to rising expectations from ICAR. Fixing up of a thus far unreached growth rate of four per cent/annum envisaged in the National Agricultural Policy is a pointer to the confidence and expectations of the nation in its agricultural research system.
- Simultaneously, with issues arising from poverty, malnutrition, unemployment, regional inequalities, and inefficient use of native resources synthetic inputs and resultant degrading quality of the natural resource base and possible shifts in climatic patterns, ICAR has to rework its research agenda on attaining the target productivity growth rates keeping sustainability concerns in sight.
- Additionally, research has to be more responsive to the new ethical issues like IPR and GM foods. Globalisation and opening of world economies call for greater vibrancy and a need to make quick adjustments in research objectives. While serving these emerging demands researchers will now become more accountable to a far larger variety of stakeholders than was ever visualised before. Also, public investments in agricultural research and education, which seldom match the real needs, will be screened and reviewed from every possible angle.

Governance Framework

- The existing governance structure of ICAR is assessed from three-dimensional perspective of governance—the internal mechanisms, performance measurement system and accountability. The internal mechanisms define the legitimacy, autonomy, decentralisation and delegation which impact the overall work culture. The two external dimensions, performance assessment and accountability, permit assessment of efficiency, effectiveness and relevance of the organisation.
- The ICAR works in a governance framework set largely within the rules and procedures emanating from the Government of India. The framework derives its legitimacy from a general body of the Society in which state governments and other departments at the centre, members of the Parliament and farmers and other interest groups are represented.
- The other main instruments of the governance framework are: the Governing Body (GB) at the ICAR headquarters, and the Institute Management Committees (IMCs, called Board of

Managementor (BOM) in DUs) at the research institute levels. The stakeholders are represented in both GB and IMCs.

- The GB is the chief executive and decisionmaking body of the Council. Societies Act of 1860 entrusts the GB the authority to manage the affairs of ICAR. One other provision of the Act provides, "whenever it shall appear to the GB of any society registered under this Act, which has been established for any purpose or purposes, that it is advisable to alter, abridge such purpose to or for other purposes within the meaning of this Act", it can do so. These two provisions of the Act are the basis of autonomous functioning of ICAR.
- Apparently, government funding should not affect autonomy of the ICAR. In practice, it exercises considerable influence, since GoI rules and procedures are applied and adopted *ipso facto.*
- *Mutatis mutandis* facility of making necessary alterations in instruments of governance remains mostly unutilised. ICAR has generally sacrificed its autonomy in favour of government rules and procedures, observed G.V.K. Rao Review Committee in 1988. In fact, subject of autonomy and a matching organisational structure at the ICAR headquarters and delegation of powers to its constituent institutes has been reviewed at least on nine occasions in the last 27 years without implementing the recommendations fully.
- Setting up of a new Review Committee, before recommendations of the last Committee were accepted/implemented, has earned ICAR the distinction of 'a most often reviewed scientific organisation in the country' and creates apprehensions in the organising capacity to deliver to the expectation of the stakeholders.
- Selective follow-up action on operationalisation of several critical recommendations has left autonomy of ICAR to look more aspirational than real. ICAR's work ethos continues to copy a government department and its institutes controlled by the headquarters on all crucial matters. Its format of governance can best be described as 'limited autonomy' and 'controlled decentralisation'.
- Other autonomous scientific organisations like CSIR have taken advantage of the *mutatis mutandis* clause to ensure a

relatively higher degree of devolution of powers to the institutes and scientists. But in ICAR real decentralisation has yet to find effect to a measurable degree in practice.

- Formal processes of performance assessment have largely been limited to the ICAR institutes and AICRIPs through quinquennial review teams (QRTs). At the individual level, a scientist's performance is monitored through annual assessment reports (AAR). The existing AAR proforma suits more to assess the work of civil servants and less of scientists. By constitution, the AAR is 'confidential in nature'. By design, average performance rating is legitimate, since a scientist is not informed to get better until the rating is below average. Also, there are no structured mechanisms to improve performance by building competencies in the deficient areas.
- What is true on performance management of scientists holds good for other functionaries also.
- The existing accountability system is aimed primarily at ensuring that the annual budget is used according to approved heads and as per prescribed rules and procedures. Since decentralisation is limited, so do the responsibility, authority and accountability. Accordingly, there is no strict accountability system based on utility of research outputs or their relevance to stakeholders. Thus, ICAR's work culture is loaded more with input accounting and auditing procedures and less with output evaluation systems.
- In summary, ICAR governance follows a centralised-decentralised format. It functions as a strong-walled trichotomous system. Majority of the scientists pursue research goals more than relevance and utility of output. The administration prefers rules to performance. The finance wing functions independently to check the level of adherence to the audit manuals and can torpedo any proposal. The synergetic interactions between the three wings are generally missing. The Council instead of full time guidance on policy-input and decisive strategy in streamlining research has perforce become another centralised bureaucratic layer. The fact that critical recommendations of several review committee reports are not operationalised is an indication of reluctance to implement change and to perpetuate stranglehold of the multi-layered

bureaucracy over the system. There is urgency to set the governance mechanisms in order, if the ICAR is to meet the challenges of the new millennium.

Recommendations

Agricultural research influences livelihoods of two-thirds of Indian workforce. Majority of them is resource poor and suffers from disguised unemployment (overcrowding with marginal productivity). Incorporating their work needs and economic aspirations while conceptualising, designing, conducting, and refining research programmmes is, thus, necessary. The unrelenting demographic pressure and social commitments made by the country in its National Agricultural Policy demand that agricultural S&T must support a minimum growth rate of four per cent/annum without damaging the quality of soil, water, vegetation and air. If past performance of ICAR system is any indication then there should be no doubt on its ability to measure up to emerging challenges and concerns. However, success will depend upon the level of government support and control, and how forcefully ICAR reengineers its programmes, governance structures and systems. Apart from assuring adequate funds, government can right track ICAR activity and productivity by continuously supporting a performance-enabling environment. On its part, ICAR must redefine its mission, programme objectives and governance framework to inculcate system efficiency, effectivity and relevance. Participating scientists had following perceptions and suggestions:

- ICAR's autonomous status, as enshrined in the Societies Act of 1860, must be reassured and reaffirmed by the ICAR leadership—both political and apolitical. For this to happen it will be necessary to steer and translate the existing provisions of autonomy into practice by supporting the implementation of the relevant recommendations made by the earlier Committees.
- Also to sustain autonomous status of ICAR, it will be essential to de-link public funding from authority to self-govern in the first instance. This recommendation is in line with other similarly placed scientific organisations. Secondly, it is necessary to redefine the initial facilitative and link role of DARE. This will eliminate the persisting confusion of DARE

exercising a command and control over ICAR's functioning. Any attempt on making DARE a superior entity over ICAR will violate latter's autonomous status granted by the constitution. Thirdly, ICAR must rewrite its rules and procedures within the ambit of *mutatis mutandis* facility available to it. The new set of rules and procedures should allow flexibility and speed in decisionmaking, remove interpretational ambiguities and build transparency in application. ICAR may consider hiring the services of a legal luminary for this purpose. Fourthly, ICAR needs to assert more as a scientific department and less as an administrative outfit. It can reach that goal by vesting, alongwith technical responsibility, greater administrative and financial authority in its subject matter divisions.

- In order to overarch economic development with technological solutions, it will be more than necessary that ICAR is steered by professionals who not only possess subject brilliance and real life knowledge but also have concern and appreciation for unique needs of scientists and scientific research. Suggestion on organisation and structure supporting scientific leadership is not new. It is being made to reconstate that it is the scientific leadership that in the past has paid rich dividends in ICAR and other autonomous scientific organisations like CSIR, DBT, ISRO and BARC. In order to strengthen technical support to GB, a Standing Sub-Committee of Professionals should be created to assist it in S&T related decisions.
- Intimately connected with the question of autonomy at the ICAR organisation level are issues related to decentralisation and delegation of powers to institutes, coordinated research projects and scientists. While autonomy requires legal sanction of sources external to ICAR, decentralisation does not. It is an internal matter. Armed with this proviso, ICAR has decentralised its functioning by disbursing a fair degree of independence to its constituent units, but not enough authority to manage in total the affairs and environs of research laboratories. Additional grant of powers recommended by the earlier Committees will strengthen the process of decentralisation further.

- Scientists who have hardly any authority in managing their research can be empowered by institutionalising the concept of project based budgeting (PBB) which is already in practice across international research centres and national institutions like CSIR. The final report on the subject of PBB is hibernating in the ICAR for the last three years. When the government has directed to follow zero based budgeting for justifying investments of public funds, implementation of PBB becomes more urgent than ever.
- So that devolved powers are used without fear and fuss, clear definition of the roles and functions of administrative and finance divisions becomes necessary. In a scientific organisation, input of administration and finance should ideally complement the technical functions. Their concurrent involvement when administrative decisions and financial commitments are made will minimise the later objections and questions.
- Convergence of technical, administrative and finance functions is necessary to reach scientific goals in a time-bound fashion and with efficiency. Structurally, these three functions must originate and concentrate at the same point i.e. subject matter divisions/institutes. In no circumstances, the in-vogue multiple approval system (re-approval of already approved items) should be allowed to perpetuate any further.
- Apart from adjustments in the internal structure, ICAR can reach high standards of scientific output and efficiency by interfacing its education, research and extension efforts with other public (SAUs and other R&D departments) and private (industry) agencies. In pursuance of forging no-delay linkages, it will be necessary to design appropriate guidelines on collaborative mechanisms and inter-organisational arrangements.
- Group recognised that good governance leading to sustainable agricultural development would not be possible without implanting a set of core values (institutional ownership, commitment to strategic vision, responsiveness, efficiency, transparency, team working, workforce diversity, consensus orientation). Institutionalising such value systems require a shift in focus from enforcing rules and procedures to ensuring

practice of principles and codes of ethics, both at the system and individual levels. In order to infuse such value systems sustained efforts and investments in human resource development will be absolutely necessary (see a later section).

- In place of annual assessment reports, participating scientists perceived to introduce a performance management system, which is non-punitive but focuses on individual building in deficient areas. It was recommended that performance should be measured in terms of quantifiable results against goals/objectives/targets set for the reporting period. The performance of a scientist should be weighed against performance of support sections also. The report should be made open to provide counselling for possible human resource development. The performance management system developed by NAARM and endorsed by the Directors of the ICAR institutes should be operationalsed expeditiously.
- So that various functionaries—both in research management and out of it—can be assessed and held accountable with respect to organisational processes (delegation of authorityforthe tasks for which an individual is responsible), compliance of accepted rules and procedures, outputs and impact and relevance and utility of the output. It was broadly recommended that accountability measurements should focus more on relevance and applicability of a functionary's output. Simultaneously, it was recognised to intertwine responsibility and authority together.
- While assessing individual's performance is important, to appreciate that research is largely inter-disciplinary and is carried out by teams of scientists. Necessary accountability systems that encourage and evaluate teamwork need to be developed. The administrative and finance personnel must also be viewed as members of the team and their performance should be measured with reference to contribution they make to the outcome and timely execution and completion of research projects.
- Since accountability and output are inseparable, so are accountability and performance. Whether it was performance management or accountability assessment, it was suggested to include the following elements: (i) self- identification of goals/

objectives and targets, (ii) delineation of responsibility and definition of clear authority, (iii) self-assessment of performance/constraints/improvement and (iv) open evaluation of feedback for improvement and capacity building.

- In order to strengthen and institutionalise good governance through organisational change, building client perspective in science and technology initiatives, fostering linkages, instilling value systems, and introducing performance management and accountability systems, it was felt necessary to build talented scientific resource supported by equally skilled technical, administrative and financial services. The future was underlined to belong to informed, motivated, creative and committed humans. Since strength and competence of each individual is pivotal to build an effective team capable of meeting emerging social challenges and economic concerns, investments for development of human knowledge and skill, outlook and behavior was recognised as the core of organisational change. ICAR must launch a comprehensive human resource development programme to serve the training needs of various functionaries. The first requirement will be to make an assessment of training needs and scope. So that training gets prominence in employment conditions, an instrument of policy on HRD may be developed and put into practice.

28

Priorities of Research and Human Resource Development in Fisheries Biotechnology

Background

In India, 16 per cent of the world population has to be sustained on 2.4 per cent of the global land area. Apparently, aquatic resources had to be given high priority for additional food production. Like agriculture, the fisheries sector is also dependent on renewable natural resources. Compared to animal production system, fisheries sector enjoys some advantages. Firstly, fish is the cheapest source of animal protein and a health food. Secondly, high fecundity (up to 1 million eggs) and fast growth rate (growth coefficient often > 1.0) of fishes have no parallel among other animal protein sources, like livestock including poultry. These biological advantages of fishes offer considerable scope for achieving nutritional security in India. Understandably, the projected annual increase in food production target for the next decade is as high as 10 per cent for fish and other aquatic products, whereas it is just 2.5 per cent for cereals.

Asia has been the centre of fishing and aquaculture activities. Among the Asian countries, India ranks second in culture and third in capture fisheries. Fish production has increased by more than 10 times and the contribution of the fisheries to the Indian GDP by more than 3 times during the last five decades; a growth arguably one of the highest within the food production sector.

However, the annual per capita fish availability (8 kg) is less than the world average (12 kg) and the quantity (11 kg) recommended by WHO for nutritional security. India's potential for increasing capture fisheries from the coastal areas is limited. Capture fisheries from rivers is also declining due to water abstraction, sedimentation, habitat destruction and pollution. Hence, India has to depend heavily

---

Policy Paper 21, December 2003.

on aquaculture. At present, the technology exists for rearing only 15 fish species, while as many as 39 and 52 species are profitably cultured by the Chinese and Koreans, respectively. Asian countries suffer an annual loss of more than $ 3,000 million owing to disease problems. Intensive efforts are in place to develop diagnosis. It is also recognised to build trained manpower in this area of fishery research and development.

Fish genetics is a rapidly growing field of science but in India it is still in its infancy. Selective breeding, a proven method for strain improvement, is known to enhance the expression of desired traits by 3 to 5 times; the super strain rohu, 'Jayanthi' is an example. However, it must be indicated that selective breeding may result in transmission of unwanted traits, due to mixing of whole genome of parents. It may also prevent transfer of desired genes from one species to another. On the other hand, transgenesis allows the introduction of novel genes or augmentation of a specific trait by just transferring the tailored copies of gene from the same or other species. In the area of trangenics also, we have just made the beginning and are, thus, far from developing disease-resistant transgenics. Research in genetics and molecular biology of fish is expected to yield probiotics and vaccines to eradicate or contain microbial diseases, which sometimes rapidly wipe out the farmed fishes, even before the disease is diagnosed. Marine plants and animals can be a very important pharmaceutical source to treat dreadful human diseases. Fish is a highly perishable commodity. Biotechnological research may lead to superior methods of preservation, production of value added products and bioactive compounds, quality control and so on. In order to conserve the valuable germ plasm of Indian fishes, there is need to develop simple and widely practicable methods of sperm preservation and technique for interspecific androgenetic cloning.

In order to discuss aforesaid issues, the National Academy of Agricultural Sciences, organised a workshop. Several aspects of human resource development, molecular markers, genomics and transgenics, aquatic health and management, conservation genetics, and bioactive compounds were covered during the deliberations of this workshop.[1]

---

1. A two-day workshop was organised by the NAAS, at Centre for Cellular and Molecular Biology, Hyderabad, on 25 and 26 August 2003.

Recommendations

1. Prioritisation of Research in Biotechnology of Fishes

Biotechnology is a fast growing field of science. The number of publications appearing on Fish Genetics, Molecular Biology and Biotechnology, especially from USA, Japan, Korea and China are phenomenal. Considering our genetic resources, requirements, expertise and infrastructure, the following areas have been tentatively identified for financial support in the area of research on biotechnology of fishes on a priority basis.

1.1 Molecular Markets

Because of the technical simplicity, protein markers like isozymes and/or allozymes have long been used as 'molecular tags' in identification of species/strain/stock; but this technical simplicity is counterbalanced by the need for fresh tissue. With the advent of PCR technology, molecular markers have become a powerful tool in many areas of fisheries research including identification of species/strain/stock and fishery product(s) of endangered or banned species, evaluation of phylogeny, delineation of stock structure, measuring levels of genetic variation in wild and cultured stocks, conservation genetics, determination of breeding strategies, linkage and gene mapping, marker assisted selection and assessment of gene manipulation techniques. DNA markers include nDNA(RFLP, microsatellite, RAPD, AFLP) and mtDNA (RFLP, PCR). Microsatellites are excellent markers in the study of population genetics whereas, RAPD have been shown to be good markers for pedigree analysis.

1.1.1 Marine Capture Fisheries

An important consideration in the management of exploited fishery resource is the identification of discrete populations or stock units, which are genetically distinct self-maintaining groups. Sex specific markers are useful to assess sex ratio at the earliest embryonic stage and to confirm the genotype of the androgenetic, gynogenetic and hormonally sex reversed progenies. In recent years, aquaculture of the clown fish *Amphiprion* and the food fish *Epinephelus* are targeted. These fishes undergo natural sex reversal from male to female in the former, and female to male in the latter. There is an urgent need to develop sex specific markers for these fishes.

Sardinella longiceps and *Rastrelliger kanagurta,* which constitute the bulk of Indian exploited marine fisheries must be subjected to an in-depth study using protein and microsatellite markers. There is a need to prepare a series of atlas for geographic distribution of these species and description of population size and structure as well as their breeding.

For mariculture, there is a need for developing markers for molecular taxonomy of the genus *Epinephelus* and *Amphiprion.*

There is also a need for developing markers for *Hilsa ilisha.*

Since aquaculture concentrates on a very small range of species and an equally narrow genetic base in these species, deliberate release of stocks for ranching can modify the genetic composition of wild resource. Hence, there is a need to monitor this activity by the concerned agencies.

1.1.2 Freshwater Capture Fisheries and Aquaculture

While genetic background is crucial for success in selective programmes, hardly any programme has been initiated in India for genetic selection of economically important traits such as body growth, disease resistance, fecundity and food conversion rate for even the major carps, which are the mainstay of freshwater fisheries and aquaculture. The vast and diverse Indian aquatic resources are known to provide among the richest genetic resources in the world. Yet, there is no recorded information describing various strains of major carps, which occur in different agro-climatic zones of India. Japanese have made a series of atlases for the distribution of selected fish species characterised by different ploidy status and there are only a couple of records on the occurrence of haploids and tetraploids in India. To promote aquaculture, Chinese have made a large scale annual production of over 100 million triploid crucian carp seedlings by crossing tetraploid and diploid Japanese crucian carp. Considering these deficiencies, the following programmes are suggested:

- Identification and preservation *of* different strains *of rohu* and *catla, Penaeus indicus* and *Macrobrachium rosenbergii.*
- Development *of* reference and/or resource family lines, genetically distinctive populations/strains of Indian major carps and identified penaeid and palaemonid species by crossing and rearing under controlled conditions.

- Isolation and characterisation of polymorphic DNA markers for rohu, tiger and giant prawns.
- Construction of genetic linkage map containing at least 500 markers each for rohu and catla.
- Development of markers, products and morphologically indistinguishable progenies/larval stages of exotic/banned fishes.

1.2 Genomics

Marker assisted selection can accelerate the selective breeding process and such markers can also be used for genome mapping. In this context, a commercially important foodfish, *Labeo rohita* and a shellfish, *Macrobrachium rosenbergii* are recommended for genome mapping through RFLP, RAPD, AFLP and microsatellite markers.

1.3 Transgenesis

Although selective breeding is an excellent method for strain improvement of crops and farm animals, it suffers from drawbacks such as: trans-species barrier, mixing of whole genomes resulting in transmission of unwanted traits and requirement of relatively longer time. Since transgenesis allows introduction of novel genes or augmentation of a specific trait by transferring tailored copies of the gene from the samc or different species, it has become a powerful tool to bio-engineer an organism to provide products and services. Fish has been regarded as the best vertebrate model system because of its oviparity and fecundity. However, non-availability of piscine vectors has been a handicap. At Madurai Kamaraj University, in order to construct transformation vectors for the indigenous fishes, growth hormone genes of rohu, *Labea rohita,* catfish, *Heteropneustes fassilis* were isolated, cloned and sequenced; their fidelity was confirmed in prokaryotic and eukaryotic systems. A vector was constructed with grass carp b-actin promoter driving the expression of r-GH. The sperm electroporation technique was standardised to ensure 25 per cent hatchling survival and 37 per cent presumptive transgenics without suffering any deformity. Genomic integration was confirmed in 15 per cent of the tested individuals (Ti) belonging to family lines 2 and 3; another 25 per cent of the juveniles (Te) were also proved transgenic but with the transgene persisting extrachromosomally for longer than 1 to 2 years. Further like drought resistant gene, cloning and sequencing of salinity tolerant gene may also prove a boon to aquaculture. Development of salinity tolerant transgenic fish and shrimp may be a long-term programme but may not be an insurmountable task.

Considering the above constraints and accomplishments, the following programmes are recommended:

- Follow-up research on transgene of growth hormone gene transmission on rohu need to be pursued further.
- Isolation, cloning, sequencing and construction of vector for genes responsible for aestivation and dormancy and their introduction into animals and plants of desirable species.
- Isolation, cloning, sequencing and construction of vector for genes responsible for salinity tolerance and their introduction into eggs to generate a salinity-resistant fish or shrimp.

Existing transgenic technology, like, microinjection, electroporation and lipofection are successful in isolated cases. Establishing methods of gonadal and embryonic stem cell isolation and manipulation for transgenesis in fishes would offer new methodology and far-reaching implications. Blastula stage cells can be transplanted successfully into another blastula by microinjection. It has been shown that donor cells colonise in the gonads of fishes like zebrafish, trout and medaka.

- Transgenesis through gonadal stem cells is to be explored.

1.4 Aquatic Medicine and Health Management

Bacterial, fungal, protozoan and viral infectious diseases are widespread among the natural fish populations. However, the fishes raised with intensive aquaculture systems are more vulnerable to diseases. Frequent outbreaks of diseases, especially those caused by viruses in recent years, are recognised as the major constraint to aquaculture. Effective control and treatment of the diseases require access to diagnostic tests that are rapid, reliable and highly sensitive. Conventional methods of bacterial identification are time-consuming and not amenable to mass scale screening. Molecular genetic tools facilitate mass screening and classification of the microbial pathogens. Highly discriminatory molecular typing methodologies allow rapid and sensitive differentiation within even a single serotype. For example, there are over 200 serotypes of *Vibrio cholerae,* but only serotypes 01 and 139 cause cholera. Following recommendations are made in the area of aquatic medicine and health management:

- It is strongly recommended that a Network Project on microbial Diseases of Fishes and Shellfishes is established.
- The proposed project may focus on two most important diseases *viz.*, white spot syndrome disease of shrimp and white muscle disease of freshwater prawn. In addition, participating institutions may take up research on diseases that are locally important in the areas like disease surveillance and reporting, indigenous kits for rapid identification of microbial pathogens, vaccines and immunostimulants for improving host-defence, novel probiotics and bioremediations, cell lines from host species, disease-resistant broodstock through marker-assisted selection, aquaculture health products for the presence of exotic pathogens, exploratory research on exotic pathogens in molluscs and other aquatic species, DNA probes and monoclonal antibodies as bio-sensors to determine fish quality, techniques for rapid identification of pathogens in aquaculture systems and chemical pollutants by biofluorescence and bioluminescence in aquaculture systems.

1.5 Conservation Genetics

Due to habitat destruction, overexploitation and chemical and biological pollutants, several species of fishes are designated at the global level as endangered/threatened. As a major agricultural country, India uses annually 18 mt of NPK fertilisers, 0.55 mt pesticides, 0.125 mt synthetic detergents; at least 25 per cent of these manmade chemicals ultimately enter the sea through rivers. Large scale earth moving and clearance of forests have also led to an unprecedented sedimentation rates of major Indian rivers. Consequently, siltation rate of reservoirs is also many orders higher than that estimated at the time of construction. Hence, there is an urgent need for researches in conservation genetics. This aspect could be explored on the following lines:

1.5.1 Sperm Preservation

Strategies adopted for *in situ* conservation requiring legislation for the protection of habitats and germ plasm are of regional importance. Conversely, *ex situ* strategies like cloning and cryopreservation of germ plasm, involving scientific and technical development, have global significance. The National Bureau of Fish Genetic Resources, Lucknow has a sperm bank facility for only about a dozen fish species: *Catla catla, Labeo rohita, L. dussumieri, Cirrhina mrigala, C. carpio, O. mykiss, Salmo trutta, Tor putitora, T. khudree, Tenualosa ilisha* and *Horabagrus brachysoma* and protocols for other commercially important species need to be developed.

- Explorative research on protocols for year round production of adequate seedlings of (i) endangered species, (ii) eggs of commercially important *Epinephelus tauvina* and sperm of the bycatch E. *merra* and (iii) migratory species like the eel using eggs of suitable species.

1.5.2 Development and Preservation of Cell Line

Barring very few exceptions, development of fish cell lines in India is highly restrictive. Keeping that in front, there is an urgent need to develop and preserve cell lines for culturably important penaids, *Penaeus monodon, P indicus* and *Macrobrachium rosenbergii* and to diagnose viral diseases, and to screen and certify imported brooders and seedlings, and exportable shrimp commodities.

1.6 Inventory of Bioactive Compounds

Aquatic organisms can play an increasingly important role in biotechnology and medicine. They are rich sources of novel and structurally divergent compounds. Some of these bioactive compounds could serve as models for new drugs, antifouling compounds, morphogens, signal substances, pheromones, neurotransmitters and antifertility compounds. Hence, genetic control of the biosynthetic pathways of selected bioactive compounds, needs to be worked out. Studies are also warranted to determine whether the genes for biosynthesis can be sequenced and cloned for practical pharmaceutical applications. Thus, large quantities of the compound can be made without disturbing the organisms, from which they are isolated. Application of DNA technology may thus offer exciting opportunities for advancing science and technology associated with bioactive products from the aquatic environment. Following suggestions are made:

- Preparation of an inventory of anti-viral and anti-bacterial compounds with potential application in aquaculture.

- Isolation, purification and evaluation of heparin from nudibranchs and algae development of fermentation technology for production of enzymes and enrichment of aquafeeds using mangrove microbes.
- Development of enzymes from specified sources with higher activity.

2. *Human Resource Development*

The programmes of trading and research are weak and hardly match with those seen in some well established institutes. Hence, it is recommended that an adequately funded Centre for Excellence in Fisheries Biotechnology is established with specific mandates to (i) offer post-graduate course in Fisheries Biotechnology to train annually a dozen students, initially to meet the Indian requirement but later to benefit the Third World countries in Asia and Africa, (ii) organise advanced level (three months) mid-summer training in fishery biotechnology at the identified units to a dozen graduate students, selected through competitive examination, who will work on cloning, sequencing and gene expression, as well as cell and tissue culture and (iii) generate manpower with competence for DNA microarray technology, 2-D gel electrophoresis, sequencing and characterisation of genome and protein. In order to make most efficient use of existing manpower and infrastructure, it was recommended to encourage multidisciplinary programmes of research and development by promoting alliances among conventional universities, national laboratories, fisheries institutes and fisheries colleges.

29

Strategies for Agricultural Research in the North-East

Introduction

The North-Eastern Region of India, comprising the states of Arunachal Pradesh, Assam, Manipur, Meghalaya, Mizoram, Nagaland, Sikkim and Tripura, lies between 21.5° N to 29.5° N latitude and 85.5° E to 97.5° E longitude. It has a total cropped area of 5.3 million hectares and a population of around 39 million. The region falls under high rainfall zone and the climate ranges from subtropical to alpine. The region is characterised by difficult terrain, wide variations in slopes, altitude, land tenure systems and diverse cultivation practices. The transport and communication system is poorly developed and as a result majority of the areas in the region remain still inaccessible. The infrastructure facilities like supply of input, marketing, institutional credit and extension services are inadequate. Majority of the population, predominantly tribal, is dependent on agriculture and land-based activities. The agricultural production system in the region is predominantly rainfed, mono-cropped at subsistence level. Slash and burn agriculture is still practised in almost all the states except Sikkim on steep slopes with reduced cycle of 2-3 years as against 10-15 years in the past. The region, once richly endowed with rich genetic diversity of plants, has been denuded due to human interference by adoption of unscientific land use system. With rapid increase in human and livestock population and the rising demand of food, feed, fuel, fodder, fibre, timber and the other developmental activities, the farmers have been forced to exploit forestland and water resources at sub-optimal level in complete defiance of the inherent potential. This has resulted in progressive decrease in forest cover, loss of biodiversity, serious soil erosion leading to depletion of plant

---

Policy Paper 9, June 2001.

nutrients, gradual degradation and decline in land productivity and its carrying capacity, silting of major river basin causing recurrent floods in the plains, and drying up of perennial steams as well as ecological imbalances. Gradual degradation of these resources is of prime concern and calls for location-specific measure to conserve, utilise and manage these resources for optimising production on sustained basis without adversely affecting its quality.

Agriculture is the main economic activity in the region and despite major impact of green revolution in the irrigated areas of the country, modernisation of agriculture has escaped this region as evidenced by poor adoption of modern technologies, low consumption of fertilisers and other indicators of growth. One of the most important reasons for this slow growth is the lack of location-specific and system-based technologies. Apart from this, the dwindling resources of soil, water, flora, fauna and increasing concern for environmental safety has drawn the attention of the planners and policy makers at regional as well national level. In view of these concerns and enormous opportunities, the National Academy of Agricultural Sciences in collaboration with Indian Society of Hills Farming deliberated on the theme "Strategies for Agricultural Research in the North-East."[1]

The following records salient features of the issues discussed and the recommendations made at the seminar.

Resources-Advantage North-East

North-East India is endowed with rich natural resources in the form of soil, water, plants and animal diversity and this provides the basic ingredients for development.

Land

About 53 per cent of the reporting area is classified as forestland though there are large variations from minimum of 8 per cent in Meghalaya to a 91 per cent in Arunachal Pradesh. The land under the forest apart, about 60 per cent of the land is classified as not available for cultivation. The net area sown is hardly 7 per cent of the total reported area in the region. The hill areas present some very special ecological and geophysical features. The terrain is mostly rugged and

1. The seminar was hosted by ICAR Research Complex for North-Eastern Hill region, Umiam, November 1999.

steep. The unit cost of infrastructure development in the region is high and the returns are relatively low. The cost of infrastructure is high largely because of rugged nature of terrain with widely dispersed population. It would, therefore, take considerable time for this region to build up an adequate resource base.

Water

The region has large surface and groundwater resources mainly because of its location in high rainfall area and the extensive river system. But all the water resources cannot be utilised because these are inaccessible or non-reversible. The availability of these resources has not been adequately documented and as such full information is not available. The total surface water potential of the region is 928873 mm^3. The region, excluding Sikkim, has a total groundwater potential of 855 mm^3. As usual all the hilly regions have low groundwater potential, the lowest being in Sikkim. The exploitable water balance is mostly concentrated in Tripura. Water application on slopes for irrigating plantation crops poses serious problem of soil erosion.

Plants

The region is a treasure house of genetic diversity and 6-7 thousand plant species constituting approximately 40 per cent of the total flora of the country are represented here. A large number of economically important plants such as food, fodder, spices, oilseeds, fruits, vegetables, medicinal and aromatic, orchids and other flowering plants grow wild in nature. The region is either primary or secondary centre of origin for many crop plants.

Livestock

Livestock in this region comprises cattle, buffalo, sheep, goat and pigs. In the hills, draught power is rarely used for tilling the soils and most of the tribal populations do not fear cattle in the usual sense. They depend on semi-wild animals like mithun and yak for meat. Cattle is generally let loose to graze and stray in the open. Basic problem of livestock rearing is shortage of feed and fodder and absence of commercialised dairy and piggery.

Agricultural Farming Systems

The agricultural practices in the region are broadly of two distinct types, *viz.*, (i) settled farming practised in the plains, valleys, foothills and terraced slopes and (ii) shifting cultivation practised on the hill slopes. In the hills, agricultural operations are carried out at a maximum elevation of 5000 m with 'slash and burn' method.

Shifting (Jhum) Cultivation

It is a primitive mode of agriculture evolved as a reflex action under the stress of the peculiar geophysical and ecological situations prevailing in the hills of the NE region. The system reflects a sort of community farming without heritable rights over the land. The practice starts with selection of forested land, clearing and burning of the forest before the onset of monsoon, planting of various crops in an intimate mixture by dibbling and harvesting. The land is abandoned after cultivation for a period of 2-3 years and cultivation is shifted to another site. The method of allotment of land varies from tribe to tribe. In most cases it is decided upon by the village councils or by village elders, and size of the plots depends on the number of working hands in respective farm families. Out of the total area of 2.2 million ha of *jhum* lands, 17.5 per cent is cultivated at anyone point of time; the rest is left for natural regeneration over a period of several years. Consequent upon population pressure, the *jhum* cycle has, however, shrunk to 3-6 years or even less against, earlier practice of a cycle of 10-15 years resulting in soil degradation and ecological imbalance. Indiscriminate destruction of forests in the *jhum* land, coupled with high rainfall, has resulted in heavy soil erosion and consequent silting of rivers causing floods in the lower reaches.

Settled Cultivation

Settled cultivation is practised in plains and valleys. Adequate precipitation ensures availability of water for the crops grown in *kharif.* However, *rabi* crop faces water deficit, which is met by irrigation. With increasing awareness of the disadvantages of shifting cultivation, the farmers have started adopting settled cultivation. This is likely to result in reduction of area under shifting cultivation.

Present Status of Agriculture

Cropping Pattern

The cropping pattern in the region, with the exception of Sikkim, is characterised by predominance of rice as the lead crop. In Sikkim, maize is the dominant crop. Food crops account for more than 80 per cent of the gross cropped area, which is suggestive of prevalence of subsistence agriculture and lack of crop diversification. About 70 per cent of the gross cropped area is accounted for by cereals alone. Within the region, however, there are exceptions. Sikkim records 9.3 per cent, and Tripura 6.2 per cent of the gross cropped area under oilseeds. These state level variations are a manifestation of the differences in the status of agricultural development prevailing in the constituent units of the region. A remarkable feature of shifting cultivation is that a wide variety of crops is grown in the *jhum* fields. *Jhum* paddy is the dominant crop and is mixed with maize, millets, beans, tapioca, sweet potato, ginger, cotton, tobacco, chillies, sesamum and vegetables. The production units are more or less self-contained, each family growing crops according to its needs. Major food crops of the region are cereals and pulses which are cultivated in 78.03 per cent of the cropped area.

Crop Intensity

The region's agriculture is characterised by low crop intensity (117 per cent). Physical constraints like weak and underdeveloped irrigation network, extremely cold weather in the higher elevations limit the crop season to summer months only. Low density of population in the hills and sociological constraints like values and attitudes, which are not wholly conducive to market oriented agriculture, have resulted in near stagnation in agricultural production in the region.

Irrigation

The available data on irrigation shows that about 24.42 per cent of the net sown area is irrigated. Of the total 302.96 thousand ha of the irrigated area, 56.89 thousand ha (18.78 per cent) is irrigated through surface flow, 5.51 thousand ha (1.82 per cent) through surface lift, 0.41 thousand ha (0.14 per cent) through groundwater lift and 240.15 thousand ha (79.29 per cent) through other irrigation methods. The normal way of irrigation in the hills is by diverting small streams. In other areas, the irrigation is provided from natural sources only during rainy months. The hydro-geomorphology of the hilly region is such that groundwater resource cannot be easily utilised for irrigation purpose. In such areas techniques for harvesting rainwater appears to be essential.

Fertiliser

Fertiliser consumption in the region works out to be as low as 23 kg/ha as against the national average of 95.6 kg/ha (1999-2000). The low level of fertiliser consumptions is attributable to heavy loss through run off of rainwater during the *kharif season* and due to slow and limited spread of fertiliser-based technology in the region.

Seed

Seed is the most important component of agricultural production system. Information available from different sources indicates that there are large gaps in the supply of this vital input. The time lag in the supply of seeds frustrates the production programmes of the region. The spread of HYV has been poor as evidenced from the fact that only 56 per cent of the rice area has been brought under HYVs in NE hill states against 74 per cent in the country. Of the total quantity of improved seeds required by the farmers, roughly 40 per cent is met from within the region leaving a gap of 60 per cent in its present requirement. This gap is very likely to widen over time, as the percentage of adoption is sure to increase in near future. Except Assam, organised seed production programme does not exist in any of the NEH states.

Agro-based Industries

The status of agro-based industries in the region is not very satisfactory because of lack of development in this area. However, there has been some growth in traditional agro-based industries like rice mills, oils mills, etc. Hindustan Paper Corporation has started a paper pulp unit in Nagaland. This paper mill uses bamboo, which is widely cultivated in the region, as the raw material. The scope of growing soybean and groundnut in large scale as agro-industrial crops has been strengthened with the establishment of *vanaspati* plant in neighbouring Assam. Research results indicate that there exists very good potential for these two crops in the region. Despite considerable potential for development of fruit processing industries in the region, there has not been much progress in this area. The North-Eastern Regional Agricultural Marketing Corporation (NERAMAC) was set up in the region for marketing and processing of fruits.

Foodgrains and Oilseeds

An analysis of area, production and productivity trends in North-Eastern region shows that the region produced a total of 5.2 million tonnes of foodgrains from an area of 3.78 million hectares during 1998-99. Thus, the average productivity of the region during the above period has been around 1377 kg/ha, which is much below the national average of 1620 kg/ha. The population growth is higher than the growth in foodgrain production in the region. The region faces a deficit of 2 million tonnes of foodgrains, which is met by imports from other parts of the country. It drains the capital out of the region. This can, however, be curbed by making the region self-sufficient in foodgrains. The grain harvest of NE region in the post-Green Revolution era has continued to rise; yet per capita production flattened because of population growth.

Crop improvement research in the country in general and NE region in particular has so far been commodity-oriented. Moreover, improvement programmes pursued elsewhere in the country do not take into account the ground realities prevailing in different farming systems of the region. Further, the unique soil and agro-climatic conditions justify breeding programmes targeted specifically for the region. A case study of rice shows that most of the varieties introduced from other states had limited success whereas the varieties developed in the region by involving indigenous land race/variety, have been highly successful. Work done in varietal development in rice and wheat through enhancement of harvest index has increased the grain yields.

Within the region, there are marked variations in the performance of the states. The per cent area under HYV is highest in Tripura and Manipur being 46 per cent and 42 per cent respectively. The use of fertiliser per unit cropped area has decreased in Arunachal Pradesh, Meghalaya, Nagaland, Sikkim and Tripura whereas it has increased in Assam, Manipur and Mizoram. Overall, productivity of foodgrains has increased substantially in Manipur from 1763 kg/ha to 2411 kg/ha (37 per cent), Mizoram from 1317 kg/ha to 1600 kg/ha (26 per cent) and Meghalaya from 1161 kg/ha to 1349 kg/ha (16 per cent) during last couple of years.

The growth scenario of pulse and oilseed production is a mixed one. The region produces 111.7 thousand tonnes of pulses and 239.1 thousand tonnes of oilseeds from 176.1 thousand ha and 424 thousand ha with an average yield of 634 kg/ha and 563 kg/ha respectively during 1998-99. Non-availability of adequate quantity of quality seeds of improved varieties and lack of use of inputs like fertiliser, weed control and disease and pests protection are major constraints in pulses and oilseeds production in the region. The total area under pulses in seven hill states in only 46,300 ha and that of oilseeds is only 93,800 ha. However, the area under pulses and oilseeds showed a substantial increase from 11,000 ha in 1993-94 to 16,400 ha in 1997-98 and 21,000 ha in 1993-94 to 28,500 ha in 1997-1998 respectively.

Horticulture

Endowed with diverse climate, the NE region offers ideal agro-climatic conditions for cultivation of a wide variety of tropical and temperate fruits in its hills and valleys. The dominant horticultural crops of the region include pineapple, citrus, banana and areca nut. The region has a total area of 35.7 thousand ha. under pineapple and 35.6 thousand ha under citrus. Banana is grown in 54.9 thousand ha with a production of 707.6 thousand tonnes. The area under papaya, coconut, areca nut, temperate fruits, apple and miscellaneous crops are 9.14, 21.06, 81.32, 2.31 and 9.50 thousand ha with a production of 129.48, 128.61,63.17,58.08,6.73 and 51.64 thousand tonnes respectively.

Orange is a unique crop of this region, however, average productivity is poor except in Assam and Meghalaya. High productivity from well-maintained orchards and trees bearing 2500-3000 fruits have been reported. Orange is a highly remunerative crop in Jampui hills of Tripura where organised efforts made by the state department has paid rich dividends. Productivity of pineapple in Assam compares well with that of all India average but it is much lower than the neighbouring West Bengal where production technologies *viz.* flower inducements, balanced nutrition and high density plantations have been adopted in North Bengal situations. These technologies need to be tried at large scale in NEH region to improve the productivity. Productivity of banana in Assam and Meghalaya (12-13 t/ha) is, however, at par with that of the neighbouring states like, West Bengal and Bihar.

Ginger and turmeric are well-established cash crops of Meghalaya. Ginger is grown in Mizoram too with high productivity. Large-scale production of planting material and marketing support has encouraged ginger cultivation. Large cardamom is a major plantation crop in Sikkim.

Potato is the most important vegetable which gives quite high average yield (17.1 t/ha) in Tripura. This state has achieved the distinction of producing true potato seed at commercial scale. Meghalaya and some other states also produce potato but productivities are low. Among vegetables, beans, radish and cauliflower are grown extensively in all the states.

Livestock and Poultry

Livestock in this region comprises cattle, buffalo, sheep, goat and pig. Poultry is yet to develop on commercial lines. Many tribal populations have no habit of rearing cattle for milk production. Therefore, animals are reared largely for meat. Feed and fodder both in quantity and quality are major constraints for any improvement in livestock and poultry production.

Fishery

Fish resources of the region reside in the tributaries of the Brahmaputra and Barak, the rivulets and streams and reservoirs, lakes, ponds and paddy fields. Opportunities for both warm and cold water fisheries are available in the region. These resources can be broadly classified into two categories namely riverine comprising rivers and streams and inland comprising reservoirs, lakes, tanks and ponds.

The region has total riverine fisheries resources of 14,648 km and inland fisheries resources of 82962 ha. The bulk of the fishery resources, however, consists of tanks ponds and lakes. The region produces annually 48227 metric tonnes of fish. The extent of exploitation of aquatic resources of the region remains quite low. Of the over 14,000 km of the riverine aquatic source very little is presently harnessed. One of the reasons for this is the sizeable length of the rivers and streams lying nearly fallow. Lack of regulatory measures for conservation of fish stock and industrial pollution are an added constraint in fisheries development.

Farm Power

Availability of farm power is a pre-requisite for timely field operations that are essential for optimum productivity in agriculture.

In farming systems dependent upon animate power sources (men and animals) about 1 kW/ha is considered sufficient. However, our farms are semi-mechanised where a farm power of about 2 kW/ha is considered essential. In NE region, the farm power availability is estimated at 0.67 kW/ha with a maximum of 1.44 kW/ha in Meghalaya and a minimum of 0.17 kW/ha in Arunachal Pradesh against the average farm power availability of 1.15 kW/ha in the country.

Studies have revealed that under rainfed farming, seed-bed preparation and sowing, harvesting and threshing, inter-culture, and weeding, and transport are the most energy intensive operations. Means will have to be found to bridge the gap between availability and requirement of power for farming operations in the NE.

NE region experiences hot and humid weather, which poses serious problems for post harvest handling. In humid weather, it becomes difficult to dry the agricultural produce to safe moisture levels and this results in huge post-harvest losses.

Agricultural Research

The research needs of the agricultural sector of the region are largely met by the ICAR Research Complex for NEH region, Barapani and its centres in Arunachal Pradesh (Basar), Manipur (Imphal), Mizoram (Kolasib), Nagaland (Jharnapani), Sikkim (Gangtok) and Tripura (Lembucherra). Two agricultural universities *viz.*, (i) Assam Agricultural University at Jorhat and (ii) Central Agricultural University at Imphal also contribute to this requirement. The former has eight regional agricultural research stations located in eight agroclimatic parts of Assam whereas the CAU has six colleges spread over all the six states and headquarters in Manipur.

The other ICAR establishments that cater to the research needs of the region are: NBPGR regional station, Shilliong; CPRI Substation, Shillong; NBSS&LUP regional station at Jorhat; Ramie research station, Sarbhog; CPCRI substation, Kahikuchi and CICFRI, Guwahati. In addition, National Research Centre on mithun, yak and orchid are located at Medziphema in Nagaland, Dirang in Arunachal Pradesh and Pakyang in Sikkim respectively.

Agricultural Education

The literacy rate of the region varies between 49.1 per cent in

Meghalaya to 82.27 per cent in Mizoram. The region is ahead of many other states of the country in literacy. However, it has lagged behind for of agricultural education mainly due to lack of infrastructure. The agricultural education of the NEH region is the responsibility of the Central Agricultural University, Imphal and Assam Agricultural University, Jorhat. Assam Agricultural University, Jorhat has six colleges: two for agriculture, two for veterinary science, one each for fishery and forestry. These colleges have played important role in imparting education in the field of agriculture and animal husbandry.

The Central Agricultural University, Imphal is the first such university of the country which was established by the ICAR at (CAU) Imphal, Manipur in 1993. The CAU envisaged establishment of seven major campuses in six states of NEH region namely Arunachal Pradesh, Manipur, Meghalaya, Mizoram, Sikkim and Tripura with provisions for establishment of other smaller institutions, especially krishi vigyan kendras and regional research units, at different places in these states. Agricultural education in the region is imparted through College of Agriculture at Imphal, Manipur; Colleges of Agriculture, Medziphema, Nagaland; College of Veterinary Science at Aizawl, Mizoram. The other colleges under the Central Agriculture University are yet to become functional.

Agricultural Extension

Agricultural extension is primarily the responsibility of the departments of the state governments of the region. The ICAR research complex conducts some extension activities through the extension division, krishi vigyan kendras and trainers' training centre located at Umiam. The extension directorates of two agricultural universities in conjunction with 12 krishi vigyan kendras existing in the region help in transfer of technology.

Policy Recommendations

Major theme-wise policy recommendations are as follows:

1. *Resource-base, Potential, Problems and Strategies for the Future*

a. Soil and Water Resources

- Soil and water resources of the region need to be properly inventorised, characterised and evaluated for appropriate uses.

- Implementation of appropriate soil conservation practices should be mandatory for any cultivation programme in the hills, where soil and water losses are beyond the permissible limits.
- Watershed approach of soil and water conservation and development of land-based production programmes in participatory mode need to be propagated.
- The surface run off should be captured and used by following appropriate irrigation methods.

b. Plant Resources

- For plant genetic resources, a collaborative approach should be followed under the Indian National Plant Genetic Resources System (IN-PGRS). Policy planning for research and development of agriculture in NER should include horticulture and forestry to simultaneously cover agrobiodiversity.
- Diversity mapping should be done along with inventorisation at micro-level for effective management, conservation and use of crop species and their relatives.
- Agroecological niches should be clearly identified and records should be maintained on prevalent crops and cropping system, the ethnic group and economic background of indigenous communities. The information collected should form the basis for designing 'on-farm' conservation.
- Regional networking should be planned/developed to enable safeguarding regional interests for genetic diversity.
- Specially designed genetic reserves, gene sanctuaries and/or genetic garden should be earmarked for wild species of food value and other economic importance. Strategies, approaches and technologies including biotechnology should be developed/standardised for gradually bringing these wild economic species under cultivation. Linkages should be strengthened with NBPGR, other government departments, universities, NGOs and communities to get desired results.
- Greater emphasis needs to be given on (a) *in situ* conservation of endangered species, *ex situ* conservation of base collections in field gene banks and *in vitro* storage and cryopreservation of important germplasm, where tissue culture protocols are

available. Germplasm screening for processing and diversified use should be encouraged.

c. Livestock Resources

- Characterisation of indigenous livestock and poultry resources and their conservation need to be emphasised as there are many valuable genes for adaptability and disease resistance available in these animals.
- The swamp buffaloes of the region and goats need to be improved either through selection or crossbreeding. Pigs of the region are required to be improved by using exotic breeds. Similarly, poultry, including ducks and other animals like rabbit, mithun and yak need to be exploited to boost farm income.
- Research on standardising improved technologies of artificial insemination, intra-uterine insemination, embryo transfer for improved reproduction of livestock is also the need of the hour.
- For animal health improvement, research on disease control/ health care, vaccine production, diagnostics, disease surveillance and monitoring are extremely important.
- Feed and fodder resource enhancement should receive urgent attention.
- Research is called for: (i) on cultivation of indigenous grasses, fodder trees for feed and fodder (ii) nutritional requirements of various categories of animals, including cross-breds and (iii) conservation of fodder for the lean period. These steps are needed for improvement in animal nutrition.
- Management system research and animal product processing and marketing research are other areas, which need attention.

d. Fisheries Resources

- Areas requiring emphasis are: standardisation of technologies suitable to varied agrarian practices and investment potentials of the region. This requires establishment of an elaborate research network across the North-Eastern states. This is also true of the species-based systems, with a high biodiversity of fish species present in this part of the country.

- Development of suitable and specific aquaculture packages for acidic soil condition, upland waters and lowland waters.
- Research programmes aiming at intensification and diversification of freshwater aquaculture in tune with the geo-morphological feature of the region *viz*., upland aquaculture and lowland aquaculture.
- Training to farmers and entrepreneurs at different levels along with better coordination between extension functionaries.
- Extension activities in terms of demonstrations, publications, farmers' meet, village adoption, production of audio and video materials for promoting scientific aquaculture.
- Extensive bio-monitoring of all rivers in time and space to assess threat perspectives in relation to biodiversity conservation.
- Regular monitoring of reproductive biology of economically important fish species so as to suggest remedial measures for depleting population.
- Conservation and rational exploitation of *beel* resources which are the back-bone of fish and fisheries in the region.
- Introduction of fish husbandry in *beel* ecosystem. It would be prudent to exploit the nutrient enriched marginal areas of the *beels* profitability with new technologies like pen culture/cage culture successfully demonstrated by CICFRI, Barrackpore. The pen-culture technology developed by CICFRI for 'fin' fish can open new vistas in the fishery activities of the region as it is less capital and labour intensive and has the added advantage of producing reasonably high fish/prawn biomass without disturbing the main fishery of the lake proper.

2. *Traditional System, Their Problems, Potentials and Strategies for Improvement*

a. Shifting Cultivation

Following course is recommended:

- Detailed survey to identify specific problems of various agro-climatic situations and ethnic groups.
- Selection of crops and varieties to suit different farming systems based on the needs of farmers. Identified crops should

have assured market and should fulfil specific needs such as fodder, etc.

- Development of different types of farming systems suited to irrigated and rainfed situation coupled with efficient management of soil and water.
- Proper use of the biomass and recycling of wastes in the various farming systems.
- Studies on development of suitable irrigation system for hill-slopes with scientific water management.
- Participatory research and extension involving farmers to bring about community empowerment by considering a number of socioeconomic issues in order to achieve success.

b. Agroforestry and Horticulture

- A greater intervention through agroforestry and horticulture is necessary to remove the unbalanced and unsustainable short cycle of *jhuming* and limited opportunity to expand arable lands and their mechanisation on the slopes.
- Traditional agroforestry practices existing in NE region need in-depth understanding for agronomic improvement.
- There is need to encourage product diversification in a unit of land through agroforestry to increase land's capacity to produce and bring the land to its full potential by linking to assured marketing channels by identifying demands and outlets or outputs.
- Collaboration on rubber-based farming system will help identification of the most suitable and economically viable farming system models by taking care of adapting low input agricultural practices, which can be followed by a smallholder. Rubber cultivation will help in soil conservation, enrich organic matters, improve soil properties and microclimate, besides being eco-friendly.
- Vast expansion of area under rubber is possible where land is denuded and is not suitable for cultivation of other crops. From the ecological restoration, socioeconomical development and rehabilitation of shifting cultivators' point of view, rubber can play a crucial role to alleviate the problems in the North-East to a large extent.

- Research on mushroom cultivation and processing of the produce need to be encouraged.

c. **Agricultural Research and Education**

- To raise productivity of existing crops, intensive research efforts are required to (i) breed/identify suitable varieties of existing crops, (ii) employ newer technologies to speed up crop improvement, (iii) evolve suitable agro-techniques (iv) develop an integrated pest and nutrient management system and (v) develop appropriate soil and water management techniques. Similarly, for improving productivity of animals besides breed upgradation, research efforts should be directed towards animal nutrition and health care. Since, agriculture is rarely practised in isolation, development of appropriate mixed farming models, based on landholding size, will be helpful in improving the productivity in the livestock as well as agriculture sector.
- Grassland improvement scheme must be introduced and cost of meat and milk production should be reduced for the enterprise to become economical.
- To bring cash in the hands of the farmers and to raise their economic status three pronged approach is required which includes: (a) identification of suitable production technology, (b) development of technologies suitable for adoption at farm level to reduce the post-harvest losses and to increase the shelf life of produce to enable transport of perishable commodities to distant consuming centres and (c) marketing support.
- There is an urgent need to document the traditional knowledge, collect different plant materials and ascertain their medicinal properties. Useful plants can be popularised for large-scale cultivation, which will help in providing employment and improving income of large number of tribal and other rural families.
- Short or medium range programme on horticulture-based cropping system; water management including micro-irrigation and fertigation, greenhouse cultivation of vegetables and flowers have already received research attention. Further refinement of the technologies and their transfer will bring

perceptible change/improvement in production of different crop commodities.

- Commercialisation of micropropagation protocols for banana, orchid, cardamom, and ginger. Refinement of TPS and micro-tuber production technology in potato. Improvement in STG techniques and cross-protection in citrus for virus elimination and control; hybrid seed production of temperate vegetables in high altitude areas.
- In order to reduce post-harvest losses at production centres low cost eco-friendly on-farm storage structures can play a crucial role. Significant advancement has been made in this direction and some small and medium sized cool chambers on the principles of evaporative cooling have been devised. Further refinement of the technology will go a long way. Also, standardisation of packing line operations and proper packaging of different commodities are of urgent need.
- Pesticide residue management and newer product development will add values to the produce.
- In crop improvement programmes, new genes need to be introduced for increasing total biomass, which should then be followed by more efficient partitioning. This will require crossing the cultivated races with wild types, many of which are not sexually crossable. This can, however, be accomplished through embryo rescue/protoplast fusion and other biotechnological approaches.
- Among abiotic stresses, cold is a limiting factor in realising full potential in most of the crops. Biotechnology can be employed for incorporating cold-tolerance through introduction of genes for relevant enzymes such as *glycerol-3 phosphate acetyltransferase* in to high yielding varieties. Research on insect-pest management need to deploy resistant genes rather than depending on chemical control in this high rainfall region. Overall, a new regional initiative is needed that focuses on the varietal improvement of cereals (rice and maize), pulses and oilseeds.
- Establishment of a College of Agricultural Engineering & Technology at Sikkim under Central Agricultural University, through teaching, research and extension function can become

hub of technology upgradation in the region for both agricultural production as well as post production management. Post-harvest engineering and technology can help farmers earn more and increase employment through value addition to their produce and by-products.

- Improvement of tea, ornamental plants, medicinal plants, forest trees through application of biotechnology should receive immediate attention.
- Judicious use of biotechnology tool in crop and livestock improvement programme/research is important as the conventional breeding methodology sometimes provides quicker results at less cost in many cases.
- Human resource to serve immediate requirement of academic and developmental activities in the region is an urgent need. It is also necessary to produce middle level technicians/ professionals to assist developmental activities.
- In the context of growing public concern for environment, globalisation, household food security and eco-regional imbalances, new directions are required in planning and transfer of technology by designing more effective linkage between scientists and farmers for technology information to meet future challenges of growing demand of food.

30

Redefining Agricultural Education and Extension System in Changed Scenario

Introduction

The 20th century witnessed radical developments in science and technology. These developments had an impact on the social and economic changes that took place in the world. Such changes gradually influence the day-to-day life of people at the grassroots levels. Agriculture related science and technology had a major impact. Agricultural education and extension have been geared to harness the modern science and technology for higher productivity and production. This substantially helped to reduce the food scarcity in India. But sustainable food production is still the primary pursuit.

Economic liberalisation and WTO regime pose a big challenge to Indian agriculture. To cope up with changed global scenario, agricultural education and extension would have to be redefined. This is a major concern for academics and policy makers. An approach to our own problems will have to be worked out. This, in turn, warrants reforms in agricultural educational systems.

Challenges and Issues

The agricultural education in India is faced with one of the biggest challenges. It has to identify its role in equipping the human resources for enhanced agricultural productivity and sustainable use of natural resources. Agricultural colleges and universities were initially assigned to disseminating scientific knowledge and skills to the farming community and to train them to use such skills for better output. As a backup for such a mission, agricultural research was encouraged to adapt the scientific knowledge to suit to the realities

---

Policy Paper 31, June 2005.

of rural societies. However, these initiatives could not keep pace with the fast changing scientific and technical improvements and gradually failed in their objective to cultivate the most modern skills and attitudes to both agricultural students and farmers. Developing a mechanism to update the curricula of agricultural education and extension, at least on a five-year regular basis, is crucial. This is very relevant to teaching, research and extension functions of the university as they form the inter-related, theoretical and practical basis of modern agricultural education in India. Moreover, the link between these three major strands is weak, particularly because extension is still not well recognised as the other two academic functions of the agricultural university.

The knowledge, generated by research, often gets obsolete by the time it is disseminated through education/teaching and finally trickles down to farmers and practitioners through extension due to big lapse of time. The universities should consider this problem seriously and should reform their academic activities to strengthen the links between research, teaching and extension.

Economic liberalisation and WTO regime demand manifold re-orientations in our agricultural education and extension system. Market-oriented agricultural education and extension along with changes in agricultural marketing policy for the national and international markets is the need of the hour. A cadre of social scientists with clear insight into agricultural related disciplines, such as agricultural economics, agri-business, marketing management, rural sociology, agricultural anthropology, agricultural ethics and politics, has to be built up to provide the backup for solving the problem of globalisation in agriculture. This will imply total re-organisation of agricultural education with emphasis on other agricultural related social sciences and effective agricultural policy research. Such a re-organisation will empower our agri-graduates and scientists to face up the challenges posed by the demand of the national and international markets.

Any individual initiative needs certain motivation for sustenance. Pursuing agriculture as a profession/job for a considerable time is also influenced by motivational aspects such as monetary gains, employment, better quality of life and social acceptance. The number of agricultural producers is declining. The population that makes a living directly from agriculture continues to fall in developing countries. People, with rural background and engaged in traditional agriculture, are reluctant to let their children and younger generation to pursue agriculture as a full time profession. Those, with rural background and genuine interest in agriculture, do not have the basic education to go in for higher education in agriculture. This prompted the rural youth to go in for non-agricultural courses and moving in of urban based agriculture graduates without any practical knowledge of agriculture. Agricultural education has to be made more practical and rural realities oriented while agricultural extensions courses should be tuned to serve the needs of the rural community through dynamic interaction with rural societies. Such extension activities would enrich the academic institutions and universities.

Developing countries lag behind in introducing educational courses on environmental and sustainable agriculture. In India, too, there is a need to provide such a curricula re-orientation to academic institutions to create an environment sensitive faculty and to help bring about attitudinal changes among rural communities.

To examine these issues and make recommendations for needed reforms in agricultural education and extension system, a workshop was organised in Delhi by the National Academy of Agricultural Sciences (NAAS).[1] These deliberations culminated in a set of recommendations after discussing the following issues in three sessions:

- Challenges facing agricultural education and their implications.
- New strategies and institutional reforms to meet the challenges.
- Quality, relevance, reach and preparedness.

Recommendations

1. Challenges Facing Agricultural Education and their Implications

- There has been a dilution in the quality of agricultural education. Agricultural universities, with inadequate infrastructure, financial support and autonomy have

1. A workshop was organised in Delhi by the NAAS under the Convernership of Dr. Panjab Singh, Vice-Chancellor, BHU and formerly Director, School of Agriculture, IGNOU, on 10th December 2004. Organised in a panel discussion mode, it brought together 30 educationists with diversified experiences. A concept paper by the Convener was sent to the participants in advance to set the agenda for discussion. Each of the three sessions had a key speaker, followed by presentations from the panellists and open house discussion.

mushroomed. To ensure academic excellence, agricultural universities should have complete autonomy coupled with accountability.

- Dilution in the quality has been, mainly, due to an imbalance in the academic staff structure of the universities. The recruitment policy, and also the policy of freezing new recruitment, needs to be reviewed as presently about 40 per cent of posts are lying vacant in these institutions.
- Centralised planning of agricultural education system and curricula is not addressing the local needs to the extent required. There is tremendous disparity in regional assessment, as quite a number of regions of our country have not been benefited by the developments. There is a need for establishing more central agricultural universities like the IITs.
- Rural students need to be encouraged to study in agricultural universities. Steps may be taken to promote such admissions as urban-based agricultural graduates are not so comfortable in rural environment.
- Specialised courses in educational technology should be developed to upgrade the teaching skills. A teachers training institute in agriculture at the national level would help.
- After research, teaching and extension, training should be the fourth function of the agricultural universities. Directorate of HRD and Entrepreneurship could be considered to steer this function.
- ICAR and VCI systems of examination and evaluation in the same agricultural university create disparity in assessing the quality of manpower generated. ICAR and the agricultural education system need an independent regulatory body like the UGC to streamline and give direction to the system.
- The agricultural education system needs to be redefined so as to equip the new graduates with subject competency, self-motivation, positive attitude, agri-business skills, knowledge of computer and information technology, and communication skills in both English and regional languages.
- A manpower planning document is needed at the central and state level for proper agricultural education management including employment.

- A clear proportion for the financial allocation for agricultural education, research and extension education needs to be formulated. At present, only 10 per cent of ICAR budget is spent on education.
- More time and energy have to be earmarked by teachers for extension education programmes and activities to be fully aware of the farmers' problems. Teachers should also help to disseminate the traditional knowhow and technologies of farmers.
- SAUs should have unified administration and complementarity of departments and multidisciplinary teamwork in the development of programmes of education, research and extension. The Model Act of the ICAR should reflect in the functioning of the state agricultural universities (SAUs) and deemed universities.
- Systems could be put in place at SAUs for quick communication of new knowledge to students in classrooms, extension workers and farmers. SAUs could have flexible course-credit system buttressed by continuous internal evaluation.
- SAUs could provide regular specialised training to rural youth, particularly school dropouts, and adults who are not eligible for enrolling for formal agricultural courses.
- An SAU could have corporate board of management with adequate powers under the University Act and have organisational and operational autonomy.
- The successful models in agriculture demonstrated by the private sector, business groups and industries involved in agriculture should be studied and adopted. The private sector should be given a bigger role in agriculture extension policies/ activities being planned/implemented by the state and central governments.
- In the liberalised educational policy scenario at the national, regional and global levels, it is essential to have a better coordination between institutions and universities to implement restructured and relevant courses. Private-public partnership, to strengthen the present system of education, research and extension in India, is need of the hour. Roping

in talented scientists of the private agri-research sector as guest teachers in agricultural universities could be a first step towards this end.

- The curriculum should be regularly updated to incorporate the changes in agricultural scenario. Specific curricula could be developed for technological and skill development of women in market driven technological enterprises and sustainable management of natural resources.
- The Planning Commission found in 2000 that the contribution of the primary sector (mostly driven by agriculture) was on the decline while the tertiary sector (driven by industry) was having a slower growth. The agricultural education system needs to be reoriented to cater to the needs of the emerging sectors and to ensure that excess manpower is not generated in slow growing sectors.
- Agricultural education should lay increased emphasis in future on topics like, alternate farming, biofertilisers; pressurised irrigation; integrated water management; integrated nutrient management; integrated pest, disease and weed management; resource optimisation; post harvest technology and value addition; and marketing.

2. *New Strategies and Institutional Reforms to Meet the Challenges*

- The basic objective in redefining agricultural education in the changing scenario is to promote development and delivery of educational programmes that would enhance the employment potential and build up an easy, accessible and cost effective knowledge intensive information system. The guiding principles, approach and strategies are as follows:

(a) Guiding Principles

- Initiatives should have long term impact.
- The focus should be on ameliorating quality, relevance and reach.
- Linkage with the market requirements and demand of stakeholders.

(b) Approach

- Performance and output oriented.
- Optimal utilisation of existing infrastructure and resource.

(c) Strategies

(i) Education

- Shaping up academic environment through total quality management at all levels.
- Entrepreneurship development and self-employment orientation in agricultural education should receive high priority through infusion of vocational courses (certificate/ diploma levels).
- Periodic review and revision of curriculum consistent with national and global scenario, market trends, self-employment avenues and industries requirements.
- Devising of qualification framework.
- Make agricultural education an independent subject at school level and integrate it with the higher education system.
- Emphasis on distance education in agriculture with extensive use of innovations of IT, digitalisation of course content, virtual universities, e-learning and video conferencing.
- ICAR needs to play a more proactive role in initiating, implementing, reviewing and monitoring reforms in education system.
- Distinguish agricultural education from training for public service, thereby, fostering training aimed at meeting the needs of the private sector and preparing students for entrepreneurship.
- The educational system today suffers from lack of accountability of the teachers and institutions. The quality of individual institutions be critically assessed, and the accreditation system should be made more stringent to enforce accountability.
- Facilitate, promote and support agreements, cooperation and coordination at inter-institutional and international levels.

- Promote integration of population, environment and sustainable development themes into agricultural education and extension programme. Include specialised training in degree programmes.
- Develop well-trained human resource for doing research in advanced areas of science to become internationally competitive.

(ii) Research

- Need a paradigm shift from single discipline orientation to multi-disciplinary approach.
- Encourage privatisation, planning, monitoring, evaluation and assessment as core component of research management process.

(iii) Extension

- Harness benefits of new scientific advances such as bio-technology, cloning, remote sensing, modelling, information and IT for farmers and rural communities. Internet aided extension networks have to be materialised.
- KVK's have to be strengthened as nerve centre for dissemination of technical knowhow.
- Extension programmes to be based on rural participatory approach.
- Formation of commodity specific extension kiosk on indigenous food, horticultural crops, livestock, poultry, rabbitry, beekeeping, mushroom, medicinal and aromatic plants, sericulture, tea and value addition enterprises.
- Distance education for farmers and small agri-entrepreneurs need to be explored and strengthened through television, radio, interactive audio and video systems, besides print and programmed learning materials which would be the distinguishing feature of extension teaching and learning process.

(d) Institutional Reforms

- Effective partnership linkages with advanced centres of education in the country and abroad.

- Greater autonomy to academic institutions, SAUs and statutory power to ICAR.

3. *Quality, Relevance, Reach and Our Preparedness*

- Though quality is a very important issue in agricultural education and extension, it was given least importance at present. A total quality management (TQM) system should be evolved to ensure quality education.
- The concept of total quality management must refer to conforming to the expectation of the stakeholders. The paradigm shift in agricultural education should include teaching students not just what is currently known but also to keep them abreast of the new knowledge essential for meeting the challenges of new economic environment and all that would contribute to socioeconomic development and public good. The education should lead to development of analytical skills, exposure to international marketing, total quality standards and comparative advantages. It should also facilitate confidence in ensuring sustainability with high productivity and quality of the produce while meeting international standards. Post-harvest technology and value-addition are also key to maximum benefits from exports.
- Total quality management in agricultural education needs to focus on improving quality of courses, instructional processes, human centred development, students support services, building strong work culture and electronic communication system. Ultimately, the quality and skills being imparted must inculcate expectations of different stakeholders.
- Universities could provide non-formal education and vocational course including technological empowerment of women and rural youth through distance education.
- The private colleges that provide high quality agricultural education should be given accreditation with the central and state government universities. These colleges/universities should also be given assistance to sustain their quality and standard.
- Training is an important component. Training institutes of high quality should be established on urgent basis to provide better training to both in-service and outside personnel in

agriculture. Distance learning methods are recommended to meet the high demand on training.

- Exchanging academics between ICAR and SAUs should be encouraged to avoid inbreeding and encourage quality of agricultural education.
- Agriculture is increasingly demanding interface with industry, and hence short-term and long-term self-supporting courses on agri-business and marketing should be developed by the central and state universities. The experience of IGNOU, and State open universities should be taken into account in this regard.
- Agricultural students and practitioners are aware of the quality aspects, but they don't put them in practice. This attitude must change. Research, training and extension agriculture should be in continuum in for achieving quality ideals. Young people should be encouraged to the system and contribute fresh ideas.
- Practical aspects that need prominence in the curriculum are absent in the present system. SAUs should run short-term courses for plus-two level students to enable them in engaging themselves in income generating activities. Suitable tie-ups with industry should also be encouraged to enhance the relevance in agricultural education.
- The recommended 10 per cent raise in funding every year is not implemented by the government and universities. This adversely affected upgrading the facilities/quality of agricultural education. Centres of Excellence should be identified and the responsibility of revamping agricultural programmes should be given to these institutions.

4. Revamping Agricultural Education

- Re-orienting and modifying the course curriculum to suit the demands of the job markets and to bridge the mismatch between manpower demand and availability in different areas should be done.
- In curriculum, post harvest technology, agro-processing value-addition, marketing and entrepreneurship develop-

ment as well as management intricacies should be given emphasis.

- Agricultural education should be made innovative to absorb futuristic trends and skill-orientation rather than based on note-memorisation of new knowledge.
- Agricultural education system should produce professionals and for that, the system has to balance between classroom teaching and practical sessions and experiential learning based on actual work-based experience.
- Establishing and creating facilities/infrastructures for in-plant experiential learning in SAUs and agricultural colleges.

5. *Revamping Agriculture Extension Education*

- Transfer of technology mode to technology application mode.
- Group approach and farmers participatory approach to agricultural extension.
- Involvement of community-based organisations.
- Associating stakeholders in extension.
- Formation and training of self-help groups and extension through them.
- Rural location-specific knowledge centre.
- Need-based, problem-solving, skill-based vocational training for self-employment for farmers, farm women and rural youth.
- Shift to commercial farming and agri-business approach in extension.
- Linkages between extension education, continuing education, non-formal education, distance education and vocational education, so that each supports and strengthens the efforts of the other and *vice versa*.

31

Stakeholders' Perceptions on Employment Oriented Agricultural Education

Introduction

India is principally an agricultural country. Seventy per cent of the Indian population lives in villages. Fifty-nine per cent of the total workforce depends on agriculture for its daily bread. Almost 25 per cent of the gross GDP is contributed by agriculture. So striking is the dependence of growth in Indian economy that it follows the ups and downs in growth rates of agriculture. One out of every eight rupees earned through exports owes its origin to agriculture. A steep surge in production since late 1960s has made country self-reliant in foodgrain production with buffer stocks full to the brim. Since Independence, 0.66 per cent growth/year in gross agricultural area combined with 2.67 per cent/year increase in foodgrain production led to a net increase in per capita foodgrain availability from 395 g/day to 475 g/day. An outstanding achievement, as it happened despite an unabated population build up at a rate of 2.15 per cent/year or 13 million more mouths to be fed every year. Not only in foodgrain production, the country has made revolutionary progress in milk, fish, eggs, fruits and vegetables production also. It is the self-sufficiency in food that has raised the international status of India. It is this exalted exuberance that has infused decisive confidence to negotiate with an incisive voice on many sticky clauses of global treaties and conventions.

Technically qualified manpower provided the base for the radical metamorphosis of agriculture—what William Doug of USA called Green Revolution. A study by IFPRI on linkages between government spending, growth in productivity and poverty in rural India showed that investment in education was among the major contributors for both poverty reduction and productivity enhancement. Right since

Policy Paper 25, May 2004.

independence, the country laid maximum emphasis on the development of its human resource through education. In support of that slowly and steadily a comprehensive system of higher agricultural education has been evolved and developed. The system consists of state agricultural universities (SAUs) and their constituent/affiliated colleges, deemed to be universities (DUs), central universities (CUs) having agricultural faculties, Central Agricultural University (CAU) and a number of public and private funded agricultural colleges. Compared to a total intake capacity of about 1,500 students per year at the time of independence; currently, some 21,000 agricultural graduates and postgraduates are produced every year. Despite this whopping rise, hardly there is an attempt that links trends of employment needs of various sectors of economy (public, private or selfentrepreneurship) and clients (students, farmers, agri-industry) against the kind and number of manpower produced by the system of higher agricultural education. With that persisting neglect, there is no visible concern on making projections on manpower demand and supply at the national level. Also, the system lacks a country level computerised manpower information system with the facility of updating, retrieval and dissemination of information to serve the policy objectives corresponding to development and utilisation of graduates and postgraduates that education system churns out every year. As it exists, the emphasis is achieving the set targets of admission capacity without matching it with the changing market demand for employment. Already there is an air of all round concern

Figure 31.1

Agricultural Universities and Admission Capacity

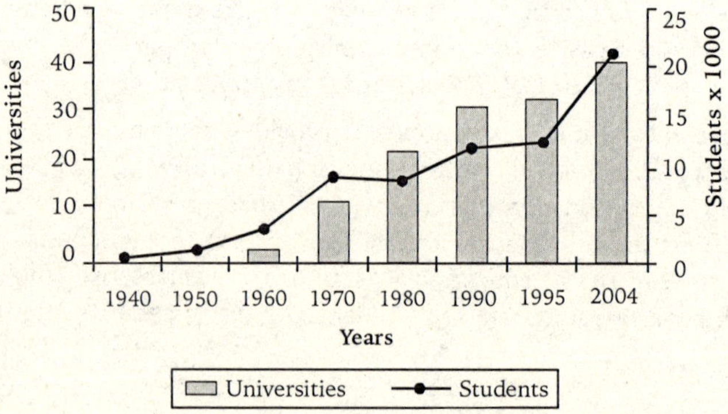

for rising unemployment among agricultural graduates and postgraduates.

In view of the above concerns, NAAS organised a brainstorming session on, Employment Oriented Agricultural Education.[1]

Salient recommendations emerging from the deliberations are given below.

Recommendations

1. General

- Indian higher agricultural education setup consists of some 260 colleges imparting education in diverse disciplines of agriculture and allied subjects, animal and veterinary sciences, fisheries, agricultural engineering and home science. These colleges are either constituents of 34 SAUs, 5 DUs, 3 CAUs, 1 CAU or are affiliated to 17 general universities. Of the 21,000 graduates and postgraduates produced annually, nearly one-fourth is girls; one half comes from rural background.
- Until early 1990s, number of graduates and postgraduates produced by the agricultural education system was not large when seen from the employability point of view. With maximum absorption of graduates and postgraduates in government jobs, problem of unemployment was hardly an issue. Public sector was absorbing more than 50 per cent of the total stock of graduates and postgraduates. This situation has changed dramatically with the spin-off effects of liberalisation and opening up of the economy becoming clearer and clearer. Currently, unemployment among graduates and postgraduates is so rampant that it can hardly go unnoticed or

---

1. NAAS organised a brainstorming session on "Employment Oriented Agricultural Education" on 9 February 2004, under the Convenership of Dr. J.C. Katyal, DDG (Education), ICAR. The purpose was to examine the status of national agricultural education system and the kind and number of manpower it generates. In order to reach this goal, discussions also focused on the employability of graduates and postgraduates *vis-à-vis* the changing market demand. What should be the possible reforms in the agricultural higher education, that would harmonise the increase in number of graduates and postgraduates with the generation of employment, was another key objective of these deliberations. Academicians, education management experts, industrialists, NGOs and students participated in the session.

The programme started with the opening remarks by Dr. V.L. Chopra, President NAAS followed by Dr. J.C. Katyal's presentation giving overview and key points for discussion. The entire group brainstormed issues flagged by Dr. Katyal for discussion.

remain unattended. Many reasons are behind this development.

- With food situation becoming comfortable, agricultural S&T is no more a priority sector of government funding. This has adversely affected the availability of jobs in public sector. Opening up of economy lays far greater emphasis on the development of private enterprise; reducing thereby the over dependence on public sector for employability. Accordingly, number of government jobs in public sector has been on the decline. Also, growth in processing industry and services sans updating of conventional course curricula and methods of its delivery to support these ventures has dichotomised the kind of marketable and produced manpower.
- Since, consequences of ongoing shifts in world economic order and associated trends in surge of sectors and activities are seen to grow in future, reengineering of agricultural education to suit their requirements for graduates and postgraduates is a must. Significant adjustments in syllabi and time for practical and practice sessions also become necessary, in that, products of our agricultural education setup are professionals who do not hunt around for jobs but can create jobs for themselves. On the one hand, self-employable individuals are answer to problem of joblessness, on the other, they are projected to play a crucial role in extending applicability of S&T for eradicating poverty by making farming cost effective and its produce more diversified, market appropriate and profitable.
- In order to introduce more time for practical classes and practice sessions, there is need for liberal investments to create appropriate physical plant and build relevant competence in teaching faculty. Against this perceived exigency, public funding for agricultural research and education, in reality, has tightened as the food situation has eased. Falling investment was viewed with concern, since, it was neither conducive to fight scourge of poverty nor for infusing greater application of R&D for sustaining growth in agricultural productivity. According to an IFPRI report, investments in agricultural R&D and education, next only to roads, are the most influential investments made for poverty

alleviation and productivity enhancement. Current share of agricultural education within the total budgetary provisions does not seem all that impressive.

- With 90 per cent of a university's budget consumed for payment of staff salary and committed establishment charges and the remaining appropriated by research, hardly much is left for effecting improvements in content and delivery of education. In fact, distinct provision for education improvement is unheard of in the existing broad heads of budgeting in the SAU system. Education is treated as a routine and casual activity involving a professor lecturing students herded in a classroom. Use of modern information and communication technologies for enhancing student-professor interactivity in teaching and learning during practical and practice sessions is either nil or very low. In fact, teacher-taught coalition is a model adopted in other parts of the world to effect intended improvements in agriculture through products of education.
- In order to unshackle education of the existing passive mode of teaching, investments on upgradation of course curricula by refreshing the content and refashioning the mode of delivery were considered necessary. The importance of overarching knowledge needs in emerging areas of frontier sciences, economic activity and environmental concerns and competence of teaching faculty needs to be appreciated. As it exists, teaching faculties neither possess adequate knowledge to launch courses in these upcoming subjects nor exude confidence to handle interactive sessions involving face-to-face traditional mode or modern open and distance learning techniques. Since, knowledge base of a teacher and his competence to deliver that knowledge are the key determinants of effective teaching and learning, investment for training and retraining of faculty was considered necessary. In order to empower teachers with knowledge and overhaul their teaching skills, it was suggested to make training part and parcel of budgeting and service conditions through launch of an instrument of HRD Policy.
- ICAR continues to support SAUs, DUs and CUs for infusing reforms in agricultural education to make it more relevant and

useful to diverse stakeholders. Some improvements like, development of model course curricula, uniform guidelines and norms for management of academics, have taken place but are not enough to link education to employability, alleviating poverty or sustaining country's economic growth without harm to health of natural resources. Nearly 12 per cent share of education, within the overall budgetary provisions of ICAR, constitutes merely 2 per cent of the total budget of SAUs. There is an urgent need for raising investments by State and ICAR to institutionalise reforms for building quality of agricultural education.

- The proportion of agri-sector contribution to GDP is going down, whereas value addition is on the rise. Accordingly, emphasis needs to shift for developing manpower in industrial agriculture than in general agriculture.
- Apart from the expanding job possibilities in private sector, there are some planned programmes of GoI for environmental conservation and employment generation. Typical examples are promotion of biodiesel involving investment of Rs. 15,000 million and propagation of bamboo at a projected cost of Rs. 10,000 million. Besides, there are focused changes in the Extension Policy of GoI. Whether these are setting up of farmer call centres, launch of agricultural channel or establishing one Krishi Vigyan Kendra in each rural district of the country the avowed purpose is to take agricultural technology to farmer's doorsteps. While revising the curriculum and programmes of agricultural education, these initiatives of GoI and attendant policies need to be taken into consideration.

2. *Regulation*

- Recent mushrooming of private colleges and universities at an unprecedented pace is alarming. These quickies entice academically mediocres for admissions into various courses and charge them fee, which is unproportional to infrastructural facilities and quality of education they offer. More worrisome features of this development are: (i) the ill-conceived legalisation of affiliation of private colleges by certain SAUs against the provisions of the ICAR supported

Model Act for State Agricultural Universities, (ii) prevailing disregard to ensure existence of a minimum physical plan before private institutions are licensed to begin imparting education and (iii) absence of an enforcing mechanism to accredit these institutions after they have been opened. Additionally, with profit as the prime motive, private institutions end up hiring substandard human resource for teaching also. There is a need for establishing a regulatory authority within ICAR/DARE to sustain quality of agricultural education on the lines of UGC and AICTE.

- ICAR/DARE in alliance with Indian Agricultural Universities Association may develop a draft policy paper for creation of a statutory but autonomous authority within former's ambit. Once this draft is vetted from legal angle, it can be converted into a bill for approval of the Parliament. Since an Accreditation Board to assess quality of agricultural education already exists, what is now required is to empower it with legislative authority to enforce quality. As it prevails, agriculture, including higher agricultural education, is a state subject. This situation contrasts with other professional courses, whose regulation of growth and quality vest constitutionally with the Union government.
- Agricultural education needs to be made compulsory at school level preferably during 5th–8th standard, as it exists in Uttar Pradesh or elsewhere in the world i.e., China, Chile and Brazil. From an optional subject status, it should be made compulsory and integrated into the existing science curriculum syllabus.
- Agricultural courses, offered as a separate stream at school level in certain states of the country, should be recognised equivalent to science stream. Students opting and passing this stream should be eligible for admissions, like science stream subjects, to graduate and postgraduate courses. Participants perceived that this shift would attract more students to join agricultural courses at senior secondary level.

3. *Reorienting Agricultural Education*

- Any attempt to reorient agricultural education for employability must be preceded by a need assessment on the kind and number of manpower projected to serve upcoming sectors of economy and country's national commitments (i.e., poverty reduction, employment generation, non-degrading use of natural resources, taking scientific information and knowledge to farmers' reach and value addition) and international obligations and happenings (various issues arising from World Trade Agreement, General Agreement on Trade in Services, ethics of Intellectual Property Rights and Genetically Modified Organisms and Foods). It is necessary to review the present skill composition of graduates and postgraduates *vis-à-vis* futuristic job profiles in consonance with needs of various stakeholders (public institutions, private enterprises, service industry, NGOs and self-employment seekers).

- In order to assure a steady flow of takers of the government sponsored agri-business and agri-clinic schemes for self-employment of graduates, the need for introducing business management, trade, marketing, cooperatives, banking and credit related subjects in the existing syllabi was recognised. Simultaneously, in-vogue exclusive emphasis on formal degree programmes was felt to be harmonised with non-degree skill oriented certificate and diploma courses to bolster village-based technical services. Introduction of lower-tier education was seen to create a cadre of para-professionals who can serve as an interface between farmers and graduate professionals. As in other fields (i.e., medicine and engineering), these lower level professionals are crucial in enabling farmers in knowledge application and giving feedback on relevance and applicability of the existing knowledge and in developing new knowledge.
- It is proposed that the proportion of time allocation for theory and practical at various levels of education can be as follows:

UG :	Theory & Practical 50-50 per cent
Diploma :	Theory & Practical 40-60 per cent
Certificate course :	Theory & Practical 20-80 per cent

- In view of the fast expanding developments in modern information and communication techniques (ICT) for education in natural and social sciences, it was realised that agricultural education cannot afford to ignore their use any more. In addition to emphasis on ICT-based curriculum delivery techniques, it was also felt to explore the potential and possibility of introducing open and distance learning (ODL) opportunities alongside traditional classroom mode of education. Tie-up with organisations/institutions which impart education through ODL was considered a preferred option to find out the extent and prospects of introducing their use in agricultural education where experiential learning demands 50 per cent or more of a student's time.

- More than formal education, exploring the use of ODL in training and retraining of faculty, graduate professionals and para-professionals to keep them abreast with the new knowledge and developments in S&T was suggested. Harnessing flexi time, space and pace potential of ODL in offering lifelong learning opportunities for farmers and farm workers, particularly those belonging to fair sex, was recommended as a key strategy to build a knowledge society. Whether the goal is to anchor necessary structural and contextual improvements to make formal agricultural education better or the mission is to build human competence through non-formal education for infusing science-driven growth of agriculture, investments for efficient and effective materials and methods of teaching and learning come in the forefront of all other investments. A refinement in knowledge base of faculty and farmers on a continuing basis is a fundamental strategy world over to enhance agricultural productivity and profitability and to alleviate rural poverty.

4. *Practice Training*

- In line with the focus of the brainstorming discussions on reorienting agricultural education to support employability, participants envisaged the need for developing professional competence and capability through integration of teaching in modern subjects with learning in real life situations. Unfailingly, the new syllabus must be relevant, responsive and sensitive to serve the needs of various stakeholders. In that endeavour, reengineered education ought to support self-employment (individuals), employability in upcoming sectors of economy (private e.g., business houses, processing industry, transfer of technology through NGOs or consortia of specialist individuals) and development of graduates and postgraduates well versed with issues and concerns arising from WTA, IPR, GATS and other conventions and treaties ratified and signed by India.

- While the need for introduction of courses and programmes in biotechnology, food processing and packaging, business management including codex standards, marketing, trade and credit/lending institutions and international agreements was recognised, one year practical training in any subject of professional learning in field, model plants, factory, business enterprise or engineering workshop was also recommended.
- Practice training will cover every aspect of a chosen vocation from beginning (say cultivation) to end (say consumption) and will be closely supervised and periodically evaluated. In view of this recommendation, participants called for redesigning RAWE (rural agricultural work experience) to accommodate hands on training in actual field or factory conditions. In-depth discussions followed to find answer to the question, "whether the one-year practice training should be continuous or spread over the four-year duration of the course?" Majority view was to have six months supervised training at institution based and managed instructional farm, model plant or workshop. During this period, teachings in theory and practice should be juxtaposed in time and space. Remaining six months training should be in the farmers' fields, a commercial farm, business house, processing industry or engineering workshop. The argument favouring this arrangement is that those of the students who are attached to private sector for training in the 3rd year and follow-up with a project writeup assigned by the private sector they have better chances of getting absorbed there. Somewhat similar arrangement is happening in engineering education and it can be adopted with appropriate refinements in agricultural courses also.
- Training for appropriate period, preferably one year, is desirable with flexibility in training type. Accepting this viewpoint, it was agreed that typically institutions with strong focus on basic subjects need to prepare students in unconventional areas like global trade and treaties with international perspective. With socioeconomic courses it was

felt that three months practice training was enough at UG level as the students have to be strong in theoretical aspects.

- PG research should be as per national priorities, institutional mandate and stakeholders' needs. Its planning, objective setting and methodology should be conceived with that perspective. Involvement of concerned individuals (farmers' representatives) and groups (processing industries) right at the point when institutional deliberations are held on research priority area identification and conceptualisation should be ensured.

5. *Subject Area Diversification versus Specialisation*

- Compartmentalisation of disciplines based on mindless multiplicity of subject areas was unhealthy. It distanced professionals from farmers who seek new knowledge on entire production system rather than that of an isolated discipline.
- Existing diversification of subjects into 11 faculties at UG level hardly serves any significant cause of specialisation. More or less syllabus coverage is similar for the first three years. In the guise of diversification, students develop a narrow focus, which is not desirable from the point of developing a wide learning canvas and overall comprehension of art and science of agriculture. Majority of the members in the house felt that at the graduate level the need is to produce generalists. Their specialisation should emanate from one year spent for practical training.
- Multiplicity of programmes at PG level needs a re-look. Currently, M.Sc. and Ph.D. degrees are awarded, respectively in 95 and 80 disciplines. It was agreed that consolidation of subject areas should be attempted keeping in front the existing ground realities and emerging economic and environmental scenes and scenarios. A specific recommendation was made that at PG level the specialisation through major and minor fields of study must rise from problem solving research on a topical theme covering crop/ animal, commodity or system. It was further urged that thesis research representing a blend of upstream basic and downstream-applied research ought to be founded on envisioned national priorities and mandated institutional

mission. Reformatting programmmes on these lines will expectedly offer a holistic view to education in building knowledge economy. This will also help in incorporating sustainability perspective, which from the prevailing viewpoint should integrate various aspects of water, energy, health, agriculture and biodiversity.

- The UG:PG ratio is very narrow in agricultural and allied sciences as compared to other professional streams like medicine and engineering. Keeping in view the increasing role of dissemination of fruits of research and development in enhancing productivity, alleviating poverty and ensuring biosphere integrity, it was recommended to consolidate, and wherever necessary, to prune PG education. The room, thus created, could be used to strengthen village-based services by reorienting UG education.
- A spin-off effect of rising number of graduates opting for postgraduate education is amassing lower quality graduates to service rural communities. There is mismatch between the existing supply of graduates in conventional agricultural subjects and their need in upcoming areas like agri-business management, agro-processing, dairy technology and veterinary services, marketing and storage, environment, biotechnology and information and communication technology. Hence, education at UG level, alongwith basic subjects on strengthening farm services, must cover syllabi in theory and practice of business management against the backdrop of produce handling, marketing and environment principles.
- Whether it was strengthening of village based services or producing graduates in emerging subject areas, admissions beyond the present numbers should not be stretched. Instead it was recommended to launch need and skill-based certificate and diploma courses to produce a cadre of para-professionals who would support graduate and postgraduate professionals in their pursuit of serving rural communities.

6. *Admissions*

- The present system and procedures of admissions, based on merit, common entrance test or merit-cum-common entrance test, were considered satisfactory.

- As the dividing line between mathematical and bio-sciences is becoming thinner and thinner and also considering the importance of bio-physics and mathematics, the students from these backgrounds could also enter agricultural courses. Likewise, it was recommended that graduates from pure science stream should also be admitted in postgraduate courses, if they accept to take add-on remedial courses to build knowledge base in agriculture.
- Suggestion to permit arts and commerce students to enter professional agriculture courses at graduate level was not supported.
- Admissions to various subjects and disciplines should be based on regular reviews and assessment of job markets and corresponding manpower needs. In order to forecast future employment, patterns, including stakeholder perspective, were of fundamental necessity. Since, shifts in job markets follow rising or falling importance of various sectors of economy, a country's national commitments and international obligations and happenings (emergence of genetically modified foods and organisms), subjects and methods of teaching and learning must be open ended with possibilities to infuse fresh ideas as and when necessary. For measuring up to these expectations, it was recommended to hold more frequent meetings of academic councils and boards of studies to refine subject contents and to introduce new disciplines. In gathering presence and input of diverse stakeholders in these discussions was considered an invaluable strategy and distinct opportunity to enhance scope and relevance of reengineered course curricula.

7. *Reorientation of Agricultural Education: Stakeholders' Viewpoint*

- Private seed sector expressed preferential interest in candidates with postgraduate qualifications to: (i) develop new varieties/hybrids, (ii) maintain quality, seed health, farm advisory services and (iii) generate related technology. Their specific demand for graduates was in the area of seed multiplication, field supervision and marketing. In order to fill this need, the seed sector wanted to focus on these aspects while rewriting the course curricula.

- NGOs spelt out the need for graduates and postgraduates well versed in organic farming, bio-tech applications in agriculture, input use knowledge, standard agronomic practices, natural resources management and ecological principles, processing, packaging, cold chain management, diagnostic techniques to identify nutritional disorders and diseases of both plants and animals, weather information and advisory management, ITK and its scientific enhancement, WTA covering IPR, rural development programmes, crop insurance, banking and credit lending institutions and markets and marketing. A new look syllabi should contain subjects and methods of delivery to build appropriate knowledge and skill base of futuristic graduates.
- In order to define the knowledge profile of para-professionals, besides building pedantic knowledge and practical skills in a particular vocation, the prospective candidates should be provided with opportunities to gain information and knowledge on village life, rural livelihood means, natural resources and native ecological principles of their use, traditional practices and beliefs *vis-à-vis* modern techniques on raising diverse crops, livestock including their varieties and breeds, necessary use and maintenance of farm implements and machinery, practical exposure to general field problems and basic expertise in employing and linking present day communication means for interpreting weather advisories and market trends.

8. *Entrepreneurship Building*

- Entrepreneurship training to prepare graduates for setting up farm advisories and services (agri-business and agri-clinics) should not be treated as a separate activity of teaching and learning. In place of current practice of imparting training after graduation, building entrepreneurship competence and spirit should be integrated into the re-scripted course curricula. In order to support service needs of diverse production systems, each SAU/institution must offer a basket of region-specific entrepreneurship courses. In pursuance of that objective, erecting necessary infrastructure for practical sessions and building faculty competence to teach refashioned courses were identified as the frontal strategies and investment initiatives.

- Specific entrepreneurship trainings for faculty should be organised at the leading institutions like NDRI in Dairying, IARI in agriculture, CIFE in fisheries, IVRI in animal health and Diagnostics, NAARM in education technology and some premier SAUs in known areas of excellence.
- With possible entry of big corporate houses into contract farming looming large, doling out of more franchises for micro-enterprises was expected. It was recommended to prepare graduates in appropriate knowledge and skills so that they are able to embrace these micro-enterprises with maximum ease and minimum investment and risk.
- Business in any enterprise is capital intensive; agriculture for that matter is no exception. Because of procedural wrangles, bank loans are not easy to come by. In many instances, investments in agricultural enterprises may not prove remunerative also. Softening loan and credit terms is necessary to attract more and more students setting up agri-business and agri-clinics for self-employment and service of rural communities.

9. *Gender Sensitisation*

- Gender stereotyping i.e., offering specially tailored soft assignments to girl students was unanimously opposed. Instead it was recommended that gender sensitisation be integrated into the curriculum i.e., uniform subject code for boys and girls. The contributions by male and female students to various aspects of agriculture can be reflected in the curriculum. More than course curricula reorientation, participating girl students wanted all out efforts to create basic facilities like boarding and lodging, which at present are deplorably poor. It was recommended that ICAR might consider giving one time special financial dispensation to build at least two hostels/SAU to accommodate girl students.

10. *Faculty and Facility Development*

- ICAR needs to enhance its support for faculty development both in terms of funding as well as encouraging greater number of training programmes such as summer/winter schools or conduct of refresher courses through centres of advanced studies. Courses and programmes building competence in upcoming areas of education and skill development should receive priority.

- Faculty may be placed for short spells as work culture observers in appropriate industry. This exposure is foreseen to help in forging and developing linkages with industry for subsequent placement of student trainees. More importantly, short-term placement of this kind will be highly useful in changing faculty's outlook in assigning graduates for real life training and research in industrial agriculture.
- Mechanisms for providing some kind of incentives to retain best students for faculty jobs should be devised.
- To complement faculty output in reorienting agricultural education for employability, the crucial role of strengthening and building appropriate facilities and infrastructure for teaching and learning was recognised. In furtherance of that cause, specific one time financial support was sought for creating state-of-the-art facilities for experiential learning in practical farming, commercial agriculture, dairying and food processing plants, soil and water testing, disease and nutritional disorder diagnostics and farm machinery maintenance and repairs.

In summation, reorientation of agricultural education within a reasonable timeframe and visible impact on employability should be launched as a mission to introduce necessary structural changes in system and setup to capture modern realities, challenges and opportunities. In fulfilment of that mission, partnership with various stakeholders, who influence and are influenced by the quality of agricultural education, ranks first among the reforms. Emphasis of futuristic agricultural education should be on introduction of a revised course curricula embracing new teaching and learning areas (typical examples; agri-business and entrepreneurship, diversification of farming, processing and value addition, natural resources and environment management, marketing, international trade and treaties and biotechnology) and delivery systems (information and communication technology), deployment of highly qualified faculty, creation of infrastructure for extensive practical sessions and linkages with private partners to support career building training.

32

Reorienting Land Grant System of Agricultural Education in India

Introduction

The establishment of the agricultural universities in India, on the pattern similar to that of the land grant colleges of the United States, made a landmark in reorganising and strengthening agricultural education system in India during the post-independence period. The contributions made by these universities, as partners of the National Agricultural Research System (NARS), are well recognised. The green revolution, with its impressive social and economic impact, would have not been possible without the significant contributions made by the agricultural universities, both in the form of trained scientific manpower and the generation of new technologies.

We can legitimately take pride in these achievements. However, of late, there has been a widespread concern about the sustainability of the system so also the gradual decline both in academic standards, and excellence in research. The agricultural universities continue to do good work, but high academic standards of earlier years seem to have suffered a setback, due to various obvious reasons. Accordingly, the need for reviewing the system has been felt to have a better understanding of the issues in order to take corrective measures for revamping the system.

The higher agricultural education system in the country, based on the concept of Land Grant system of education, adopted in various states, (education being a state subject) consists of 33 agricultural universities i.e. 25 state agricultural universities (SAUs), one state level university of horticulture and forestry, two state level universities of veterinary and animal sciences, one central agricultural university, and four national research institutes (one each in agriculture, dairying,

---

Policy Paper 6, December 1999.

veterinary and fishery) enjoying the deemed to be university status. Besides, agricultural education is also being imparted by a number of private/government colleges and faculties affiliated to the general universities, not following the land grant system.

The task of coordinating and supporting the existing agricultural education system in the country had been assigned by the Government of India to the Indian Council of Agricultural Research (ICAR) during sixties on the pattern similar to that of the University Grants Commission without conferring any statutory authority. The ICAR, therefore, rightly undertook this task in an advisory/participatory mode and developed a Model Act in 1966 for adoption by State Agricultural Universities with such changes as were deemed necessary. The Model Act was revised twice, once in 1984 and again in 1994. The SAU Acts in some cases, however, do not conform to the required provisions of the Model Act, especially in terms of governance. Ten states now have more than one agricultural university, each having jurisdiction over a part of the state territory. In the states of Tamil Nadu and West Bengal, the initially established agricultural universities have spawned off veterinary and animal sciences universities, and in Himachal Pradesh a horticultural and forestry university. Most of the universities have constituent colleges at more than one location in the area of their jurisdiction, some due to incorporation of distantly located pre-existing colleges, while others due to proliferation and dispersal of teaching campuses to satisfy regional/political demands, despite the fact that the dispersal of the teaching function under multi-location situations lowers the quality of education even in universities having adequate financial and appropriate managerial resources.

In agricultural universities, teaching, research and extension education functions are integrated to mutually re-inforce each other. Teachers are directly involved in deciding the course content, methods of teaching and also the evaluation mechanism. The system provides for quick communication of new knowledge to students, extension workers and farmers. Teachers and students are made to be responsive to the changing needs that agricultural development poses in terms of requirement for technical manpower. Agricultural universities have been enjoying fair amount of organisational and operational autonomy. It has proved beyond doubt, that semester system is better than the traditional system. Agricultural universities have served as pivotal heads of new knowledge, and instruments for increasing agricultural production and transforming the rural life.

During the course of implementation and execution of new agricultural education system in India, some pitfalls, discrepancies and errors have crept in with regard to uniformity of system of education, admission of students, curriculum, quality of teaching and teachers, practical training including work experience, modernisation of laboratories and equipments, etc. However, the ICAR has made some efforts to remove many of these problems through constructive recommendations of high level committees on various aspects of agricultural education which will eventually involve development of sound policies in order to bring effective change in meeting the challenges of the next millennium.

It is also a fact that despite four decades of existence of these institutions it has not been possible to ensure a stable funding mechanism commensurate with their requirements. As a result, most of them face acute financial hardship frequently. Often, the flow of funds is irregular and a major percentage of it is diverted for payment of salaries, leaving very little for research, teaching and extension education activities.

Initially, most of the SAUs had adopted a progressive policy, in recruitment of teachers as well as admission of students which enabled cross fertilisation of ideas and enhancement of academic standards. Gradually, the process of inbreeding and the lack of proper linkages among various academic institutions has resulted in deterioration in quality of teaching and research.

The adopted system has proved beyond doubt extremely beneficial and various faculties *viz.* Agriculture, Veterinary and Animal Sciences, Engineering and Technology, Home Science, etc. have all come under a common umbrella. In addition, integration of teaching, research and extension functions has, by and large, been achieved. As already mentioned the Land Grant pattern of education in India has met with problems during the course of implementation. Hence, the need to reorient the system has become imperative.

Policy Recommendations

1. Policy Issues

1. The Government of India should ensure that the ICAR has the required statutory power to promote, support and coordinate higher agricultural education in India on par with that of University Grants Commission.

2. The agricultural universities may also redefine their mission, taking into consideration the present state of Indian agriculture, as a result of the tremendous progress made since independence. The new mission statement should recognise that Indian agriculture now has expanded clientele including NGOs and the private sector, and above all, a faster-growing middle class population of more than 200 million people with increased purchasing power.
3. The Government of India should constitute a Standing Committee for each state with Secretary, Department of Agricultural Research and Education (DARE) as Chairman; respective State Agricultural Production Commissioners as Vice-Chairmen with respective Vice-Chancellors, Secretaries of Departments of Agriculture, Horticulture, Animal Husbandry, Fisheries and Forestry as members and Deputy Director General (Education), ICAR as Member-Secretary for periodic monitoring of the functioning of agricultural universities and for providing timely guidance and support for solving their problems and making mid-course corrections, as considered necessary.
4. Effective mechanisms for impact assessment and evaluation of agricultural universities are required to be put in place, including accountability for the administrative, financial and academic achievements. At the same time, they be given proper funding support and freedom to perform.
5. The present practice of proliferation of agricultural universities through creation of new universities, campuses, colleges and through bifurcation of existing universities is indeed counterproductive and not in the interest of quality education. This practice needs to be stopped at the state level. On the other hand, there is an obvious need for the consolidation and downsizing of existing institutions for efficient utilisation of limited resources.
6. Agricultural universities must ensure adherence and effective implementation of the concept of each faculty member getting involved in at least two of the three basic functions *viz.*, teaching, research and extension education.
7. Both at the level of staff recruitment and student admission, the practice of inbreeding adversely affects the teaching and

research standards. While responding to the specific regional problems of the states, agricultural universities must lay emphasis in maintaining greater national perspective in faculty building.

8. Students should be discouraged to take more than two degrees from the same university in order to curb inbreeding which often has deleterious impact on quality education.
9. Agricultural universities must encourage admission of students on an all-India basis in order to attract the best available talent. The current practice of admission of 15 per cent of the undergraduate and 25 per cent of the postgraduate through all-India entrance test conducted by the ICAR should continue and the proportion be gradually increased.

2. *Strengthening of Agricultural Education*

10. It is absolutely essential to bring private/government colleges and faculties in general universities offering programmes in higher agricultural education into the fold of agricultural universities and the ICAR for improving the quality of education after careful evaluation and screening.
11. The central and the state governments should strengthen the existing agricultural universities in order to equip them for making India a developed nation and to meet the challenges of the next century. India is expected to not only have sustainable food security but also become a leading exporter of agricultural products and services.
12. To strengthen the agricultural education system, the ICAR be encouraged to conduct an overall review of agricultural universities by involving eminent educationists, scientists and administrators. The proposed review must focus on important aspects, such as, university governance, financial sustainability, personnel policies, teaching, research and extension education as also their linkages with other institutions and organisations in the public and private sector.

3. *Funding of Agricultural Universities*

13. The state governments must ensure proper financial support to the agricultural universities by allocating at least 15 per cent of the total budget of the departments of agriculture,

animal husbandry, fishery, horticulture, forestry and any other related to agriculture.

14. The central and state governments may devise a mechanism to provide, to respective agricultural universities, a lumpsum grant as core fund to be used in future, exclusively for the maintenance and renewal of existing infrastructural facilities on the campus. This will mitigate the effects of uncertain funding.
15. The development grant provided by the ICAR to agricultural universities under plan allocation should be reviewed and adequately enhanced.

4. Course Curricula

16. The Land Grant pattern of academic institutions permits a great deal of flexibility to the students, especially at the postgraduate level so as to plan their course of study according to their requirements and capability. This provision is however, not uniformly availed by students for various reasons. This needs to be implemented uniformly.
17. There is a need for rationalisation of policies with regard to admission, duration of the programmes, requirement of credits, evaluation and grading and updating as well as revision of curricula at regular intervals for various degree programmes in the agricultural universities.
18. Complete transparency is required to be introduced in the internal evaluation system at the undergraduate and postgraduate levels with an inbuilt system of checks and balances that ensure proper record keeping of the performance of the students with the concerned teachers for future reference.
19. With social, economic and scientific changes, several new disciplines such as agri-business management, biotechnology, molecular taxonomy, environmental sciences, plant and animal genetic resources, intellectual property rights in relation to World Trade Agreement, Biodiversity Convention and other international commitments have become important. These disciplines and emerging areas must find a place in the teaching programmes of the agricultural universities.

20. With the technological revolution, which is presently getting a momentum through effective use of genetic engineering and molecular biology, basic research is assuming pivotal importance to provide a sound base for competitive technological advances. Basic sciences, therefore, need to be given adequate importance by agricultural universities both in terms of resource allocation and faculty involvement.
21. Social and behavioural science components need to be given specific emphasis in view of our understanding of the interdisciplinary character of the problems of the farming community.

5. *Human Resource Development*

22. A suitable mechanism is required for periodic assessment of the scientific and technical manpower requirement for agricultural R&D institutions in the country. This will help maintain a reasonable balance between the manpower generated and opportunities for their gainful employment.
23. To attract talent and avoid inbreeding, admission for postgraduate programmes in the agricultural universities should be made open to students from basic sciences and engineering in the related disciplines. Such students should, however, be given remedial courses in the discipline of agriculture.
24. To ensure technological empowerment of women the home science education should be made broad-based by redefining the goals, changing the nomenclature of the programme, introducing market-driven enterprises and sustainable management of natural resources.
25. A system of recognising outstanding performance of teachers, linked with appropriate incentives and rewards, is required to be put in place by all the agricultural universities.
26. More scientists and teachers in the agricultural universities should be trained through summer-institutes/refresher courses so that they are better equipped for the assigned tasks. Facilities available for training in advanced countries through bilateral and multilateral agreements should be made use of.

6. Linkages

27. There is need to provide proper linkages between agricultural universities, The ICAR institutions and agro-based industries. There should be an effective system of exchange of teachers and researchers for short durations through linkups with the universities and academic centres of good standing. Joint programmes of research should be developed with foreign universities and research institutions.

28. For better appreciation of the needs of the industry for technological development, it would be advantageous to implement fellowship programmes for movement of the university faculty to the industry and *vice-versa*.

29. There is a need to develop proper linkage with general universities and the UGC for improvement of educational standards in affiliated colleges imparting agricultural education.

33

Guidelines for Improving the Quality of Indian Journals and Professional Societies in Agriculture and Allied Sciences

Introduction

The publication of journals provides an easy window for widespread sharing and access to the scientific knowledge and new discoveries. It serves as the main gateway for rapid reach to the correct and well-edited version of scientific investigations, research output and conclusions drawn. However, there is considerable variation in the scientific quality depending on accuracy, originality and manner of presentation of such information in different journals published by different scientific professional societies in the country. This leads to varying credibility/reliability/acceptability of the scientific knowledge communicated through various journals. It is in this context that NAAS organised a brainstorming session on improving the quality of Indian journals in agricultural research, education and development.[1]

The deliberations in the various sessions culminated in the

1. A session on Improving the Quality of Indian Journals in Agricultural Research, Education and Development was organised on September 12-13, 2005 at NAAS office by bringing together a group of about 40 selected experts comprising distinguished fellows of the Academy and presidents/editors of Delhi based professional societies publishing leading journals in agricultural sciences, in a panel discussion mode. The discussions covered various topics such as research contents of the articles submitted, editorial board—its constitution and functioning, peer review of the articles, publication process, functioning of the scientific societies, financial incentives and resource mobilisation.

   Dr. Mangala Rai, Secretary DARE & DG, ICAR, in his inaugural address, expressed concern over the generally low impact factor of our research journals and emphasised the need of having competent peer reviewers and professional editors for improving the quality and impact factor of at least one scientific journal, in each of agricultural disciplines. In this connection, he also stressed the need to improve the quality of research. The infrastructure created and funding enhanced for research programmes through NATP earlier and NAIP now under ICAR are likely to provide the needed impetus for quality research.

Policy Paper 34, March 2006.

following recommendations as guidelines for improving the quality of Indian journals and professional societies in agriculture and allied sciences:

Recommendations

1. Quality of Journal

1.1. Editorial Board—Its Constitution and Functioning

Since the Editorial Board plays a key role in processing the articles, its proper constitution and functioning become main determinants in the scientific quality, contents and relevance of the journal.

(a) Constitution

- The editorial job being extremely scientific, the choice of editors can not be decided upon by a majority election. Hence, the Editorial Board should be constituted by nomination by the Executive Council.
- There should be some sort of pragmatic mechanism to constitute the Editorial Board. One of the mechanisms suggested was that each member of the Executive Council should suggest 10-15 peer scientists in his/her area of specialisation. After thorough scrutiny of the proposed names, the Executive Council should select at least one name for each major sub-theme area.
- There should be a broad uniformity in the constitution of the Editorial Board of all the Indian journals in agriculture and allied sciences to the extent possible.
- Editorial Board should consist of one Chief Editor/Editor possessing distinguished broad expertise of the entire focused theme of the journal and a minimum of five members of proven competence representing different sub-theme areas, and should have a tenure of two to three years as per constitution. The Chief Editor/Editor should not necessarily be confined to the location where the society's headoffice is housed. In order to enhance the visibility, the Editorial Board may also include one or more reputed foreign scientists.
- Chief Editor/Editor should be a member of the Executive Council.

(b) Functioning

- The Editorial Board should develop precise and comprehensive instructions for preparation of manuscripts of articles by the authors. These instructions should be published in an issue (volume) of the concerned journal preferably once in a year. However, if the instructions are modified/changed earlier, the modified version should be published immediately in the next ensuing issue (volume). The instructions should elaborate specific criteria indicating which article can be accepted as full-length and which as short communication.
- The Chief Editor on receipt of an article will process it for peer review. For this purpose, the Editorial Board should also develop guidelines for the peer reviewers to enable them to undertake the objective and detailed review of the articles and also to offer clear-cut recommendations regarding their suitability or otherwise. The articles with contents of routine/repetitive nature should be discouraged, and scientific merit of contents should be the chief criterion to be taken into account while evaluating the article.
- The Editorial Board should meet preferably before publication of each issue. However, if for some reasons, it is not possible, the board's meeting must be held at least once a year.

1.2 Peer Review Process

- A list of about 8-10 peer reviewers in each sub-theme areas be prepared by the Editorial Board and Executive Council of the society. The selection of reviewers should be based on their own record of high level scientific publications in journals of repute. The members suggested the desirability of including some Fellows of NAAS among the list of peer reviewers.
- Chief Editor/Editor in consultation with the concerned Editor/Board Member should first himself/herself judge the suitability of the article for consideration to be published in the journal, and then only the choice of appropriate reviewer(s) needs to be made.
- At least two peer reviewers should be involved for vetting each article. It was felt that while sending the article to the reviewers, the identity/affiliation of the authors need not be kept secret.

- In view of a general complaint of delays on the part of the reviewers, it was recommended that the commitment of the reviewer(s) should be sough first before the article is sent to him or her for review. The Board should also have an alternative reviewer identified beforehand.
- The society should cover all cost of stationery and postage in communication of the articles and not the reviewers/editors. Since, the cost involved is generally small, it would be better if self-stamped envelopes are sent to the reviewers/editors. Use of electronic media in the entire publication process will reduce the cost drastically, beside accelerating the pace of process. This was strongly recommended.
- Societies should provide the comprehensive list of the actual reviewers at the end of the last issue of each volume, irrespective of the fact whether they rejected or accepted the articles.
- The reviewers should not be over burdened with number of articles. It was suggested to put a limit of maximum two articles per volume for any reviewer.

1.3 Quality of Contents

- The declining quality of journals reflects the falling standard of research in our institutions. The poor research planning and fast declining research support were considered as the main causes of declining quality of research in the country. This needs serious thinking and remedial action plan. The participants called upon the ICAR to initiate this process to stop further decay of science in the country.
- A good journal should carefully apportion its space between the basic, strategic and applied research articles. Care should especially be taken in selecting the articles in the applied areas. The article must carry some new message *viz.* new techniques, methodology or process description, etc.
- Short communication is a kind of 'Hot Scientific News' related to new techniques, methods or a product. Only the articles received as 'short communication' should be considered for publication in this category. Most often a full-length article when found unsuitable is recommended or publication as a

short communication. This practice needs to be stopped.

- Inordinate long delay in publication is one of the reasons for scientists not to submit their articles in the Indian journals. The society should take all possible measures to ensure completion of the peer review process within a period of maximum three months.
- The society should consider publishing a few relatively high quality articles with novel findings out of turn, on priority. This will act as an incentive to attract good quality articles.
- Sometimes even a good piece of work is rejected due to poor language and presentation. The Editorial Board should help in rewriting and, if needed, even reanalysing the data on behalf of the author(s). This is a common practice being followed by some foreign journals. This practice will help accommodating such deficient authors, but good researchers.
- The journal should represent today's science, and not just any report or repetition of the research work done earlier/ elsewhere in or outside the country. Quality of science-originality, strategic nature and relevance of work-should be emphasised. The journal *Current Science* can be taken as an example.
- Review of the ongoing scientific research in the country should be one of the important roles of the society. To meet this goal, the journal should include one good review article in each of its volumes which should be the 'state-of-art report on scientific research'.
- Almost every university publishes its own journal that includes articles based on the work done mostly within the university. However, there is wide variation in the scientific quality of the journals. Many journals are of low standard and lack in rigorous peer review and editing. The members suggested that such journals should be renamed as 'Research Bulletins' rather than research journals. Nevertheless, the university may give due recognition to publications in these bulletins while evaluating the performance of scientists/faculty members for promotions or any such incentives.
- The universities which are maintaining high scientific standard of their accredited journals, should be provided necessary

financial support by ICAR to further enhance the quality.

- ICAR also publishes a few multi-disciplinary journals. There is a need to upgrade the scientific quality of these journals, so that these journals serve as model to be followed by the professional societies.
- The societies/journals need to be accredited by the ICAR through a credible agency like NAAS to set the standard. The accredition needs to done regularly on the basis of discrete quantitative criteria at certain intervals. The ICAR should provide the resources and such support needed for the accreditation work. The members felt that the ICAR may also consider whether the articles published in non-accredited journals should be considered as valid publications for performance assessment.
- It should be made mandatory that only the accredited journals/societies will be provided financial support by the ICAR. This will be a genuine discouragement to the ever proliferating and poor quality journals being published both in public and private sectors.

1.4 Publication Process

- After an article is finally approved by the Chief Editor/Editor, he/she will put down the date of acceptance on the manuscript and will inform the concerned/corresponding author(s) regarding its acceptance for publication. He will also inform the concerned corresponding author(s) about rejection of the article as soon as the same is finally disapproved.
- Each article when finally accepted, should be thoroughly and meticulously edited by a well-known professional for correcting all errors, grammatic mistakes, etc., for improving flow, readability and coherence of language, for indicating the size and legibility of tables, figures and charts as well as their proper placement, and for specifying which word(s) and sub-heads should be bold/italics, etc. On the whole, editing should ensure uniformity of the style as also overall excellent get-up of the journal.
- The Chief Editor/Editor will arrange the accepted full-length articles sub-theme wise and priority wise for publication in the issue (volume) of the journal. These will be followed by

articles of short communications nature. The material will be passed on to the Secretary of the Society for publication in double columns in a reasonably good quality A-4 size paper.

- Periodicity, timeliness of publication, and regularity of journals are extremely important. The publication should not be delayed beyond two months. At least two issues of a volume of the journal should be brought out each year.
- Generally, every society at the time of its seminar/symposium/ conference/convention publishes the abstracts of the articles received as a separate volume. The members felt that the society should publish 'Extended Summaries' instead of abstracts. In this way, the interest of those members whose articles are generally not published in the journals is also served.

2. *Improving Functioning of Societies*

- Proper functioning of the society is crucial for ensuring the quality and punctuality of the journal published by it, which in turn depends on the persons chosen to fill the positions in the Executive Council (EC). The EC should consist of persons of eminence whose contributions in the discipline are well known. Therefore, the constitution of EC should go by the strength of science and not merely by democracy. It was suggested that a search committee can be authorised to pick up names, a few of them could, for example, be fellow of reputed academies like NAAS, INSA, etc., which could be then ratified by the general body.
- It was strongly felt that the Secretary and the Chief Editor/ Editor should work independently.
- The Constitution of the society is a determining factor for meeting its goals, functions, growth and future development. Society should, therefore, take the help of professionals to draft its constitution. The copy of the Constitution should be made available to all the members for ensuring transparency and implementation of the same in letter and spirit. The Constitution should regularly be revised and updated based on society's own experiences as well as the lessons learnt from other societies in the country.

- The society should be well organised having its office furnished with modern communications and other facilities. Help of management expertise could be sought to organise and develop office procedures to make it more efficient. As an incentive, the ICAR should constitute an award to the best functioning society.
- Use of computers and electronic media for communication should be practised. The time has come for each society to have its own website, which will help bringing the members of the society closer. The website will include, among other details, the database of the membership including the details of addresses, status, designations, subject-interest, etc. The website will be upgraded on regular basis, so that the members remain abreast of all the forthcoming activities and new developments.
- Traditionally, society as a forum form members to come together and exchange information, present their work, etc. Therefore, the society should be proactive in promoting modern/advanced sciences by keeping the scientists updated with latest advancements in scientific research and modernisation of teaching.
- Although, a journal is an official organ of a society to disseminate its quality, content and extent of reach in the subject, yet it should not be the only focus/target of the society. To be visible, the society should organise regular national and or international activities, and be in communication with its members. The role of societies should, therefore, include some of the following activities.

    a) Improving teaching in the respective disciplines should be one of the responsibilities of the societies, which will help upgrade the science itself, and its practice. Hence, modernisation of the course curricula should be one of the major activities. Since the societies have a wide membership of qualified people, they can organise such activities quite objectively.

    b) Societies should also review the status of scientific research going on in the country from time to time and should come out with reviews that can be shared among its members for their use and critical appraisal. The review articles should be well structured and invited by the Executive Council on a topical subject.

c) Training programmes for young scientists particularly with regard to scientific writing and research proposal formulation, planning, conduct as well as result analysis would go a long way in raising the standard of science and published literature. The society can organise these trainings by utilising the services of the expert members who come to participate in the annual general body meetings, seminars, symposia and conferences, etc. NAAS would be in a better position to organise such training programmes. It will be desirable if ICAR could support such long-term HRD initiative.

d) The society should encourage the scientists of repute to write standard books. The authors should first get the synopsis of the proposed book(s) vetted by the peers in the discipline. The society should provide the required logistics including even seed money to carry out this project. The final format of the contents should have approval of the concerned society.

e) In recognition of the outstanding contributions made in research/teaching, the society should offer fellowship to the distinguished members on a regular interval, who are selected based on a set of transparent criteria developed by the society and not on any *ad hoc* basis, and without any age discrimination.

f) Organising at least one international seminar/symposium/conference once in six years would be a very stimulating and scientifically rewarding activity to keep the science of the disciplines vibrant and shared among its members. The society should also organise a national level seminar once in two to three years. This will keep the society scientifically functional and active.

g) Outcome of the society's symposia and conferences often does not meet the purpose, particularly due to delayed publication and poor circulation. The proceedings and recommendations should be published preferably by the well known professional publishers which can go a long way in reaching the clientele for whose benefit the symposia/conferences are organised. The proceedings must be published within a year and not later.

h) The society should project itself as the final word on the subject to the policymaking bodies of the country. Brainstorming on issues critical to national interests needs to be organised from time to time, and policy papers published and distributed to the users.

i) Incentives need to be given by awarding not only its own members, but also the non-member scientists. These incentives should be the recognitions based mainly on the contributions in the discipline/sub-discipline.

j) Prompt response to the queries of the members is essential. Therefore, the society secretariat should remain up-to-date in information and quick in response.

- The minutes of all society meetings including that of the General Body should be faithfully recorded and the proceedings circulated in time giving opportunity to each member to respond, if necessary.

3. Financial Incentive and Resource Mobilisation

Quality of journals is very much related with the financial health of the societies. The financial position of majority of our societies is far from satisfactory. The following measures were suggested to remedy this malady:

- Publication of journal is the single most important economic activity of any society. Hence, the ever increasing cost of publication is a major concern for the society. It was suggested that the societies should explore the possibility of contracting well known private commercial publishers from within and outside the country to manage publication as well as distribution of journals, while keeping the invite for articles, their reviewing and editing to themselves.
- The practice of providing free copies of the journals to the society's life members needs a relook. It involves not only the

publication cost of the free copies, but also a large amount that is needed for meeting the postage and stationery charges. Members felt that some suitable criteria should be evolved to minimise these costs. One of the suggestions put forward was to stop posting free copies to the life members after 10 years of his or her retirement. Consensus of the members on this issue may be obtained for its smooth implementation.

- It was strongly felt that each professional society should develop a corpus fund and the ICAR should contribute one-time grant towards the corpus of each accredited society, beside enhancing the annual financial assistance for publication of journal which is quite low presently.
- The society should look for mobilising its resources also through non-conventional sources i.e. private companies and organisations having similar interest. Like IITs, the societies should also approach its distinguished alumni world over to donate to their corpus fund.
- The experience shows that to deliver 1000 copies of a journal, each of 100 pages, there should at least be 700 paid subscribers. A 10 per cent outside paid subscription will be desirable to further ease the situation. Therefore, the societies should consistently work towards increasing the paid subscription.
- Currently, there appear to be about 100 agricultural research journals being published in the country, as per the ICAR records. The members strongly felt that all attempts should be made to reduce this to a reasonable number by amalgamating/combining journals of similar scope and mandates. Although it will not be an easy job, but ICAR can play a positive role in this direction by organising dialogue with the societies.

4. General Recommendations

- Accreditation of the society/journal should be made mandatory to grant recognition. The funding of the society by ICAR should be linked to this. The ICAR may also consider giving higher weightage to the articles, published in the accredited journals for purpose of scientists's assessment. The ICAR should identify a suitable nodal agency to perform the accreditation work based on well-defined quantitative criteria on a regular basis. The NAAS was thought to be the most ideal agency for doing this job.

- Need for hiring a professional editor and paid by each society was highly emphasised. In this context it was suggested that the ICAR should be approached for providing financial support to implement this recommendation.
- Members recognised the potential of electronic media for publication of journals, such as on-line submission and processing of articles as well as on-line distribution of journals to foreign members, and also for the day-to-day management of the society's functioning. It will both hasten the process and reduce the society expenses drastically. It was, therefore, suggested that the ICAR should provide one-time grant to the accredited societies to install the IT facilities on a priority basis.
- While it was appreciated that considerable homework was done by NAAS in assigning impact factor to different research journals, members did point out some of the lacunae in the existing practice. A consensus decision was taken to organise another round of discussions and have a relook on the whole issue including development of quantifiable precise parameters, as it affects not only the societies, but also the scientists who publish articles in these journals.
- In some fields, there are more than one journal with more or less similar scope and objectives. The ICAR should start a dialogue to discuss possibility of merging such journals into one good journal. Financial support to such journals should then be enhanced.
- Periodical review (say every five years) of the quality of journals and functioning of societies be organised by NAAS with financial assistance of ICAR. Attempts should also be made to remedy the deficiencies wherever existing, with the chief aim to motivate and strengthen the scientific competence of such societies.

Section VI

Globalisation and Agri-Exports

34

WTO and Indian Agriculture

Implications for Policy and R&D

Introduction

Agriculture will continue to remain at the centre stage of socio-economic development in India. Notwithstanding its outstanding performance in making the country self-sufficient in foodgrains, deceleration of its performance beginning with mid-nineties is of serious national concern. Two major developments impacting Indian agriculture during nineties have been: (1) the Agreement on Agriculture implemented from 1st January, 1995 under WTO, and (2) emergence of highly volatile price regime thereafter. Several recent studies, on the causes of poor performance of agriculture, often relate to WTO agreement on agriculture and its likely adverse implications against promise.

It is also a fact that WTO agreements are now a reality and these agreements can only be modified. The global scenario in future is going to become more competitive and the pressure for liberalisation of domestic market would also grow. In this context, India needs to follow two pronged strategy. One, based on the post WTO experience of last 10 years India should continue taking active part in negotiating agreement to its advantage with sound arguments. This would require objective understanding of the implications of the changed trade regime, promoted and planned under the auspices of WTO. Such an understanding is of critical importance to play an effective role in future negotiations as well as to plan adequately for designing our policies and economic activities including agricultural R&D programmes which are basic to accelerated sustainable agricultural development. Two, in order to effectively operate in the WTO driven

---

Policy Paper 38, December 2006.

environment, the member countries need to devise appropriate domestic policies and strategies. The crux of these policies should be to identify weaknesses and strengths of domestic produce *vis-à-vis* major competing players, and to improve competitive attributes of our produce, involving production, processing, marketing, trading practices and other processes from the farm to the final destination. Upgradation of competitive strength requires improvement in policies, infrastructure, institutions and technology. Out of these, the major role lies in technology and its favourable interplay with institutions and policies. India's agricultural research system has stood several tests successfully in the past and has helped the country to tide over formidable food crises and other challenges. To address WTO related challenges, research system should know what is precisely needed from it. Specifically, some of the questions that arise in this regard are: (i) what is the nature of challenges in different enterprises, products and locations that agricultural research system should address to impart competitive strength to Indian agriculture comprising dominantly small and marginal farmers; (ii) what are the desirable attributes to make our products competitive; and (iii) what are the institutional and policy imperatives? Recognising the need to find answers to these questions, the NAAS organised a one-day workshop.[1] The deliberations of the workshop were organised into two technical sessions followed by detailed discussion. This Policy Paper brings out some of the critical views and recommendations of this dialogue for larger public debate and policy action.

Uruguay Round (UR) Experience and Implementation

Agriculture was kept outside the purview of GATT till 1995. However, UR has succeeded in bringing agriculture on the main track of GATT and agriculture trade is now firmly within the multilateral trading system. All the member countries of WTO are committed to follow set of rules embodied in WTO Agreement on Agriculture (AoA) which covers: (i) domestic support, (ii) Market access i.e., tariffs, and restrictions on imports and exports, and, (iii) export subsidies. The agreement sought reduction in trade distorting domestic policies like price interventions and subsidies; reduction in

1. Under the Convenership of Dr. Mruthyunjaya and Co-convenership of Dr. Ramesh Chand to discuss the above issues involving experts, policy makers, economists, leaders and other stakeholders in the national agricultural innovation system at New Delhi on December 23, 2005.

export subsidies; replacing quantitative restrictions on trade with tariffs and reduction in tariffs to encourage more and freer trade. It was projected that trade liberalisation and implementation of AoA would bring large benefits to the developing countries through improved access to the developed countries' markets, increased trade and better pricing environment for tropical and other products of interest to the developing countries.

The Uruguay Round AoA was first step towards the reforms in agriculture trade. The Article XX of the Uruguay Round agreement on agriculture required WTO members to review the agreement after about five years, i.e. by the end of 1999 or beginning of 2000, for continuing the reforms started with the Uruguay Round. This provided opportunity to review the effect of implementation of UR AoA, and, in the light of this experience, move further towards establishing free, fair, and market-oriented agricultural trading system. Negotiations for the next round of AoA were started in March 2000 and have passed through several phases. Presently, there is a complete stalemate as the WTO members could not come to an agreement. The reason for the delay and stalemate in concluding the new round is the sharp differences among the members on various aspects of AoA. Implementation of UR commitments has been a tough task for several member countries and it has exposed vulnerability of various segments of agriculture to global market forces. In most of the cases expectations placed on UR AoA or promises related to this did not materialise. The promise was that trade liberalisation and implementation of AoA would bring large benefits to the developing countries through improved access to the developed countries' markets, increased trade and better pricing environment for tropical and other products of interest to the developing countries. However, there was a distinction between reality and the promise.

The biggest challenge to the developing countries' agriculture in the post WTO period was posed by unprecedented and unforeseen decline in international agricultural prices. The prices of cereals, fish, sugar, cotton and beverage started declining after 1996 and reached historically almost the low level during last 25 years around year 2000 (Table 34.1). Though there is some recovery in the price cycle in the recent years, yet the current level of prices of above mentioned commodities is 15 to 44 per cent lower than the prices prevailing in the beginning of WTO. Because of this decline in prices in post WTO period, developing countries' exports were badly hit and several countries like India were taken aback by import influx of commodities, in which they thought they had strong competitive edge. This caused adverse impact on farmers' income, employment and livelihood security.

Table 34.1

Export Prices of Primary Agricultural Commodities in Post WTO Period, (Base 1995=100)

	1995	*1996*	*2000*	*2005*
Cereals	100	119	67	85
Fish	100	86	76	85
Shrimp	100	97	113	63
Sugar	100	92	66	81
Beverages	100	85	65	83
Cotton	100	82	60	56

Source: WTO International Trade Statistics, 2006.

The decline in international prices happened contrary to the projections that implementation of WTO AoA would result in reduction in subsidies and thus, increase in cost of production and prices. Developed countries responded to decline in global agriculture prices by providing huge support to their farmers to protect their incomes. However, the developing countries neither had mechanism nor resources to protect their agriculture and farmers against such adverse trading and pricing environment. As per the Uruguay agreement, the developed countries were committed to reduce base level of support (taken as average of years 1986-1988) to agriculture by 10 per cent by the year 2000. In reality, OECD countries on the whole, and EU and USA in particular, provided much higher level of subsidies to their farmers during most of the post WTO period as compared to the already very high base level (Table 34.2). Further, more than 60 per cent of this support was in trade distorting market price support.

There is a feeling that the developing countries did not bargain properly in the UR discussion, while the developed countries secured the balance of AoA in their favour. Besides being discriminatory, the agreement is said to be ably manipulated by the developed countries to benefit their agriculture at the cost of developing countries.

Because of all these reasons, the developing countries have turned highly conscious and are very careful about the minute details of various provisions of any future AoA. This has led to hardening of the positions, particularly relating to the excessive support and export subsidies in OECD countries, access to developed countries' market, and special and differential treatment.

Table 34.2

Farm Support in OECD Countries before and after WTO

Value of Output and PSE: $ Billion	*1986-1988*	*1995-1997*	*2002-2004*
1. Total Value of Production (at Farm Gate) OECD Total	597.00	769.00	740.00
2. Producer Support Estimate (PSE) OECD Total	243.00	252.00	254.00
3. Market Price Support: OECD Total	188.00	170.00	156.00
Percentage PSE	37.43	29.62	30.35
PSE in EU $ Billion	101.70	117.60	114.27
PSE in USA $ Billion	36.40	26.30	40.41

Source: Agricultural Policies in OECD Countries, OECD, 2005.

Post WTO Trade Regime: Lessons for India

Wide-ranging economic reforms introduced in India during 1991 boosted the agricultural trade (exports as well as imports), and the net trade surplus in agriculture increased from $ 2 billion during 1992-93 to $ 4.33 billion during 1995-96. The trade got a fillip with WTO agreement during 1995, thereby resulting in the net trade surplus reaching $ 6.8 billion in 1996-97. However, the problems of downword trend in exports, increase in imports, sharp year to year fluctuations in net trade, erosion of self-reliance, etc., started soon thereafter, despite further liberalisation of trade. These trends raise questions about further liberalisation of trade in general and specific commodities in particular, and the implications of present agreement including ongoing negotiations on AoA in WTO, etc.

An analysis of the global trends after WTO indicates unprecedented decline in the global prices of agricultural commodities, severely hitting earnings from agricultural exports of the countries. India too could not gain much from WTO and global liberalisation, though India generally performs better than the other countries when the international price situation is favourable. As the exports are becoming increasingly competitive, it is not possible to promote exports without improving produce quality, efficiency and cost reduction. From the analysis of the trade scene and suggested future strategy by commodities (Table 34.3), it is evident that: India is facing challenges in traditional export items; the challenge is not from the developed countries, but from the developing countries; major import of vegetable oils are from the developing countries (Malaysia and Indonesia); and India has done well in export of high value products to the developed countries.

Export of raw products like cotton, wheat, sugar, coffee and tea is likely to become very competitive and India would be required to relook into the benefit of promoting export of these commodities. Export prospects are brighter with soybeans, oilseeds, oil meal and cake, fruits and vegetables, and fruit preparations. Thus, high export prospects are seen with high value products, horticultural products, processed products, marine products and rice, provided global competitiveness in costs and quality is maintained in domestic production, marketing and supply.

India has to learn lessons from these trends and other experiences. The AoA created an environment of trade reforms and initiated trade liberalisation in agriculture. The other lessons include: (i) WTO seeks to remove the barriers of 'our own market' and 'their own markets', (ii) India need not be extremely defensive and inward looking, as our agriculture has demonstrated strength which needs to be appropriately used to compete in the global market, otherwise it will become a case of missed opportunity, (iii) as a net exporting country, India stands to gain from increase in international prices, which is possible only if we relentlessly fight for reduction of domestic subsidies and support and export subsidies in the developed countries (once this is done, the possibility of import threat for our key export products like dairy, wheat, cotton, sugar, etc. will disappear), (iv) import threat can be surmounted by general provision of raising tariff to bound level and of asking for special treatments/packages like S&D treatment, sensitive product, etc. and (v) to comply with quality standards to effectively compete in the global market, we must demand as a first step that all countries should notify their quality requirements clearly on WTO website.

Table 34.3

Post-WTO Trade Scenario for Major Commodities and Implications for Future Negotiations and Strategy

Product(s)	*Trade Scene*	*Main Factor*	*Future Policy and Strategy*
Rice	Export adversely affected; Import threat.	Increased competition from developing countries like Vietnam, Thailand.	Improved competitiveness of domestic production.
Wheat	Export adversely affected; Import threat.	Low prices and subsidies and support in EU and US.	Seek elimination of export subsidies and domestic support in OECD.
Oilcake	Export adversely affected.	East Asia crisis and GM varieties in USA, Argentina and Brazil; Subsidies in USA.	Improved varieties of oilseeds particularly soyabean; Seek reduction in subsidies in USA*.
Sugar	Export adversely affected.	Subsidies in EU and USA.	Seek elimination of export subsidies and domestic support in EU and USA*.
Cotton	Export adversely affected; Imports increased.	Decline in domestic production and subsidies in USA.	Seek elimination of domestic support in USA; Technologies to compete with Bt cotton*.
Tea, Coffee, Spices	Export adversely affected.	Competition from Vietnam, Indonesia, Sri Lanka and other developing countries.	Improve competitiveness of domestic production; Do not seek too much protection for developing countries.
Horticultural Products	Exports increased; More scope.	Rising demand for high value and processed food.	Seek improved market access; Improve processing, packaging and transport of produce.
Meat and Meat Products	Export increased; More scope.	Preference for low cost and safe products.	Seek reduction in subsidies in US and North America*.

Contd...

...contd...

Product(s)	*Trade Scene*	*Main Factor*	*Future Policy and Strategy*
Dairy Products	Imports possible; Checked through tariffs.	Subsidies in EU, USA and Canada.	Elimination of export subsidies and domestic support in EU, USA and Canada*.
Soyabean Oil and Other Vegetable Oils	Serious import threat.	Superior technology in other major producing countries.	Improvement in domestic production technologies and processing.
Palm Oil	Very sharp rise in import which meets 40 per cent domestic demand.	Very cheap price and close substitution between different vegetable oils.	Consumer awareness about quality; Prohibit blending; Upgradation of technology of oilseeds in the country.
Wood and Wood Products	Sharp rise in import; Depressed domestic prices.	Low duty.	Raise duty.

Note * Developed countries are not going to totally dismantle domestic subsidies of different class in the near future. They will only agree for marginal reduction. Hence, our competition in exportable items has to reckon this reality in addition to the market dynamics across products. By and large, the market price fluctuations in primary produce is far higher than in processed products. Our competitiveness should incorporate this component.

Recent Round of Negotiations (13-18 December, 2005) at Hong Kong

It reaffirmed the declarations and decisions adopted at Doha as well as by the General Council in 2004 for full commitment to give effect to them. There is a general realisation that agriculture remains the most distorted and difficult sector in the WTO. On trade distorting domestic support, rationalisation was attempted by forming three bands with higher linear cuts in higher bands. The other decisions included developing discipline to achieve effective cuts in TDS (overall cut in TDS to be greater than the sum of individual cuts), exempting developing country members with no AMS commitments from reduction in *de minimis* and the overall cut in trade distorting domestic support, review of green box criteria to cover the interest of developing country members, parallel elimination of all forms of export subsidies by 2013, elimination of distorting trade practices of State Trading Enterprises, provision of special products and special safeguard mechanism to the developing member countries, elimination of export subsidies on cotton by 2006, and duty-free and quota-free access to cotton exports from LDC's.

The broad lessons from the Hong Kong negotiations are:

- Single country has no power in negotiations but a group of countries have.
- Safety nets are important to India.
- There is a need to reorient and reprioritise R&D according to the global trends.
- Domestic reforms in Indian Agriculture should have been started much earlier to improve competitiveness.

Enabling Policy and Institutional Mechanism to Safeguard Agriculture Under WTO Regime

Intellectual Property Rights (IPR) Issues

The technology must focus on the poor marginal and small farmers, who own nearly 80 per cent of the farm holdings being less than two ha. Several national acts have been passed and amended, and international treaties concluded over the years regarding IPR related issues. The scope of IPR covers seed, agri-chemicals, tools, input delivery, controlled systems, post-harvest agriculture, food products, non-food products, etc. Now, prior approval of National Biodiversity

Authority has been made mandatory for seeking IPR on innovations using national bio-resources and associated traditional knowledge. Patenting is highly pervasive and all embracing, including plants and animals, gene sequence, single nucleotide sequences, etc. in the developed countries. The impact of such patenting is likely to cripple research on the patented biodiversity in the developing countries. Patenting has to go along with the socioeconomic development as well as standard of science and technology development of a country. The patents have to benefit the public, and the patented products should be available to the public. While planning and protecting innovation, lot of protection, maintenance of IPR, evaluation of IPR assets, handling IPR infringements, full package *versus* unitary package of technology, etc. become critical issues to the institutions.

Technical Barriers to Trade (TBT) and Sanitary and Phytosanitary (SPS) Issues in Food Sector

The main objective is to protect risks to human beings, animals and plant life. The guiding principles should be transparency and avoidance of arbitrary SPS standards. In the negotiations, implementation issues are not adequately addressed. These mainly include reasonable interval/longer time frame for the developing countries to comply with other countries, new SPS measures, review of agreement at frequent intervals, participation of developing countries in setting SPS standards, technical and financial assistance to establish mechanisms, follow-up procedures, etc. The time framework for compliance for the developing countries should be raised from the present 6 months to 12 months. Similarly, 60 days period to react to revisions is a quite short for the developing countries. Equivalence is an important step for smooth implementation, but it is arbitrary at the moment. Equivalence issue should be a regular agenda in the SPS Committee, and agreement should be reached on the recognition of equivalence in accordance with the agreed procedure. There is lack of experts/specialists and lawyers knowledgeable about international law and science, and also non-availability of technology. Major steps should be to review SPS agreement at regular intervals, monitor the use of international standards, provide technical assistance to needy member countries and address the concerns of member countries in respect of special and differential treatment, regionalisation, specific trade concerns, use of *ad hoc* consultations, and to liaise with the Codex, OIE and IIPC.

The TBT like labeling, packaging, and specific nutritional attribute claims, are the other measures to protect consumer interests, and these are not restricted to food alone. In the US, there are about 26 TBTs, but no significant progress has been made in the recent negotiation, though at Doha it was decided to reduce and even eliminate TBTs. The major constraints in meeting food safety standards compliance include: poor raw material quality, costly laboratory tests, lack of hygiene awareness, lack of product and process standards, delay in cargo handling at the ports, etc. India may have to learn from success stories of other developing countries like Thailand and Malaysia, and soon gear up to address such issues.

State Level WTO Issues

The states are important stakeholders in trade and therefore, should be regularly consulted in trade policy matters. The experience and views on trade were discussed with special reference to Karnataka, which has made a mark in the export of silk products, Bangalore rose, onion, and gherkins. The State, however, faces the usual constraints of quality raw material shortage, infrastructure bottlenecks, high interest rates for agricultural investments, in addition to inadequate availability of export finance, small and fragmented holdings, poor power supply, inefficient Agricultural Produce Market Committees (APMCs), lack of market intelligence, lack of international airports and good seaports, and poor co-ordination with the Centre. The State has developed clear-cut short-term, medium-term and long-term strategies to enhance the export performance. Keeping these strategies in view, the State has also outlined the specific initiatives required for exporting its different agricultural products. Active participation of the states in policy discussions at the Centre is required. There is a need to establish a mandatory Nodal Agency (WTO Cell) with a two-tier structure, one at the apex level with the Ministry of Agriculture, Government of India and another at each of the State Headquarters linking the export houses of the State with the Agriculture Ministry. The other action points are:

- SWOT analysis and prioritisation of targeted export commodities.
- Establishing Commodity Boards for the targeted export commodities.

- State canalisation in niche commodities.
- Establishing export processing zones.
- Bringing the agriculture sector under Concurrent list.
- Strengthening the required homework and other detailed preparations both at Central and State level, before joining the negotiations at WTO.

Response and Reorientation of the R&D System

Response of R&D system to the changing trade scenario over the years was slow and less comprehensive. Recent experiences unequivocally suggest targeted approach to derive benefit from WTO. For the re-orientation of R&D systems, following features of our R&D system should be kept in view:

- Size and spread of National Agricultural Research System (NARS).
- Inability to fully and effectively mainstream the demonstrated benefits through externally aided projects like National Agricultural Research Project (NARP), National Agricultural Technology Project (NATP), Agricultural Human Resource Development Project (AHRDP), etc.
- Poor research–extension–farmer-market–policy linkage.
- State R&D system collapsing on account of sharp decline in State financial support.
- Lack of co-ordination between departments and ministries dealing with agriculture both at the Centre and the State level.
- Lack of effective public-private sector partnership.

Suggested broad directions for the reorientation are: (i) system related efforts, and (ii) programme related efforts. System related efforts to retool and re-energise our NARS include wide range of organisation and management reforms covering human resource development, providing further functional autonomy to our national institutes, infusing greater coordination among all development departments, research prioritisation, monitoring and impact assessment, IPR management, etc. These efforts will impart efficiency and accountability. Programme (Science) related efforts include priority attention to targeted export commodities, pursuing research on production to consumption system, strengthening basic and strategic research in frontiers of agricultural sciences, etc.

Policy Recommendations

WTO is receiving the deepest indulgence of everyone, as it is affecting the major sectors of Indian economy and agriculture in particular now, and more intensively in the coming years. A major concern growing with the increasing impact of WTO is, as to how the small and marginal farmers' who dominate the Indian agriculture, depend heavily on agriculture for their livelihood, have small marketable surplus and operate under heavy constraints to be competitive in a subsidised agriculture production and trade regime, could benefit from WTO. The concern more often swings to the other side that the spreading tentacle of WTO with reduced tariff regime and increased access to Indian market for the products from subsidised agriculture could severally damage the agriculture based livelihood of majority of Indian farmers. The challenge to policy makers is how to protect Indian agriculture from the impending WTO threat, enhance the competitiveness of Indian farming and make farming a viable and self sustaining enterprise to improve and ensure livelihood security of the farmers. A strategy to address this challenge shall necessarily involve reorientation and injection of market linked dynamism in Indian agricultural R&D, strengthening of supportive institutions to serve the resource poor farmers, and steering fast the change with appropriate policies and trained humanware. The deliberations of the workshop suggested the following policy initiatives and action points:

- India needs to devise appropriate domestic policies (extensive domestic market reforms, heavy investment in building and maintaining infrastructure, etc.) to improve efficiency and competitiveness of domestic produce.
- It should continue to play leadership role in negotiating agreements with sound analytical basis and support of other developing countries with similar interest. A dedicated group of about 100 experts, on full time basis, should work on the WTO issues to provide analytical basis for negotiations and to help in planning appropriate strategies to strengthen Indian agriculture to face increasing trade liberalisation and globalisation.

- Export of high value products, horticulture products, processed products, marine products and rice (details given in Table 34.1) should be promoted.
- India has to counter the challenges in the export of traditional items from the developing countries. In this regard, prioritisation, enhancing production and processing efficiency, marketing and transport infrastructure, maintaining quality, stable supply, etc. need immediate attention.
- There is a need to go whole hog in reforming domestic market as has been done by China. The constraints of multiplicity of laws in agricultural marketing, processing, storage, transport should be immediately addressed to impart simplicity, reduce transaction costs and attract private investments in post harvest management. Further, enforcement of Model APMC Act which encourages direct marketing and contract farming should be implemented in true spirit.
- There should be continuous insistence for reduction in domestic and export subsidies in the developed countries. Developing countries should put up a unified approach to ensure substantive reduction in subsides by the developed countries.
- All countries should notify their quality requirements of agricultural produce clearly on the WTO website.
- There is a need to assess the priorities of our national projects including the R&D programmes, and to develop and maintain the quality of infrastructure at the highest level. There should be strict monitoring of the national projects to ensure expected output as per the time schedule.
- Since IPR is becoming an important issue, a clear-cut policy on IPR, its protection, maintenance, evaluation, handling of IPR infringements, etc. should receive priority attention at the Central, State and institutional levels.
- As regards SPS issues, India should demand reasonable internal/long time frame to comply with other country's new SPS measures, review of agreement at frequent intervals, providing technical and financial assistance to establish SPS

testing/certification mechanisms, correctly defining the equivalence, capacity building for developing expertise in the area, etc.

- A special campaign is required to create awareness for appreciating quality aspects of farm produce among the farming community, traders, consumers and exporters.
- India has to learn from the experience of other developing countries like Thailand and Malaysia regarding compliance to food safety standards.
- The Centre should consult states and receive their active support for trade policy formulation, WTO negotiation, etc.
- There is necessity to establish Export Processing Zones and Commodity Boards for targeted commodities.
- Agriculture needs to be brought under the Concurrent list of the Constitution.
- There should be a merger of several related departments like irrigation, fertiliser, food, agriculture, etc. for better coordination and synergy.
- Public investment in agriculture has to be raised, particularly in R&D including extension.
- The system related efforts towards reorientation of R&D system will include general moratorium on establishment of new institutions, development of first rate human resource through quality agricultural education, need-based training in India and abroad, coordination and convergence of all development departmental efforts, harnessing ICT for rural development, adequate funding of research programmes, project based funding/budgeting, rigorous monitoring and impact assessment, strengthening social science skills, promoting public-private sector partnership, building leadership skills, reforms in financial and procurement management with full decentralisation, strengthening policy analysis and vision-oriented market-led intelligence analysis skills, strengthening agri-business development and IPR management, and campaigning for better awareness and compliance in respect of produce quality by the farmers, traders, exporters, importers and the general public.

- The programme related efforts towards reorientation of the R&D system will include priority attention to targeted export commodities particularly high value processed products, strengthening basic strategic and anticipatory research at ICAR and downstream research at SAUs where 4/5 of the scientists of NARS work, agro-ecological targets and functioning rather than national targets and functioning by the R&D system, profitability besides productivity as indicator of success, linking production with processing, marketing and consumption with focus on small and marginal farmers and farm workers, establishment of quality testing/referral labs, research on minimal non-renewable natural resource use, developing high yielding varieties which must combine high yields, high protein content and other characteristics demanded by the importing countries such as freedom from aflatoxins.

Summing Up

Post-WTO trade liberalisation helped India to achieve small increase in agricultural exports, whereas it resulted in sharp and continuous increase in imports. This has adversely affected self reliance in agriculture. The foremost reason for this adverse effect is unprecedented decline in international prices, which in turn was caused by attempts by almost all the countries to push exports and continuation of high level of domestic support and export subsidies by OECD countries. As a net exporting country India stands to gain from increase in international prices.

The last Ministerial meeting to discuss and finalise modalities of WTO agreements was held from 13-18 December, 2005 in Hong Kong. However, no consensus could be reached once again, and the meeting ended up with a resolve to complete Doha Work Programmes and to conclude negotiations successfully in year 2006. Again, year 2006 is coming to an end and there is no sight of any progress in the negotiations. Serious differences persist between three main groups of WTO members, namely, developing countries led by G20, EU and USA on almost all the issues like reduction in domestic support and export subsidies in OECD countries, formula for tariff reduction and special and differential treatment for developing countries in almost all the commitments. Therefore, there is a stalemate which implies that imbalances of UR agreement would continue. This stalemate has given freedom to the developing countries not to further liberalise their markets. On the other hand, the delay in concluding new agreement is helping the developed countries in several ways. With postponement of negotiations, the dates for implementing fresh commitments to reduce domestic support and export subsidies are automatically getting shifted to distant future. Thus, countries like USA and EU, are free to follow trade distorting policies in the absence of fresh commitments. For instance, the offer of EU in year 2000 to phase out export subsidy in next 10 years and the consequent failure to reach any agreement in 2000, provided EU, freedom to continue with these subsidies. Further, the postponed negotiations, say in year 2007, equipped EU with better bargaining position, as it can now offer to phase out same subsidies in a much shorter periods and still have a better deal compared to the offer in year 2000. While it is highly desirable to eliminate export subsidies, phase out domestic support and have separate set of commitments for developed and developing countries, it is also important to recognise that the deadlock is not helping anyone to achieve these goals. India and other developing countries need to put pressure on the developed countries to have some interim agreement to at least reduce domestic support and export subsidies as per their initial offers.

35

Globalisation of Agriculture: R&D in India

Introduction

Globalisation of the national economies which seek to reform the trade sector beset with various trade barriers arising from market restrictions such as tariffs, preferential quotas, quantitative limits, and non-tariff measures including phytosanitary stipulations, etc., is highly distorted currently. Competitiveness of countries in individual products/commodities is expected to play a major role in the international trade. Through globalisation, as trade barriers are reduced and unfair competition is disciplined, freer trade is expected to take place for many commodities.

India is a signatory to the Uruguay Round Agreement of the General Agreement on Tariffs and Trade (GATT). This makes it mandatory for India and all member countries to open up their economies to the world market. In this process, agriculture will be a key sector. For the first time in world trade history, the World Trade Agreement of 1994 brought agriculture within its policy framework. With the signing of WTO Agreement on Agriculture in 1994, India, alongwith 119 other member countries, will have to implement the agreement. The obligations and disciplines incorporated in the agreement which seek to reform trade in agriculture and provide the basis for market-oriented policies on agriculture, relate to the aspects of market access, domestic support, export competition/subsidies, sanitary and phytosanitary measures, and Trade Related Intellectual Property Rights (TRIPS).

Provisions on market access in the agreement primarily envisage tariffication of all non-tariff barriers and progressive reduction of tariff levels within a stipulated time (36 per cent over 6 years in the case

---

Policy Paper 10, July 2001.

of developed countries and 24 per cent in 10 years in the case of developing countries, with the least developed countries being excluded) as well as removal of quantitative restrictions. Domestic support measures that have a minimal impact on trade are excluded from reduction commitments. By quantifying domestic support measures termed as Aggregate Measure of Support (AMS) and its progressive reduction, the provisions in the agreement aim at disciplining trade-distorting support to farmers. To facilitate monitoring, these measures are defined and identified into three categories and stylised as boxes: Green, Blue and Amber on the basis of their impact on trade on an increasing scale and quantified at different base periods. Green box measures refer to services or benefits to agriculture or to the rural community including research and extension services with no time stipulation; Blue box to production limiting programmes to be pegged to the 1992 level; and the Amber box to measures that distort market prices such as price support, direct payment and non-exempt subsidy, pegged to 1986-1988 level and subjected to reduction progressively. Provisions in the agreement establish ceiling on the value as well as volume of subsidised exports. Invoking sanitary and phytosanitary measures shall be based on international standards, guidelines and recommendations to avoid arbitrariness and/or unjustifiable discrimination in their application.

In terms of exports, India is not a major player in the world agricultural commodity market except in the case of a few commodities such as rice, spices, soya meal, cashew, tea and coffee. As there is persistently increasing trade deficit for the country as a whole, increasing net exports of agriculture is a significant contribution to the country's economy. Hence, in the coming years, India has to focus on its agriculture not only to meet the domestic demand but also to increase its export to help the rest of the economy. The internal market is not adequate enough to absorb all the farm commodities that this country is capable of producing. We, therefore have to look for external markets. Value addition by far is the major opportunity for employment in the rural sector as biophysical resources are limited and access to which is denied to many, on account of the iniquitous distribution in ownership. The value-added products so generated would also have to seek external markets to get them absorbed. Massive entry of India's agricultural products and commodities in the world market is hence imperative for making our agriculture the engine of growth and enhance the livelihood security of the dependent population who are substantially resource poor.

Six years have lapsed since the World Trade Agreement (WTA) on Agriculture has become operative. The experience hitherto is under review by the WTO. Proposals for changes in the Agreement are being considered this year. Proposals for changes should have been sent by December 31, 2000. It is inevitable that opening up of India's agricultural sector to the world market will have implications on prices, incentive structures and various forms of protection. It is time that we take stock of our experience and performance so that we can prepare adequately to meet the challenges ahead. The workshop[1] reviewed the experience so far this country had in the implementation of the various provisions of the WTA; examined the impact of globalisation of agriculture on agricultural production, changes in input efficiencies and influence on livelihood security of the dependent population; identified the challenges and opportunities, more specifically, examined and appraised the role of R&D in agriculture in the changing scenario to take advantage from the globalisation process; and evolved strategies and proposed measures to maximise the opportunities and minimise the impact especially on the livelihood security of the vulnerable sections of the society, and came out with the following recommendation.

Recommendations

- Experience of the last six years has shown that the expected share of the markets of developed countries is yet to be realised by the developing countries. Market access continues to remain restrained on account of tariff barriers and stipulation of sanitary and phytosanitary measures. Through mechanisms such as the Green and Blue boxes as well as through very high tariff barriers (like Japan's 2000 per cent in the case of rice, it is clear that the industrial nations have succeeded in ensuring that the hope for enlarged market access to agricultural commodities from developing countries does not materialise. In contrast, our markets have been

1. The National Academy of Agricultural Sciences (NAAS) sponsored a workshop on "Globalisation of Agriculture: R&D in India" which was held at the Kerala Agricultural University, Vellanikkara, Thrissur, Kerala between 2-3 February, 2001 and deliberated on the emerging concerns.

opened up. Quantitative restrictions on the import of agricultural products will have to be removed with effect from April 1, 2001. The extent of domestic support provided by India to farming families is far below the prescribed ceiling. We also do not provide export subsidies (in contrast, total farm support in OECD countries increased by 8 per cent to USD 363 billion in 1998).

- Consciously or unconsciously, India entered into an unequal and unjust trade bargain in 1994. There is no use looking back. While being fully conscious of the various implications of the nation's participation in the WTO, the deliberations focused on charting out the preparedness of the nation to take advantage of the opportunities offered by the WTO and to meet the emerging challenges.
- Various concerns, which would have implications on sustainable agriculture and ensuring the livelihood security of the farmers, included factors that enhanced obligatory market access such as tariffication of non-tariff regulations, removal of technical barriers to trade, transparency in sanitary and phytosanitary regulations; minimising domestic support without eroding the livelihood security of the vulnerable groups; promoted export competitiveness; IPR regime that safeguards the interests of the farmers and the conservators of biodiversity. Every measure should be taken to safeguard the livelihood security of the farmers, the vast majority of whom are resource-poor operating small and marginal holdings.
- In the short run, India's competitiveness is good. We have to enhance it by providing adequate support as the input prices are rising faster than the output prices.
- Seventy per cent of our population derive their livelihood from agriculture and 80 per cent of the farmers are small and marginal category. Our post-harvest and sanitary infrastructure is very poor and massive investment in this area will be essential if we are to ensure a level-playing field in external trade in agriculture. The subsidies and other support measure should be continued by designating these under the Livelihood box to tide over the initial problems but could be revoked at a subsequent time, say, after 10-15 years. Restraints on subsidy to farmers can be overcome by canalising support

through Commodity Boards as is being done in the developed countries. But such support is granted not through direct subsidies to facilitating sustainable production through building infrastructure.

- To make use of the provisions under the Green box, an adjustment fund of USD 2.5 billion be created for investment in research technology support, infrastructure building, marketing etc. so that Indian agriculture will be competitive and a level-playing field is provided.
- Provision to bound tariff should be made use of as a special safeguard.
- That the nation's biophysical resource have already been stretched to the limits, and that future agricultural growth is to be sourced from intensification and technological progress, systematic and continuous efforts are necessary to identify and determine the R&D support required in each of the five areas referred earlier in terms of, (a) enhancing competitiveness through cost-effective productivity increases and improvement of quality in the domestic and foreign markets, (b) standardisation of phytosanitary measures, and (c) strengthening R&D institutions and their capabilities. Further, that such R&D support should be mission oriented, technology research has to be integrated with financing mechanisms.
- India's competitive advantage primarily arises from the relatively less expensive labour. By adopting capital-intensive high technology, this advantage will be lost. Therefore, caution should be exercised in choosing the technology.
- Considering the enormous variation in biophysical resource endowment in the country, diversification was seen as a means to improve the trade competitiveness. Small farming systems have to be integrated with diversification process which require work with communities. There is need for policy integration between land, water and crop development factors and financing.
- Decisions related to WTO are being taken and negotiations are being done by the central government. Even though the brunt of the problems emanating from such decisions are to be borne by the state governments, they are not consulted or

taken into confidence. A mechanism be created so that the issues concerning WTO are discussed and differences thrashed out in order to arrive at a consensus and further, that no decision is taken without proper consultation and agreement with the state governments.

- There is need to develop information systems with which Indian agriculture can respond quickly in a globalising phase and that the knowledge gap with respect to the opportunities and liabilities of entering into the WTO is enormous. There must be a cooperative effort between the State, the Centre and various institutions including NGOs and a platform should be created so that the issues identified can be thrashed out regularly.
- There should be stronger interface with trade in the country, who are more aware of the market conditions, demand as well as consumer preference and their services should be made use of.
- There is acute shortage of information with respect to the various implications of the WTO regulations. In the absence of adequate data, state governments are not in a position to take advantage of the various provisions that would safeguard the interests of the farmers and the vulnerable-groups among them. It was, therefore, recommended that WTO Cells be created in the state governments and in the agricultural universities so that relevant information is collected on a continuous basis.
- It is important to prepare a checklist of various elements which constitute the regulatory provisions under the WTO rules so that the states are aware of the kind of information to be generated and made available at the time of negotiations.
- Scientists should play an important role in the negotiations where science is involved. Issues such as, specification of phytosanitary measures, gender empowerment, sustainability etc., should be taken out of the trade and should be left to the decisions of the scientist-professionals.
- Planning of agricultural research should take into consideration the opportunities which have now become available for India to emerge as a major exporter of diverse

> commodities and products. The foremost requirement in this regard will be greater emphasis on quality of our products. Agricultural scientists for understandable reasons have been preoccupied in the past with the development of technologies which will help to increase production in the interest of food security and self-sufficiency, in general. The high yielding varieties of future must combine high yields with traits like high protein content and other characteristics demanded by the importing countries such as freedom from aflatoxins.

The Workshop realised that trade policy is an instrument of national economic development policy and it requires structured responses. We have to prepare a blueprint for the future to take advantage of the process of globalisation for which the nation is capable of. And it is time to begin now. We need a vision of the directions in which we should move. Centrepiece of that vision is that we build a strong Indian agriculture, mainly driven by a growing and diversifying domestic market and larger trade volumes in the external market. While doing so, the livelihood security of the dependent population is not only safeguarded but enhanced taking globalisation as an uncommon opportunity.

36

Sanitary and Phytosanitary Agreement of the World Trade Organization—Advantage India

Introduction

Sanitary and Phytosanitary (SPS) Agreement of the World Trade Organization (WTO) is aimed at developing measures which would ensure protection of human, animal or plant life or health, and equivalence and transparency in global agricultural trade. SPS measures are not new to global agricultural trade. However, a specific 'Agreement on the Application' of SPS measures was negotiated during the Uruguay Round for giving impetus to agricultural growth through SPS measures, derived from scientific principles, instead of SPS being used as disguised restrictions to global agricultural trade.

SPS measures are applied to both domestically produced and imported goods to protect (i) humans from animal and plant-carried diseases, (ii) plants and animals from pests or diseases, and (iii) territory of a country from spread of a pest or disease. To achieve these objectives, SPS measures must have adequate quarantine regulations and monitoring and certification system regarding how the goods are produced, processed, stored and transported for ensuring the quality of final products. The perception that crops, genetically modified through biotechnological interventions may negatively impact human and animal health and environment, adds another dimension to be addressed through SPS measures.

There is a growing concern that certain SPS measures impede the export of agricultural produce from developing countries because the latter are not adequately equipped to address various issues related to SPS Agreement. Factors that put developing countries in a disadvantageous position are:

Policy Paper 12, September 2001.

- Lack of information on aspects that affect exports.
- Lack of infrastructure for scientific research, testing, conformity assessment and equivalence.
- Non-technical persons influencing international standard-setting process.
- Inability to benefit from the provision of transparency and adaptation to regional conditions due to inadequate infrastructure and technical support.
- Inadequacy to utilise the provisions relating to special and differential treatment for developing countries. The provisions in practice remain of theoretical interest.

This paper is the outcome of the deliberations of a seminar organised on, 'Sanitary and Phytosanitary Issues', by the National Academy of Agricultural Sciences. It suggests approaches that would overcome the above deficiencies, in general and improve India's ability to use the SPS Agreement to its advantage in the matter of global agricultural trade.

It is essential for India to develop infrastructure and policies to respond to the SPS measures in their target markets as well as expectations of the consumers, by providing good quality and safe products at reasonable prices. To achieve this objective the country must ensure that: farming remains competitive and profitable; export policy is aggressive; domestic capabilities are strengthened for implementing SPS measures; products which we do not wish to enter our markets, because of the potential negative impact on health, agriculture or environment, are identified proactively. All this will be necessary to ensure free and safe trade in agriculture.

Landmarks in the Development of the Agreement on the Application of Sanitary and Phytosanitary Measures

- 1974-1979: Inclusion of SPS concerns for food safety and health of humans, animals and plants in the Agreement on Technical Barriers to Trade (TBT) during the Tokyo Round.
- 1986: Call for increased discipline in agricultural trade including sanitary and phytosanitary measures as per the *Punta del Este* Declaration, which launched the Uruguay Round in September 1986.

- 1988: Mid-Term Review of the Uruguay Round in which the following priority areas related to SPS measures were identified: (a) harmonisation of the SPS measures on the basis of international standards, (b) transparency through effective notification of national regulations, and (c) improvement in the dispute settlement process including bilateral resolution of disputes and scientific support.
- 1990: Formation of the Working Group on Sanitary and Phytosanitary Regulations. SPS measures included in a separate draft agreement.
- 1991: The Director General of the General Agreement on Tariffs and Trade (GATT) issued "Dunkel Draft". The draft included sanitary and phytosanitary issues. The final text of the Agreement on the Application of Sanitary and Phytosanitary Measures which was approved at the end of the Uruguay Round was largely based on the Dunkel text. It fulfils the general objectives of the *Punta del Este* Declaration in this area.
- 1993: Establishment of the Secretariat of the International Plant Protection Convention (IPPC) and start of standard-setting process related to the SPS measures.
- 1995: Establishment of a Dispute Settlement Mechanism under the WTO which provides for mutually acceptable solutions or adoption of a panel/Appellate Body ruling by the Dispute Settlement Body (DSB).
- 2000: SPS Committee completes draft on risk consistency. The draft provides guidelines (not legally binding) on levels of health protection.
- 2000: Finalisation of United Nations Agreement concerning trade in genetically modified organisms (GMOs).

The Goal and Main Objective of the SPS Agreement

The Agreement requires the member countries to develop domestic SPS measures that: (a) are technically sound and economically feasible to implement, (b) are based on geographical, climatic and epidemiological conditions prevailing in different regions of the country, (c) do not discriminate in favour of domestic producers, (d) recognise disease-free areas of other countries and allow import of products originating from such areas, (e) permit introduction of measures providing a higher level of protection than achieved by measures based on international standards, if these are scientifically warranted and are based on risk assessment, and (f) allow provisional adoption of SPS measures as a precautionary step in cases where scientific evidence is insufficient but probable risk of the spread of pests or diseases exists in the importing country.

International Standards

Article 3 of the SPS Agreement, envisages application of international standards as basis for SPS regulations of the member countries. Application of this Article is essential as divergence of standards and regulations adds cost to international trade. However, divergence arising from legitimate differences in societal preferences, technological developments, environmental and health conditions, may be justified. In such cases, mutual recognition of standards is a desirable option. On the other hand, where divergences are not justified, harmonisation of international standards should provide appropriate solution. In either case, the crucial requirement is the efficiency and fairness of the process of international standard-development. Benefits of harmonisation may not reach equally to all the members if the process is guided by motives that exclude other market participants or if the procedure is not adequately transparent. International standards developed for food safety by the Codex Alimentarius Commission, for animal health by the International Office of Epizootics and for plant protection by the Secretariat of the International Plant Protection Convention are recognised by the SPS Agreement. For areas not covered by these organisations, the SPS Committee recognises standards developed by other relevant international organisations.

Codex Alimentarius Commission

The Codex Alimentarius Commission, a joint FAO/WHO Commission, with a membership of over 160 countries, has developed the Codex Alimentarius, a collection of international food standards for all principal food products. The Codex Alimentarius includes nearly 5000 standards, aimed at protecting the health of consumers and for ensuring fair practices in the food trade.

India is a member of the Codex Alimentarius Commission (CAC) and the Ministry of Health is the Nodal Agency. The country has passively adopted the Codex Alimentarius and India does not have any process for developing internal standards, although a National Codex Committee and a Central Committee for Food Standards exist. These Committees should be armed with infrastructure for testing to ensure that international standards are suitable to the Indian situation if not to develop appropriate alternative standards. There is also a need to establish a network of laboratories to certify that the products meet international/national standards. The key issues are the chemical/pesticides residue limit and detection of food-borne pathogens. The National Codex Secretariat (NCS), housed in the Ministry of Health, is a non-entity. Considering the new challenges, the NCS should be provided with appropriate technical manpower and facilities to effectively coordinate and facilitate agricultural trade related to CAC requirements.

India has a great strength in producing organically grown food-using organic manures, soil-enriching cropping sequences and biological management of pests. Organic farming should be further strengthened by providing alternative energy sources to the farmers so that organic matter burnt as fuel becomes available for soil enrichment. The country has encouraged alternative energy sources like biogas and solar energy but more aggressive efforts are required through government and non-government extension agencies, to highlight the benefits that would accrue to the farmers. With increasing awareness of environmental concerns and health consciousness, national and international demand for organically grown food would grow and farmers would get higher returns for their produce. To take advantage of the emerging opportunities, the country must be supported by guidelines for the production, processing, labelling and marketing of organically grown food and putting in place a 'certification system' to assure organic production of a commodity. Certification system has to be based on continuous objective inspection, quality auditing of the inputs as well as the produce.

The International Plant Protection Convention (IPPC)

The IPPC Secretariat was established in 1993, for phytosanitary standard-setting and harmonisation of phytosanitary measures affecting trade. The nodal agency for setting phytosanitary standards in India, is the Directorate of Plant Protection, Ministry of Agriculture. The main objective of IPPC is to, "prevent the spread and introduction of pests of plants and plant products and to promote measures for their control." Special emphasis is on the 'quarantine pests' which are either not present in the country or are not widely distributed and have the potential of causing economic losses. The pests include 'invasive species' like weeds, which not only reduce yields of crop plants and add to the cost of cultivation but also pose health hazards. *Parthenium hysterophorus,* commonly known as 'congress grass', is an apt example of such introduction. It was inadvertently introduced in the country in early sixties through wheat consignments. It first established itself in areas adjoining the ports of entry but soon spread to all parts of the country causing serious human health problems, particularly asthma and skin diseases.

According to the present mandates, IPPC has responsibility for genetically modified organisms (GMOs) also, because, GMOs are included in the category of invasive species. Since, GMOs do not fall in the category of invasive species, a review of mandate allocation is called for.

The National Plant Protection Organisation i.e. the Directorate of Plant Protection has major responsibility of not only preventing introduction of exotic plant pests but also to ensure and certify that plants and plant products exported from the country are free from pests. It is a tall order and requires a very efficient and competent system to perform the following functions as per the IPPC guidelines:

i. inspection of growing plants, of areas under cultivation (including fields, plantations, nurseries, gardens and greenhouses), and of plants and plant products in storage or in transportation, particularly with the object of reporting the existence, outbreak and spread of plant pests and controlling those pests;

ii. inspection of consignments of plants and plant products moving in international traffic and, where appropriate, inspection of consignments of other articles or commodities moving in international traffic under conditions where they may act incidentally as carriers of pests of plants and plant products, and the inspection and supervision of storage and transportation facilities of all kinds involved in international traffic whether of plants and plant products or of other commodities, particularly with the object of preventing dissemination of pests of plants and plant products across national boundaries;

iii. disinfestation or disinfection of consignments of plants and plant products moving in international traffic, and their containers (including packing material or matter of any kind accompanying plants or plant products), storage places, or transportation facilities of all kinds employed;

iv. issuance of 'phytosanitary certificates' relating to phytosanitary conditions and origin of consignments of plants and plant products;

v. determine the distribution of information within the country regarding pests of plants and plant products and the means of their prevention and control;

vi. conduct research and investigation in the field of plant protection;

vii. report on the existence, outbreak and spread of economically important pests of plants and plant products which may be of immediate or potential danger; and

viii. dissemination of information on means found to be effective in controlling the pests of plants and plant products.

The Directorate of Plant Protection, which has been charged with the above responsibilities' initiated activities, started mainly for plant quarantine purposes, at the major airports and seaports. Gradually, particularly with the ratification of WTO treaty, their responsibilities increased but infrastructural facilities did not improve commensurate with the additional responsibilities. Recently, four state-of-the-art facilities have been established at Delhi, Chennai, Calcutta and Amritsar with the UNDP/FAO assistance, but these laboratories lack adequately qualified technical manpower and the resources to effectively perform all the assigned functions. For a country of our size having more than 50 ports of entry, many more such facilities are required. We also do not have complete understanding of the SPS measures of our target markets, mainly due to the lack of direct technical contact with the concerned authorities in those countries. This gap can be bridged if the country posts a technically qualified agricultural *attaché* at our diplomatic missions in the countries of interest.

The Office of the International des Epizootics (OIE)

The objectives and functions of OIE are, harmonisation of health requirements and adoption of international standards for international trade in animals and animal products. It is unfortunate that international standards are developed with very limited participation of developing countries. As a consequence, international standards are often inappropriate for use, as basis for domestic regulations, in developing countries. Also, developing countries face problems when they have to meet standards imposed by the importing markets of the developed countries.

India has vast wealth of animals and fishes which could become an important component of international trade if efficient and transparent mechanism of meeting international standards are operated in respect of animal health. Our present system does not inspire confidence. Even though a large number of animal diseases have been reported during the last five years, no information has been given about the number of animals destroyed, slaughtered or vaccinated to prevent spread of diseases. It is necessary to undertake pest risk assessment, identify areas/regions free from such pests, take appropriate measures to maintain disease-free status and develop eradication programme for diseases which could hamper international trade. Similarly, the reports on zoonoses do not give the number of cases but simply indicate'+' suggesting that a disease has been reported. Such an approach cannot build confidence in markets looking for animal products from India. The mechanism of issuing SPS certificates for the export of meat and fish products in various parts of the country requires harmonisation. Laboratories accredited to test the products meant for export should have adequate infrastructure to meet international standards.

India is advantageously placed for international trade in meat products as it produces lean meat, low in cholesterol—free of growth promoters and chemicals. The country should take measures to promote meat industry by encouraging animal farming to produce quality meat based on scientific principles.

The Standard-setting Process

The procedures for formulating standards vary among the international standard-setting organisations. Therefore, harmonisation of procedures is required for a coherent, transparent and effective system of international standardisation. The procedure should ensure meaningful participation of developing countries while developing and adopting standards taking into account the conditions prevailing in different countries. India can play an important role in debating and suggesting approaches to harmonisation of procedures, by sending technical representatives to the international meetings related to WTO/SPS, so that international standards do not become disguised barriers to trade. The process of setting international standards should be transparent to ensure reconciliation of conflicting opinions. This would require suitable modification of Article 3, so that the international standards are developed through a fair process based on consensus where all countries are effectively represented.

The safety measures and SPS standards must be based on science. India, therefore, should embark upon developing scientifically sound SPS measures, and be prepared to challenge the SPS measures of trade partners on scientific basis when required. Some countries may have difficulty but India has the advantage of having well-trained human resource which can undertake scientific pests risk assessment (PRA). Our technical resources can also be made available to other countries under suitable framework of international cooperation.

Equivalence

Article 4.1 of the SPS Agreement contains the important provision of accepting as equivalent the SPS measures of other members, if the exporting country demonstrates that its measures achieve the importing member's appropriate level of sanitary and phytosanitary protection. Some countries interpret this provision as 'sameness', instead of equivalence of measures and deprive Article 4.1 of its function, of recognising that different measures can achieve similar levels of SPS protection so that the member countries can adopt suitable SPS measures to ensure adequate SPS protection. There is a need to clearly define 'equivalence' in the SPS Agreement.

Equivalence is the ideal option when international standards are lacking or are inappropriate. India has diverse climatic, developmental and technological conditions. If its SPS measures are accepted as equivalent to those of importing countries, it will be an important instrument of enhancing market access for its agricultural products. India can also strive to develop regional or sub-regional agreements related to equivalence in SPS measures. Gradually, such measures could be harmonised within and outside the region. A major difficulty in utilising this clause of the Agreement is the non-acceptance of conformity assessment certificates by some member countries. Therefore, it would be useful to include the provision of the internationally supported accredited institutions for certification.

Mutual Recognition Agreements

Article 4.2, encourages member countries to develop Mutual Recognition Agreements (MRAs). The MRAs could either be limited to purposes like the testing methods, or they can cover all aspects including the standards. MRAs play an important role in building confidence and form a base for developing broader MRAs. The MRAs reduce costs by avoiding duplicative testing and eliminating delays. These also build valuable databases through intensive exchange of information and close contacts between the concerned laboratories in different countries. Confidence building leads to MRAs which consider the domestic requirements of the parties as equivalent so that a commodity which can be legally sold in one country may be legally sold in the other(s). India should utilise the provision of MRAs and accreditate laboratories assigned testing functions. Such laboratories should develop linkages with similar laboratories of other countries with comparable agroclimatic conditions. This will ultimately lead to regional trade agreement aimed at pooling human resources for research as well as development and establishment of regional or sub-regional laboratories, certification bodies and accreditation institutions.

Transparency

Transparency is becoming an international requirement for establishing credibility as a basis for promoting trade. SPS measures, crafted on sound science can be given wide circulation through electronic media for notification. India has built an efficient Internet-based infrastructure, which will be useful for networking of relevant public and private institutions and publishing relevant information on the world wide web. India should also take advantage of the SPS Committee for highlighting difficulties experienced by it, if any, relating to SPS measures of other member countries.

Adaptation to Regional Conditions

Article 6 of the SPS Agreement, deals with the adaptation to regional conditions, including recognition of pest- or disease-free areas or areas of low pest or disease prevalence. It is an important clause for countries like India, which have a vast geographical area and diverse agroclimatic conditions. To benefit from this clause for agricultural export, the country should identify and maintain some areas free of plant and animal pests of significance to international trade. Where feasible, cost-effective programme for eradication of a particular pest or disease can also be taken up. If required, external expert assistance can also be sought, although, considering the expertise available in the country, such assistance may not be large.

Declaring an area free of pest or disease requires scientific documentation for its acceptance by the relevant international organisations. Elaborate procedures have been developed for such recognitions. It is important that we identify export products, for example, wheat and mango, which face export restrictions due to diseases like 'Karnal bunt' and pest like 'stone weevil', respectively. Export of such items can be made from the pest- or disease-free areas. India should also seek inclusion of a clause in Article 6 to the effect that once an area is recognised as pest- or disease-free all the member countries will accept it without need for additional assessment.

Special and Differential Treatment

Article 10 of the SPS Agreement provides for special and differential treatment for the developing and least developed countries. India should prepare a list of major agricultural export products, identify the main constraints these products face in the countries of destination and request these countries (and/or the relevant international organisations) to facilitate export of the listed products. As per the provisions of Article 10, assistance could be sought for, eradicating a disease or pest, improving packaging and transportation, developing good manufacturing practices for individual agricultural products and training of personnel. So far, limited use has been made of these provisions.

Technical Cooperation

To help in the effective implementation of SPS Agreement, Article 9.1 provides for assistance to developing countries either on a bilateral basis or through appropriate international organisation. The main purpose of this provision is to enable developing countries to establish necessary infrastructural and other conditions required to promote international trade in agriculture. Technical cooperation could be used for building capacity of the officials; upgrade the technical skill of personnel working in laboratories, certification bodies and accreditation institutions, and risk assessment for a particular pest.

Enquiry Point

All member countries are required to establish an enquiry point to receive and respond to requests for information, regarding domestic SPS measures, including new or existing regulations and decisions based on risk assessment. The enquiry point can also be the nodal point for notifying the WTO Secretariat. Since, India does not have a single window enquiry point, it should utilise provisions of SPS Agreement to establish an enquiry point. Major functions of the enquiry point would be to: (a) serve as notification authority, (b) coordinate with SPS related Committees, (c) represent country's interests in international fora, and (e) identify and resolve technical issues related to agricultural trade.

Summary of Recommendations

- SPS Agreement envisages use of international standards as a basis for SPS regulations of the member countries. At present, India passively adopts the standards developed by other countries. There is an urgent need to establish referral laboratories for testing the international standards for their suitability to the Indian situation and developing standards, which would be in the interest of country's agriculture. The key issues are the chemical/pesticides residue limit and detection of food-borne pathogens.
- India has a great strength in producing organically grown food. To take advantage of the emerging opportunities the country must strengthen organic farming and develop guidelines for the production, processing, labelling and marketing of organically grown food, and put in place a 'certification system' to assure organic production of a commodity.
- The Directorate of Plant Protection, which has the responsibility of "preventing the spread and introduction of pests of plants and plant products and to promote measures for their control", started mainly for the purposes of plant

quarantine at the major airports and seaports. With the ratification of WTO treaty, their responsibilities increased but not the infrastructural facilities commensurate with additional responsibilities. Recently, four state-of-the-art facilities have been established, but these laboratories do not have adequate qualified senior technical manpower and resources required to effectively perform all the functions related to SPS measures. Moreover, considering the vast size of the country having more than 50 ports of entry, many more such facilities are required. The Department requires strengthening and complete reorganisation so that it is able to address various SPS related issues, effectively.

- The present situation is not satisfactory due to overburdening of limited resources and lack of general concerns about domestic as well as the international SPS measures. For example, the Immigration Declaration Form at the Indian ports of entry does not have any clause for declaring plant parts or food products of plant and/or animal origin being carried by the passengers. At the arrival lounge, there is no proper space and trained personnel to meet the functional needs of plant quarantine; both need correction.
- India has a vast wealth of animals and fishes which could become an important component of our international trade, if we put in place an efficient and transparent mechanism of meeting international standards in the field of animal health.
- India has an advantage of international trade in meat products as it produces lean meat which is low in cholesterol and is free of growth promoters and chemicals. However, it is necessary to take measures to promote meat industry by encouraging animal farming to produce quality meat, based on scientific principles.
- The mechanism of issuing SPS certificates for the export of meat and fish products in various parts of the country requires harmonisation, and the laboratories accredited to test the products meant for export should have adequate infrastructure to meet international standards.
- In principle, the safety measures and SPS standards must be based on science. The country, therefore, has to prepare itself to develop SPS measures which should be scientifically sound,

and when required should be able to challenge the SPS measures of trade partners on scientific basis.

- We must undertake pest risk assessment (PRA), identify areas/regions free from such pests both of plants and animals, take appropriate measures to maintain disease-free status and develop eradication programme for diseases, which could hamper international trade. Some countries may have difficulty in scientific PRA but India has the advantage of having well-trained human resource which can play a significant role in the process of sound domestic PRA. Our technical resources can also be made available to other countries under suitable framework of international cooperation.
- Transparency of the SPS measures based on sound science, is important for promoting international trade. This can be best achieved through the use of electronic media.
- SPS Agreement has the important provision of accepting as equivalent the SPS measures of other members. Some countries interpret this provision as 'sameness', instead of equivalence of measures. There is a need to clearly define 'equivalence' in the SPS Agreement. Equivalence is the ideal option when international standards are lacking or are inappropriate.
- Article 4.2, encourages member countries to develop Mutual Recognition Agreements (MRAs). The MRAs reduce costs by avoiding duplicative testing and eliminating delays. India needs to play a proactive role in utilising the provision of MRAs.
- Article 6 of the SPS Agreement, deals with the adaptation to regional conditions, including the recognition of pest- or disease-free areas or areas of low pest or disease prevalence. It is an important clause for countries like India, which have a vast geographical area and diverse agroclimatic zones. For utilising this clause for the benefit of agricultural export, the country must undertake programmes to identify and maintain some areas free of plant and animal pests of significance to international trade. The country must identify the priority export items and take measures to promote export from the pest- or disease-free areas. India should also seek inclusion of

a clause in Article 6, to the effect that once an area is recognised as pest- or disease-free all the member countries should accept it, without the need for additional assessment process.

- To facilitate effective implementation of SPS Agreement, Article 9.1 provides for assistance to developing countries either on a bilateral basis or through appropriate international organisation. These provisions can be utilised to great advantage. India can also extend special technical assistance to countries.
- Article 10 of the SPS Agreement, provides for special and differential treatment for the developing and least developed countries. India should prepare a shortlist of major agricultural export products, identify the main constraints these products face in the countries of destination and request these countries and/or the relevant international organisations to provide assistance to facilitate the export of the listed products.
- According to the International Plant Protection Convention (IPPC) pests include the 'invasive species' like weeds as well as the genetically modified organisms (GMOs). The status of GMOs needs to be redefined in the light of recent debates in the country and elsewhere. All GMOs do not fall in the category of invasive species.
- India can play an important role in debating and suggesting approaches to harmonisation of procedures, by sending technical representatives to the international meetings related to WTO/SPS, so that international standards do not become disguised barriers to trade.
- We do not have complete understanding of the SPS measures of our target markets, mainly due to the lack of direct technical contact with the concerned authorities in those countries. This gap can be effectively bridged if the country posts a technically sound agricultural *attaché* at our diplomatic missions in the countries of interest.
- All the member countries are required to establish an Enquiry Point to receive and respond to requests for information regarding domestic SPS measures, including new or existing

regulations and decisions based on risk assessment. The enquiry point can also be the nodal point for notifying the WTO Secretariat. India does not have a single window enquiry point. It must be developed for effective utilisation of the provisions of SPS Agreement. Major functions of the enquiry point would be to: (a) serve as notification authority, (b) coordinate with SPS related Committees, (c) represent country's interests in international fora, and (e) identify and resolve technical issues related to agricultural trade.

37

Intellectual Property Rights in Agriculture

The Advent of WTO

The recognition of agriculture as a rule-bound enterprise of investment and profit making became obvious with its inclusion in the intergovernmental negotiations for the General Agreement on Tariffs and Trade (GATT) for the first time in the Uruguay Round (1986-1994). This round led to the establishment of the World Trade Organization (WTO) in January 1995. Now, the WTO has at least half a dozen intergovernmental agreements that directly affect agriculture. These are, Agreements on Agriculture (AoA), Applications of Sanitary and Phytosanitary Measures (SPS), Technical Barriers to Trade (TBT), Anti-Dumping, Subsidies and Countervailing Measures, Safeguards, and Trade Related Aspects of Intellectual Property Rights (TRIPs).

An understanding of the implications and the application of these agreements, particularly the TRIPs, has become more important than ever before at every stage of planning, research, upscaling and commercialisation of agricultural technologies. The TRIPs Agreement is covered in an elaborate document—comprising 73 articles in 7 parts, namely, (i) General provisions and basic principles, (ii) Standards concerning availability, scope, and use of IPRs (iii) Enforcement of IPRs, (iv) Acquisition and maintenance of IPRs and related *inter partes* procedures, (v) Dispute prevention and settlement, (vi) Transitional arrangements, and (vii) Institutional arrangements.

There are seven forms of intellectual property rights recognised in the TRIPs Agreement. These include, Copyright and related rights, Trademarks, Geographical Indications, Industrial Designs, Patents, Layout-Designs (topographies) of integrated circuits, and protection of undisclosed information. This agreement also covers provisions related

---

Policy Paper 19, February 2003.

to control of anti-competitive practices in contractual licences, although, it does not directly relate to IPR. In days to come, when application of various forms of IPR in different areas of agriculture is put to practice, we may face serious problems unless timely remedial measures are taken, awareness is brought out and also due emphasis is given on IPR literacy, higher education and capacity building in the country.

Following establishment of the international institutional mechanisms, such as, the Convention on Biological Diversity (CBD) and the WTO, and further, signing of International Treaty on Plant Genetic Resources for Food and Agriculture (ITPGRFA), the growing importance and the global scope of IPR in agriculture are well realised and recognised. The IPR, after long debate, is recognised as an asset and means of rewarding and harvesting the fruit of agricultural research and development. Recognition of intellectual property rights provides an effective means of protecting and rewarding innovators. This acts as a catalyst in technological and economic development. The essence of regulation of IPR by law is to balance private and public interests. At the same time, equitable benefit sharing is, although, agreed upon under the CBD, is yet to be realised in effective terms.

Global and National Scope of IPR

Broadly, protection of all forms of IPR may be relevant in agriculture but its application has to be limited to the relevant domestic Acts in vogue. Hybrids in plants and animals may be protected *de facto* by not disclosing the parents, whereas protection for plant varieties may be availed by a *sui generis* system. The provision for Plant Variety Protection (PVP) made under the TRIPs Article 27.3(b), allows countries to provide such protection either through patent, or an effective *sui generis* PVP system or any combination of the two. Patents, in India, are so far available to new processes but not to all products *per se*. In agriculture, patents may be obtained for processes related to agrochemicals, growth promoters and regulators, vaccines, drugs, hides and wool, dairy technology, food technology, fuel and biogas production, bioreactors, standardisation of various laboratory protocols, environment management, etc. Copyrights and related rights, on the other hand, may be registered for databases, bioinformatics, genes and gene sequences, amino acid sequences, antibodies, etc. Application of industrial designs and the topographies of integrated circuits would be relevant, particularly in agricultural engineering. Nevertheless, in the days to come, IPR is likely to dominate the agricultural scenario irrespective of whether the technology in question is conventional or modern—biotechnology or information technology.

Countries are required to enact/amend their domestic laws in accordance with the TRIPs Agreement and the between-country disputes have to be resolved at the WTO platform, according to its dispute settlement procedures. In this context, it is important to have in place well enacted laws corresponding to the different forms of IPR that not only keep in view the basic needs of the country but are also capable of tackling complexities, which might arise at the international level.

In India, the Patents Act, 1970, constituted the basic Principal Act on the subject. This Act hardly included innovations in agriculture under the patentable subject matter. In particular, it excluded methods of agriculture and horticulture as well as all innovations in the areas of treatment and protection of plants and animals from pestilence or those aimed at increasing their productivity and value of their produce. This broad exclusion had historical impact and implications in respect of IPR protection in agriculture in the country.

India is bound by all the provisions of TRIPs Agreement, which oblige the country to enact/amend relevant domestic laws. Further, with such shifts in legal provisions and also national policies, increased private participation in agricultural R&D and far more public-private relationships, including both competition and cooperation in relevant areas, are imminent. Several legislative and institutional adjustments are being made in the country to gear up and face the challenges of globalisation. These include enactment of new legislations on Protection of Plant Varieties and Farmers' Rights Act, 2001 and Geographical Indications of Goods (Registration and Protection) Act, 1999, and amendment of Patents Act, 1970 in 1999 and 2002. The Biological Diversity Bill, 2000 is in the process of enactment and revision of the Seeds Act, 1966, is also receiving attention. The need to provide for protection in the areas specific to farm animal sector is also being realised.

Effective implementation of IPR related legislations in place and those in the offing is expected to have significant impact on the course of agricultural R&D in the country. Therefore, it is considered important to identify and develop various national policy options for addressing the emerging areas of IPR in agriculture, including the access to various protected technologies to the Indian farmers, entrepreneurs and users. It is high time that a critical analysis of the system is undertaken for its strengths, weaknesses, opportunities and threats (SWOT), to convert threats into opportunities and mitigate weaknesses through timely action.

Creating an Enabling Environment

Recognising the inherently complex nature of institutional development, prime consideration should be given to genuine requirement of resources and building an enabling environment to capitalise on the strengths. Opportunity for IPR protection in agriculture and allied sectors should be improved alongwith mechanisms for enforcement, access to resources and technology, benefit sharing, equity and justice in order to give durable effect to the national agricultural policy and the inherent basic principles of our constitution. A long lasting national commitment should be made in respect of effective institutional mechanisms and reforms, including the administrative, regulatory, legislative and judicial reforms at all levels of government functioning. Short and medium term fiscal plans must include elements of these reforms by providing resources to help meet the costs of adjustment. Resources should be tied to commitments by successive central and state governments with much needed incentive to innovators commensurate with the invention.

It is important to understand that developed economies are likely to benefit greatly from an organised IPR system due to their inherent capabilities to capitalise on such opportunities. Realisation of the gains, principles of equity and the need for a level-playing field is a real challenge. Nevertheless, in keeping with the spirit of the intergovernmental agreements, application of IPR and also maintenance of equity and social justice must be effectively addressed at the national level.

Enhanced competitiveness together with increased production should be the target for various agricultural commodities having export prospects. These include high value commercial crops, animal breeds, spices, medicinal and aromatic plants, and products like milk, meat, fish, leather and wool. Reduction in the cost of production at small farms should also be aimed at so that Indian exports become more competitive. Market-driven quality consciousness should be applied to lay far greater R&D emphasis and efforts to produce quality products that may fetch increased monetary returns per unit area, input and time.

NAAS Round Table on IPR

The National Academy of Agricultural Sciences (NAAS), recognising the importance of the above aspects, organised a one-day Round Table (under the Convenership of Dr. Mangala Rai), to deliberate on various aspects of IPR in agriculture, in order to develop a national policy framework and a road map to achieve the goals. About 40 eminent scientists, administrators, practicing attorneys, lawyers and others concerned with the agricultural sector participated and shared their views on the existing and emerging scenario in the area of IPR.

The technical sessions covered three broad contemporary issues relating to IPR in agriculture, namely, (i) Protectable subject matter in agriculture, (ii) Technical opportunities in agriculture, and (iii) Enabling environment for accelerated R&D and global competitiveness in Indian agriculture.

IPR: Protectable subject matter in agriculture: Patentable subject matter in agriculture and alternative forms of IPR were considered. It also covered what is not patentable and where protection can or cannot be granted under copyrights, designs, geographical indications of goods, trademarks, undisclosed information (trade secrets), plant variety protection, etc. The coverage also included some case studies and a few comparisons with prevailing scenario in other countries.

IPR: Technical opportunities in agriculture: IPR implications were observed on plant varieties, farmers' rights, biodiversity and environment. It also covered the biotechnological opportunities from IPR protection. In addition, coverage was made in respect of technology transfer, biosafety, institutional capacity building, human resource development and related matters.

IPR: Enabling environment for accelerated R&D and global competitiveness in Indian agriculture: A broad range of issues and concerns related to the enabling environment were discussed and deliberated upon for steering India through the existing and emerging scenario on IPR.

This included commercialisation, competitiveness, safeguards, information management, indigenous and traditional knowledge (ITK), and orientation of research and development for technology development, transfer, trade, monitoring and management in the national and international context. The recommendations that emerged as a result of these deliberations are given below.

Recommendations

1. Harmonisation of IPR System

- Recognising that the capital intensive frontier areas of technology generation require high investment and at times long gestation periods, and that IP protection is one of the important means of resource generation aimed at further enhancing the R&D, a high priority should be given to generation, evaluation, protection and effective commercial utilisation of tangible products of intellectual property in agriculture.
- A dynamic and rational approach should be followed for IPR protection and portfolio management. Protection should be availed for the intellectual property involved in inventing new technologies using one or more than one form of protection in conjunction. Choice of any form of protection should be based on its relevance, enforcement mechanism, scope, and jurisprudence. Use of trademarks for brand development of Indian agricultural products should be encouraged as safety net in agribusiness. Remedies like 'passing off' should be availed of in jurisdictions where Common Law Jurisprudence is effective.
- Realising the emergence and importance of several new tools for growth in farm sector—biotechnology, hybrid technology, biocontrol agents, biofertilisers, vaccines, diagnostics, improved implements and machinery—and also that IPR regime is bound to affect development and use of these tools, future technological options in agriculture should be fully harnessed from the knowledge, the art and the strength to realise the IPR opportunities. Core competence should be developed through appropriate means, mechanisms and systems to harness the best of the intellectual property generated.

- Recognising the need to capitalise on our national resources and capabilities to attain and sustain IPR advantages locally, regionally and globally with timely and effective action, the area of IPR in agriculture should be addressed in conjunction with traditional rights and indigenous knowledge. Access to genetic resources in the new regime is likely to be facilitated but it will certainly be regulated. Rights to equitable sharing of benefits must be suitably balanced with the rights to IPR protection wherever applicable.
- Acknowledging that the issues of IP protection by third parties based on our indigenous traditional knowledge (ITK) are sensitive and important, a high priority and liberal financial allocation should be made to the projects that may lead to development and strengthening of traditional knowledge and resource databases in order to discourage such protection by third parties.
- Appreciating that in accordance with the intergovernmental commitment by developing countries to grant product patents in all fields of technology earliest by 1 January 2005, high priority must be accorded to the development of competitive products, particularly in agrochemicals and biotechnology, in Indian agriculture, besides, further making suitable amendments in the Patents (Amendment) Act, 2002.
- Recognising the available strengths for animal genetic resources and generation of competitive technology in farm animals, poultry and fish in the country, and also realising that appropriate IP protection laws in this area are lacking, steps should be initiated on the analogy of Protection of Plant Varieties and Farmers' Rights Act, 2001 so that in future animal and fish breeds/strains and also farmers' rights on these genetic resources are protected by law.

2. *Awareness Generation and Literacy in IPR*

- Realising that awareness generation is important for confidence building in order to accept and apply IPR in agriculture and to naturalise the IPR culture, an intensive campaign should be launched to this effect, at all levels and for all relevant sections of the society. Increased general awareness should be brought out in public to enable them to respond to various opportunities, challenges and threats. Elaborate awareness tools—compact discs (CDs), documentary films, newspaper features and advertisements should be developed and widely disseminated in all languages through mass media.

- Issues and concerns, scope of application of IP protection in one form or the other, or in conjunction, various exceptions and exemptions, procedures and rules in the Indian and global contexts in easy to understand, simple language and comparisons with other countries on case-to-case basis must be analysed and presented for public appraisals. Recognising the absence or paucity of case laws, simple illustrations should be made for FAQs like what, where, why, how, have and have-nots. Potential benefits should be explained and, at least, hypothetical examples made in relation to facilitated access to genetic resources and benefit sharing, judicious application of legislative, regulatory and administrative provisions related to IPR laws, and monetary rewards or sharing of licence fee and royalty for saleable intellectual property generated by the employees in the course of R&D.
- In order to help increase the IPR literacy in agriculture and allied sectors, compendia on IPR protection and technology transfer should be published for wide distribution. Such compendia should cover rules, procedures, forms, guidelines, other important tips and selected case studies on various provisions, admissibility and application, infringement and remedies for various forms of IPR protection in accordance with different domestic laws and also in comparison with other country laws.
- Recognising that the IPR management in agriculture requires a broad portfolio management that includes the fundamental need to link IPR protection with licensing, technology transfer, upscaling, commercialisation and safeguards, all concerned institutions/organisations should generate, publish and widely disseminate relevant information and common literature on IPR in agriculture in the form of brochures and technical bulletins, etc.

3. IPR Education, Training and Human Resource Development

- Emphasising on the need to educate children—potential inventors and innovators of future years, it is time to think of developing suitable curriculum right from the, school level. Based on short stories, poems, letters, essays and short plays, these curricula should be developed in simple language and in interactive and illustrative modes. Concerned government departments and agencies should invite contributions to this effect and announce suitable and impressive awards for the selected entries. ICAR may take lead and recommend the awarded entries for inclusion in the syllabi of the Central Board of Secondary Education and the Boards of Education in various states.
- In order to enhance the level of higher education in the country for IPR in general and IPR in relation to agriculture in particular, there must be at least one compulsory course at the undergraduate and postgraduate levels in all agricultural universities and deemed universities, and also in the law colleges all over the country. Further, an LL.M. degree programme should be started in 'IPR laws in relation to Agriculture' at various law colleges in the country.
- Summer and winter schools and periodic training programmes should be conducted in the country for teachers, scientists and technical staff in order to enhance national competence to appropriately address the area of IPR in agriculture and allied sectors. Appropriate modules should also be developed for foundation level training and advanced orientation of concerned scientists at selected institutions on regional basis and at other related Centres of Advanced Studies at the ICAR institutes and the SAUs.
- Human resources in the ICAR institutes and the SAUs should be developed and strengthened in order to help efficient application of IPR in agriculture and allied sectors. Focused attention should be given in the national agricultural research system for the in-country on-job training for skill upliftment and also need-based exposure of Indian scientists to the relevant scenario in other countries. Adequate funding should be provided at the central and state levels to ascertain the much-needed promotion of HRD.

4. *Strengthening the Institutional Mechanism—Legal, Regulatory and Administrative*

- Recognising that it is important to establish an IP regime that would provide confidence in and workability for the protection of IPR in relation to agriculture and allied sectors in the country, high priority should be accorded to the process of completing the required legislative provisions and also the notification, functioning and strengthening of national institutional mechanisms corresponding to various Acts, such as the respective Controllers, National Authorities, Tribunals, Registries, etc. Further recognising that the IPR Acts mainly relate to techno-legal matters, their governance should be controlled by eminent scientists with wide experience in relevant fields and the Tribunals should also have technical members. The National Authority on Protection of Plant Varieties and Farmers' Rights Act, 2001 (PPV&FR Act, 2001), should have an eminent plant breeder as its chairperson.
- Enforcement of new Acts and Amendments related to IPR in agriculture should be speeded up. This requires finalisation of Rules and Procedures for the PPV&FR Act, 2001 and the Geographical Indications (Registration and Protection) Act, 1999 (GI Act, 1999). Government should take note of the recommendations made by the NMS on the implementation of PPV&FR, 2001 and also the draft Rules and Regulations developed as a result of the FAO-MSSRF Consultation. The final draft rules and procedures for the GI Act, 1999, should be circulated for expert opinion on areas concerning agriculture and the allied sectors. Enforcement of Patents (Amendment) Act, 2002, should be done early to protect *inter alia* the wealth of agriculturally important microorganisms in the country. The designated repository should be equipped well and strengthened as per international standards. Similarly, enforcement of Amendment Acts related to Copyrights and Trademarks should be accorded a high priority to help derive the best benefits.
- Recognising that the protection of undisclosed information is the only form of IPR listed in the TRIPs Agreement for which there is no corresponding direct law in the country, and further reiterating such intergovernmental commitment, legal

consultation process should be initiated to firm up the contextual position and decide the course of action. Development of related laws, such as, enactment of Biological Diversity Bill, 2000, should also receive attention. Appropriate legal instruments related to conservation, maintenance, trade and sustainable utilisation of animal genetic resources should be brought about.

- Simplified regulatory procedures for relevant application of IP protection and also for seeking any prior informed consent (PIC) on mutually agreed terms (MAT) for access to genetic resources and equitable sharing of benefits should be developed.
- It is recommended that parallel laws like the Seeds Act should be strengthened as they help in better application and enforcement of particular IP laws, such as the PPV&FR Act, 2001, in order to support effective implementation of *sui generis* system of protection. Similarly, Contract Law should be reviewed to strengthen the law on Trade Secret, and the law related to land ownership of small farmholders should also be strengthened to judiciously implement the farmers' rights.
- Recognising that the institutional development and strengthening is inherently complex in nature and also that it requires time, resources and will to develop institutional culture, short and medium term fiscal plans should include provision for resources that would help in meeting the costs of adjustment. Commitments by successive central and state governments should ensure availability of precommitted resources in the techno-legal area.
- Management and Information Services should be strengthened in the ICAR institutions and SAUs in order to change their basic approach to research and IPR protection. Facilities should be established and strengthened for identification of relevant research areas through patent search, literature survey, UPOV database search etc. Early and conflict-resolving information services should be set up in the broader context.
- Inventors and innovators should be provided with their share commensurate with the worth of a commercialised invention whereas incentive should be given to all inventions whether

processes or products in order to ensure a viable, dynamic and effective national institutional mechanism of IP management.

- Elaborate Clearing House Mechanism (CHM) should be developed and strengthened in relation to IPR in agriculture, encompassing all possible information on basics, thematic areas, related treaties, conventions and agreements, historical to current events and future activities. It should also have copies of all Indian Acts related to various forms of IPR, their rules and procedures, forms, guidelines and other important tips. A site on the Internet should be dedicated to this CHM and various notifications, case studies, with periodic updation of other relevant information.

5. *Strengthening the Policy Area*

- Recognising that the principal policy area related to protection of IPR in agriculture and allied sectors is the competitive commercialisation of technologies, attention should be given to further liberalisation of agricultural markets, promotion of private sector investment and more efficient technology systems.
- Codes and procedures for rewarding the concerned partners and stakeholder scientists should be developed in the ICAR, the SAUs and other concerned institutions to bring IP culture in the NARS. This may be commensurate with the gains accrued. Alternatively, a fixed proportion, at least 40 per cent of the earnings, should be given to the scientist concerned or shared among the research partners as is presently being followed in the Council of Scientific and Industrial Research (CSIR) institutes.
- Recognising that high priority should be given to strengthening of support services in farm enterprises, extension, training, research and quality control, public interventions in agriculture should focus on market intelligence, technology forecasting and early warning systems. A centre for forecasting market trends and the status of the national and international markets should be established to enhance the prospects and sustainability of competitive Indian agriculture. Market-led technologies should be developed, protected and commercialised to harness greater returns on the investments made.

- There is a strategic need to increase growth-enhancing public investment, besides capital formation in agriculture, and promoting private sector activities and resource contributions. Essential ingredients must be put in place to bring much needed commerce in Indian agriculture.
- Whereas agriculture is deregulated as a result of the ongoing reform process, the lowest income groups should be continuously protected in accordance with clearly defined policy and directives by direct and indirect support programmes.
- Recognising that in the absence of proper legal framework, misuse, abuse, overexploitation and non-judicious utilisation of animal genetic resources is rampant, particular attention should be given at the national and global levels. There should be intergovernmental negotiations to address issues like the trusteeship/ownership of animal genetic resources in various genebanks and the legal frameworks for the databanks, including acquisition of the classified data on animal genetic resources.
- In order to avail of maximum IPR-linked opportunities in competitive agriculture, India must continue to contribute towards development of a level-playing field at the intergovernmental platform between the developing and the developed economies. In the ongoing negotiations at the World Intellectual Property Organization (WIPO) for IPR in relation to genetic resources, traditional knowledge and folklore, NARS experts can play a vital role which needs to be capitalised by the government.

6. Harnessing IP-linked Technical Opportunities in Agriculture

- Trademarks should be extensively used for brand development in agriculture. Genes and gene sequences, amino acid sequences, antibodies, etc., should be protected by copyrights until there is opportunity to patent and commercialise these products. Judicious application of other forms of protection should be done as and where applicable. Protection of IPR in all cases should be essentially linked to commercialisation, sharing of royalty and other benefits, and further enhancement of relevant R&D.

- Appreciating that the agricultural research community should create/innovate, protect, and commercialise their new technologies on continuous and incremental basis, other important national responsibilities, like sustainable development, empowerment of economically weak farmers, and protection of their traditional resources and knowledge should also be prompted on high priority. Quick action should be taken to record and document farmers' varieties in the country as available over space and time and the traditional knowledge associated with their use.
- IP linked technical opportunities in agriculture may be extended to applied management of genetic resources including microorganisms. Biotechnological advances should be integrated with genetic resource management where feasible to identify, copyright and document unique genes or gene sequences. Recognising that the germ plasm registration of PGR is in practice, specifications and guidelines should also be developed for breed registration of farm livestock.
- At least five per cent of the research budget in agriculture should be allocated to protect the public sector R&D for sustainable IPR portfolio management, and technology development and mobilisation in agriculture. Where certain technologies are considered important for food security and well being but significant avenues do not exist for IPR protection and commercialisation, development and deployment of public goods must continue to be done by the public sector R&D.
- As the IP protection is likely to be far more stringent in the years to come, agricultural markets should be constantly monitored and suitably reorganised at an appropriate time. Timely, corrective steps should be taken based on critical gaps, including the kind of IP scenario likely to emerge in future.
- Competitive funding schemes should be encouraged to develop research links between profit-making and non-profit-making research institutions and to build bridges between the use of propriety and public domain resources and technology.

7. Linkages and Cooperation

- Mutually supported testing of technologies should be encouraged by a change in attitude and mindset in public-public, public-private or private-private partnerships to address high proportionate initial costs and risks, particularly that of the biotechnological R&D. Active partnerships should be further encouraged in exploring the new tools of applied genomics to understand and improve the biological systems in public interest.
- In order to provide encouragement for the public-private partnerships in true spirit, minimal codes of procedures should be developed and applied in different key areas of partnership. On selective basis, corporate culture should be brought about in some public sector institutions.
- Confidence building should be accelerated in cross-sectoral partnerships. Feeling of uncertainty in partnership calls from across the public and private sectors should be minimised. More opportunities should be provided for frequent interaction among the agricultural scientists, research institutions, agricultural industrial sector and entrepreneurs. The private sector should also complement the basic and strategic research by the public sector through appropriate funding and resource sharing.
- Voluntary or concessional legal advice may be provided in partnership deals of strategic importance to enhance competitiveness of Indian agriculture and to attend to the problems of uneven-playing field among the resource-rich and resource-poor potential partners. A common platform should be provided on sustainable basis to seek assistance from the attorneys and lawyers having reasonable agricultural R&D background. Besides, outsourcing for legal advice on case-to-case basis in order to competently address the techno-legal area of IPR protection in agriculture, the ICAR and SAU set ups should appoint law officers in their IPR Cells in order to strengthen their institutional mechanism for IP protection.
- Realising the importance of jurisdictional limits in respect of the application of IPR laws and the situations concerning enforcement and discipline, control of agribusiness abroad

should be addressed by all concerned in a national spirit. Agencies like APEDA, FICCI and CII should earmark resources and funds to meet the contingent needs for relevant transnational IPR cases involving the Indian agricultural sector and to provide emergent support on case-to-case basis.

38

Sustainable Agricultural Export

Introduction

Agriculture, a core sector of the Indian economy, accounts for 30 per cent of the country's GDP, 20 per cent of total export earnings, two thirds of country's workforce and livelihood for 70 per cent of the total population. The past accomplishments of this sector are a great strength to face the current problems and future challenges in the areas of greater efficiency (competitiveness), sustainability, poverty alleviation and continued food self-sufficiency. With trade liberalisation, agricultural exports have also become an important national goal.

The new economic regime, initiated since early nineties, has led to resetting of the goals of Indian agriculture towards global competitiveness and export orientation without compromising the basic premise of self-reliance. The emergence of the concept of sustainability of agricultural production has made the task more difficult for all those who are associated with agricultural production systems in the country. The present goals of Indian agriculture warrant reformation of strategies and action plans.

Agricultural exports increased from about 600 million US dollars in 1960-61 to 3520 million US dollars in 1990-91. During post-economic reforms period, the value of agricultural exports has nearly doubled. The share of agri-exports in total exports, however, has remained more or less stable around 20 per cent, though the share of exports in agricultural GDP has been rising. Commodities such as marine products, oilmeals, rice, coffee, tea, spices, cashew, tobacco, castor oil, groundnut, sesame, fresh fruits, vegetables, pulses etc., are important export earners and are being exported to more than 110 countries.

Policy Paper 5, September 1999.

The encouraging results of goal-oriented Green Revolution, White Revolution, Yellow Revolution etc. enthuse the agricultural fraternity of the country to set a new goal for 'Agri-Export Revolution' which is not only the need of the hour but also a compulsion to strengthen and revitalise the economy of the country. While India holds an important position in the export market for a set of traditional agricultural commodities, new areas and new commodities are likely to emerge such as live animals and animal products, fruits, vegetables, floriculture, medicinal plants and processed agricultural products. In order to understand the emerging challenges and to identify ways and means of placing our country as a major player in the global export market for agricultural products, the National Academy of Agricultural Sciences (NAAS) in collaboration with the Rajasthan Agricultural University, Bikaner and the State Government of Rajasthan deliberated on the theme "Sustainable Agricultural Export" during the Fourth Agricultural Science Congress held at Jaipur in February, 1999.

Agricultural Exports—Advantage India

Since independence, India has made rapid strides in agricultural production. Salient attainments include the following:

- Food production has increased from a mere 50 million tonnes in 1947 to over 199.30 million tonnes during 1996-97, and per capita net availability of foodgrains has grown from 395 gm per day in early 50s to 529 gm in 1996-97.
- The country is now the largest producer of milk and the second largest producer of fruits and vegetables in the world.
- Oilseeds production has increased from 5.2 million tonnes in 1950 to 22.24 million tonnes in 1997.
- Fish and marine food production has risen from 7.5 to over 50 million tonnes, at present.
- Quality seed production which was negligible in 1947, has shown a quantum jump and is currently placed at 7 million quintals. A large number of private seed companies have also come up.
- Both the irrigated and the potentially irrigated areas have increased since Independence. Efficient methods like sprinkler and drip irrigation have become popular in arid and semi-arid areas.

- The total fertiliser consumption has gone up and India now ranks fourth in the world, although the per unit area consumption is still low.
- India is also the largest reservoir of trained manpower in the world.
- Despite agriculture sector being not-taxed, prices of several agricultural commodities are below international prices, thus conferring trade advantages.

Despite impressive strides that the country has made in agricultural production, India has not yet become a major player in the international arena as far as exports are concerned.

The broad export strategy for Indian agriculture would, therefore, be to strengthen and widen the export market for established 'commercial commodities' like tea, coffee, spices, cotton, jute, sugar, oilmeals etc., and also to create and capture new export market for 'dynamic commodities' like meat, dairy products, poultry, fishery products, vegetables, fruits, floriculture etc., whose demand in the international market is buoyant. India has a comparative advantage in many of these commodities due to availability of varied agroclimatic conditions, diversified commodity mix and low wage rates leading to lower cost of production etc. The major plank of our foreign trade strategy must be on finding a niche for exports of the above mentioned non-conventional and dynamic commodities. This has to be achieved in the context of stricter control processes under Sanitary and Phytosanitary (SPS) Agreement and other non-tariff barriers *viz.*, Technical Barriers to Trade (TBT), etc.

The strategies and action plan for increasing agricultural exports which emerged out of the deliberations are summarised below:

Cereals

- Since compromise on food security/self-sufficiency is undesirable, the export of strategic cereals should be planned very carefully.
- Wheat being a major commodity of agricultural exports from developed countries, India will have to face a tough competition from them. Future projections suggest continued upsurge in domestic wheat consumption and relatively small surpluses.

- While continuing the export of *basmati* rice targeted mostly to Middle East and to NRIs in Europe and USA, the limited scope for accelerating the export of *basmati* and other scented varieties will have to be kept in mind.
- The potential market for rice lies in the Southeast Asian countries like Indonesia, Malaysia and the Philippines and in East Asian countries like Japan and South Korea and also in Sub-Saharan countries. Rice exports to these countries from Thailand have reached a plateau and India has the potential to enter these markets provided we succeed in developing and growing varieties preferred by consumers in these countries and make vigorous efforts to capture the market.
- Crops like barley and maize hold export potential either as cattle feed or as industrial raw material. The competitive edge can be further improved if the productivity of these crops is raised.

Horticultural Produce

- Opportunities of extended production of various fruits, vegetables and floriculture crops and exporting them to newer markets in Central and South Asian countries and Japan in addition to the traditional markets of Europe, USA, Russia and some countries in Africa, need to be tapped on a priority basis.
- Several spices, condiments and medicinal crops can be marketed abroad at a fancied price.
- Efforts are needed for identification of potential markets, choice of cost-effective, eco-friendly and sustainable technologies for product development, value addition and for devising innovative trade practices with emphasis on quality, transparency and timely delivery.
- Policy interventions and incentive packages leading to regional specialisation in production and processing of tradable horticultural products, foreign participation in investment, transfer of technology and buy-back arrangements are also advocated.

Commercial Crops

- The dynamic demand for coffee in the world market opens up splendid opportunities for our country to emerge as a major player.
- Despite stagnation in international market of tea both in terms of prices and quantities and the stiff competition that Indian tea faces, India should continue to work and aspire for leadership in this crop.
- The inter-year instability in production coupled with low yields of cotton stand as major constraints for cotton to emerge as a stable exportable commodity. The possibility to keep our presence in international cotton market lies in the increased productivity and enhanced stability in production.
- India holds a great potential to take advantage of the surge in export of tobacco, especially by improving its productivity.

Oilseeds

- The mission mode approach implemented through Technology Mission on Oilseeds (TMO) not only narrowed down the large, domestic demand-supply gap for vegetable oil but also paved the way for devising separate strategies for import substitution and export of oilseeds. Exportable oilseeds include the production of HPS groundnut free from aflatoxin content, developing varieties of mustard with low glucosinilate, organising production of white sesamum without pesticide residues and rice bran extraction with low silica content. Castor oil has also emerged as an important export item. The strategy for export of oilseeds must include enhanced productivity, development of needed infrastructure, pre- and post-harvest measures to upgrade the quality and popularisation of quality oilseeds production through contract farming in potential places. Improving production efficiency would be most critical in the oilseeds sector.

Dairy Products

- The growth rate of 4.5 per cent per annum in milk production in the country against the world average growth of 1.48 per cent, and its static growth in most milk producing countries indicates the potential of India to emerge as a major player in the export of milk products like whole milk powder, skimmed milk powder, butter, cheese etc., in the rising markets for these products in Southeast Asian countries and the Middle East. The emergence of the country from a net importer to an exporter of milk products since early nineties is an indication of pushing the sector from an occasional exporter to a regular exporter in the years to come.

- The immediate need is to push up the production of milk to a level of making adequate surplus for export after meeting the steadily rising internal demand. To develop dairy products of international standard for export, adequate thrust on quality and clean milk production is required.
- There is a need to segmentise the milk production systems for catering to different market segments including export market which in turn will demand investment on mechanised milking, chilling systems, quality packaging etc., with strict adherence to Codex Alimentarius Standards and developing branded market for consumers.

Meat and Meat Products

- Meat and meat products from buffalo, sheep and goat have a major export potential. Some private state-of-art abbattoir-cum-meat processing plants have established a niche in international markets covering the Philippines, Malaysia, Iran, Egypt, Gulf and Middle East countries. The meat export can be further accelerated by modernising more meat plants with world class sanitary and phytosanitary conditions, by establishing disease-free zones for meat animals and by vertically integrating production of meat animals with meat processing.
- The approach of lifting the ban on the export of culled cattle may lead to better economic returns to the farmers and to the country.

Marine Products

- The high growth rate in the volume and value of marine products, in general and frozen fish and shrimp, in particular during recent years is indicative of the export potential of marine products from India. The Marine Products Export Development Authority (MPEDA) should undertake massive advertising and marketing campaign in overseas food markets to diversify the market and products. There is an urgent need for India to modernise processing facilities to maintain international quality standards and meet the stringent quality standards of importing countries.

- Providing on-line information about international market trends, prices etc., to seafood exporters of our country will further help to boost this sector.

Exports from Arid Agriculture

The hot and cold arid zones together cover about 12 per cent of the geographical area of the country. The hot region with 32 million hectares is spread over the states of Rajasthan, Gujarat, Punjab, Haryana, Andhra Pradesh and Karnataka. Despite poor resource base in this region, low density of human population with limited requirements for ensured food security makes it a good candidate to generate surplus agricultural production leading to export options under sustainable integrated farming systems with food crops, horticultural crops, medicinal plants, forestry plants, livestock, etc. The arid region fruits like citrus, guava, *ber*, pomegranate, aonla; vegetables like chillies, garlic, okra, tomato, cucurbits; and seed spices like cumin, fennel, fenugreek etc., hold lot of export potential. Plants like *isabgol, henna, guggul,* senna, *ashwagandha, gwarpatha, safed musli, bantulsi,* etc. also hold promise. Arid regions minor forest produce like gum arabica and oloogum resins also have export potential. *Neem* is an important tree species of arid region which can be used for developing medicinal products, biopesticides, etc. The diversified livestock mix of the region also opens up a number of avenues for export earnings from livestock products.

- The export strategies for seed spices, medicinal plants and other products of plant origin include germplasm collection, conservation, evaluation, cataloguing, crop improvement, mutation breeding, etc. for quality improvement. The expanded use of biotechnology can also play a crucial role.
- Agroprocessing industries based on raw materials available in the arid region provide expanded markets for farmers, add value to the products, create additional employment

opportunities and open up channels for sustainable exports.

- Added efforts are needed to increase earnings through exports of commodities like guar gum, guar powder and splits, deoiled cakes, spices, spice-oils, cotton, wool, meat, *isabgol*, castor, etc.

Developing Effective Interface among Various Agencies

- The most critical parameter determining exportability is efficiency. The challenge for the research system is to shift its focus from increased production to globally efficient production.
- Reorientation of research agenda for export involves, augmenting production of varieties, meeting quality requirements of importing countries, strengthening of research on non-conventional crops/plants like minor fruits, medicinal plants, seed spices etc., having export potential, establishment of proper linkages with processing industries and export organisations and identification of exportable commodities and also matching of their potential areas of production in terms of agroclimatic conditions and entrepreneurial support are advocated through effective interface of various agencies from primary production to final exports.

Investment in Physical Infrastructure

- Export of agricultural commodities can be viable and competitive, if required level of investment for creation of physical facilities like transport, cold chain, packaging, port facilities etc., is made.
- Upgradation of domestic airports into international airports in potential areas like Jaipur. will greatly spur production of fruits, vegetables, flowers etc., for export purposes.
- Special incentives may be offered to encourage both public and private sector investment in various activities related to development of commodity specific physical infrastructure for export purposes. Cold storage facilities, transportation and cold chain linkages for maintaining freshness of the produce are important for items like horticultural products. Efforts be made to encourage foreign capital and technology marketing tie-ups including joint ventures for sustainable export of agricultural products.

Export Promotion

- The efforts being made for agri-exports by various agencies like APEDA, MPEDA, Spices Board, Coffee Board, Tea Board etc., need to be augmented. Along the lines of these agencies, a Meat Development Board also needs to be established.
- Contract farming on the lines of the Pepsi Foods in Punjab and replicating the same in Gujarat and Karnataka for commodities like tomato, chillies, groundnut, garlic etc., may be beneficial to the farmers and may facilitate export promotion as foreign buyers get vendor assurance and quality for the items supplied by such agencies.

Food Safety Requirements

- Commodity Boards e.g. Spices Board, Tea Board, Coffee Board, National Dairy Development Board (NDDB), Ministries of Agriculture, Food Processing, and Commerce and organisations like Bureau of Indian Standards, APEDA, MPEDA should initiate awareness programmes in a coordinated manner on Food Hygiene and Hazard Analysis Critical Control Point (HACCP) and other safety measures for export oriented food items so as to avoid rejection and detention of consignments not conforming to quality specifications of the importing countries on the ground of Sanitary and Phytosanitary (SPS) Agreement and also Technical Barrier to Trade (TBT) Agreement which came into force with effect from January 1, 1995.
- Efforts are also needed to develop HACCP and other safety measures for each of the potential fresh/processed food items separately, since single system is not universally applicable.
- Creation of quality testing and certification protocols of international standards would be necessary to fully exploit trade opportunities.

Agri-Export Information System

- In order to develop a common perspective on the constraints, opportunities and other facets of agri-exports by the various stakeholders like producers, research organisations, extension agencies, agroprocessing industries, policy makers and exporters, an easily accessible information system including database on aspects such as volume of trade, quality specifications, demand-supply scenario, border prices regulations governing certification and exports of commodities and countries along with inventory of available technologies for meeting the quality is advocated.

Developing Managerial Capability

- As the agribusiness is being enlarged from domestic to international, and multinational to global level, the demand for trained manpower to manage it is becoming more and more explicit and is increasing. Therefore, for managing agri-business, leadership and managerial ability need to be developed in areas such as agricultural production, procurement, processing, quality management, technology management, international trade, marketing, financial management, logistic management, human resource management and also strategic management. The implementation and adherence to quality assurance measures such as ISO 9000, ISO 15000, HACCP, CODES, etc. is possible only through trained manpower. Therefore, long-term educational programmes and short-term trainings are required for developing manpower for sustainable agricultural exports.

Long-term Export Policy

- In order that agriculture sector contributes to full potential of India's exports, a long-term policy leading to export orientation by differentiating the exportable items keeping in view the domestic needs, food security, price elasticity, responsiveness to R&D efforts and export competitiveness, assumes great significance.

39

Export Potential of Dairy Products

Introduction

Dairy development in India has been acknowledged the world over as one of the highly successful programmes. Dairy sector is the single largest contributor of agricultural sector to India's gross domestic product, with its annual value exceeding Rs. 100,000 crores. India has emerged as the largest milk producer in the world, with a record level of 84.6 mt during 2001-02, while the world's total estimated milk production during the same period is 584 mt. The milk production in India, accounts for more than 14 per cent of the world and 57 per cent of Asia's total production. According to FAO estimates, the world milk production has declined by 2 per cent in the last three years, while milk production in India has increased by 4 per cent. Having attained self-sufficiency in foodgrain production, there is a need to plan for strategic diversification of Indian agriculture to ensure sustainability and nutritional adequacy. Diversification of Indian agriculture focused on dairying and export of dairy products represents excellent potential leading to prosperity of the farming community representing 65 per cent of the population.

With 198 million cattle and 86 million buffaloes, India has the largest population of milch animals in the world. These livestock constitute more than 50 per cent of the buffaloes and 20 per cent of the cattle in the world. Although number of livestock is large, average milk production is far below their genetic potential. With a well-crafted strategic approach, this huge animal wealth could be utilised in right perspective for enhancement of milk production without many incremental inputs. In contrast, a small rise in milk production requires intensive inputs and crossing of genetic barriers in advanced countries of the world, where milch animals are utilised to produce

---

Policy Paper 23, December 2003.

milk at their maximum potential. With a modest effort towards managing increased milk production, India would match/exceed the milk production of major players of international trade in dairy commodities. It is this potential that can catapult India as a major dairy exporting country on global basis. A nation-wide programme for prevention and control of animal epidemics, and creation of disease-free zones coupled with efficient delivery of artificial insemination network will have tremendous impact on improving the productivity of milch animals. This, in turn, would strengthen India's entry into the global milk products market, as well as improve the quality and viability of the entire Indian dairy industry.

Unlike other milk producing countries, 54 per cent of India's milk comes from buffaloes, which is endowed with unique processing qualities. Buffalo milk is creamier, whiter, and richer in fat, SNF, minerals like Ca and P, besides possessing relatively higher levels of physiologically significant bio-immune and extranutritional factors. These unique processing attributes render buffalo milk especially suitable for commercially important dairy products such as Mozzarella cheese, cream, butter, dairy whiteners, *paneer, khoya*, etc. India can emerge as the leading supplier of these products in the world market. Buffalo milk fat is superior with regard to less cholesterol and has more tocopherol, which is a natural antioxidant. Indian dairy industry has to derive maximum benefits of the uniqueness and positive virtues of buffalo milk to obtain the competitive edge in the global dairy market.

Indian agriculture is essentially a 'crop-livestock production system', where crop residues are fed to the milch animals. Feeding cost being the main determinant of the cost of milk production, India is highly competitive against the industrially advanced countries, where stall-feeding is practiced. Besides this, dairy farmers utilise family labour available in their own household for milk production activities. Due to low labour cost, the cost of milk production is significantly lower. The most important strength of Indian dairy industry is its cost competitiveness. Today, India either matches or surpasses countries like New Zealand, Australia and Argentina as the cheapest producer of milk on global basis. Overall, 70 per cent of world's milk is more expensive than Indian milk.

India is located amidst perennially milk deficit countries in Asia and Africa. Major importers of milk and milk products are Bangladesh, China, Hong Kong, Singapore, Thailand, Malaysia, Philippines, Japan,

UAE, Oman, and other Gulf countries located in close proximity to India. China, India and Indonesia alone account for more than 40 per cent of the world's six billion-plus population. Economic growth and changes in dietary preferences in the Southeast Asian countries have stimulated consumption of dairy products, even though it is not a part of their traditional diets. Growth of quick service restaurants (QSRs), particularly pizza chains, and growing bakeries and other food processing industries also increase the demand for cheese and other dairy products. India, thus, enjoys a strategically advantageous geographical location in terms of international trade of dairy products.

Global opportunities available to the Indian dairy industry arise primarily out of the availability of a large quantity of competitively priced milk. As the Indian dairy sector produces milk without any subsidies, the country stands to gain from the fair implementation of WTO agreements. India needs to prepare itself to access the markets of developed countries such as the European Union, USA and Japan, and has already an edge with regard to the deficit regions of South and Southeast Asia, Middle East and Africa for supply of dairy products at competitive price. However, quality of milk produced in India falls below the internationally accepted standards. Development of awareness, mindset and commitment on improving the quality of milk is necessary. Intensive efforts are needed to meet the WTO's Sanitary and Phytosanitary (SPS) and Technical Barriers of Trade (TBT) agreements and Codex Alimentarius Commission guidelines on quality and safety. In order to propel the export of milk and milk products, there is an urgent need to discuss and institutionalise the various provisions of SPS and TBT agreements along with international standards of Codex as reference points.

Export Potential of Dairy Products: Format and Theme Setting

The National Academy of Agricultural Sciences sponsored a two-day Session[1] on 'Export Potential of Dairy Products' to discuss the

1. Session held at the National Academy of Agricultural Research Management (NAARM), Hyderabad under the Convenership of Dr. B.N. Mathur, Director, NAARM, Hyderabad, August 22-23, 2003. About 30 participants representing research institutes, technical consultants, industry, and public sector organisations participated in the discussion with the major objective of developing a policy document to sensitise various administrators and planners in the country to the export potential of dairy products as well as to the need for developing a suitable infrastructure and appropriate human resource for promoting the exports.

issues and challenges confronting Indian Dairy Industry and came out with some tangible recommendations for use of the planners and policy makers of our country.

The session was organised under the following four[2] major themes:

1. Export of dairy products under WTO regime—prospects and strategies.
2. Economic and policy considerations for infrastructure development and international marketing.
3. R&D, product diversification and HRD in promoting export of dairy products.
4. Quality management for global market.

Some of the major recommendations emerging from intensive discussions that need to be considered by the related Government Departments and the Indian Dairy Industry to fully exploit the vast export potential available for the dairy products under the WTO regime are summarised below.

Recommendations

1. Policy Support

1.1 International Trade

In order to cope with the trade distortion policies of developed countries within the provisions of the WTO compulsions, India needs to intensify its efforts to develop national policy for encountering the effects of subsidies, reduction in import tariffs, and adopting selective mechanisms to reduce support on items.

Recommendation: Setting up of a high powered monitoring and steering group, which is empowered to fund, and support research

2. Theme papers providing the background materials and highlighting the issues that need to be addressed were presented in four Technical Sessions by experts having wide experience and insights into the potentials and problems of dairy sector. Under each technical session, the presentations were followed by brainstorming various issues associated with the particular theme taken up for discussion. The programme format provided ample opportunities for the delegates from the academic, research, production, processing, marketing, and export promotion institutions in the public and private sector of Indian dairy industry to raise relevant issues and come out with meaningful suggestions, in the form of recommendations, to promote export of dairy products. The major issues that require immediate attention of all those agencies concerned with the export promotion of dairy products are consolidated and presented systematically in this document.

studies for policy interventions. This agency would monitor trade developments and suggest timely interventions by the government to offset any negative impact on livelihoods of dairy farmers in India. This should also focus on issues of total or gradual withdrawal of subsidies in individual countries for exports from India, tariffication, preshipment and anti-dumping provisions, and on modalities for negotiations at WTO meetings and also for preparing cases at WTO Dispute Settlement Panel (DSP).

1.2 Demand Generation

In the competitive environment, initiatives need to be taken to create awareness regarding the unique functional and therapeutic attributes of buffalo milk products as well as interest in the traditional Indian dairy products with the objective of export promotion.

Recommendation: Through aggressive media campaigns/food festivals/market intelligence, promote interest and create market for traditional Indian dairy products abroad.

1.3 Strategic Alliance

In order to compete globally for the export of dairy products, it would be strategically advantageous to establish synergistic alliances among the exporting/importing countries (on the lines of Australia and New Zealand, who fiercely compete with each other but enter the world market as strategic partners).

Recommendation: Diplomatic initiation should be made for forging strategic partnerships at the international level for the export of dairy products. In this context, India could play an active role with the dialogue initiated by it with the 22 member countries of the Austral asian Animal Productivity Society (of which India is also a member) for strategic alliances towards export of animal products to importing countries in Asia, Oceania, Africa, and South America, as well as SAARC countries.

1.4 Marketing

Strategic marketing of dairy products needs to be supported through innovative policies that encourage both multilateral and bilateral negotiations with importing countries, and thus, help to regulate and monitor production conforming to international standards. Effective strategy will also require availability of national and international databases to inform on assessments of production, prices, production conditions, consumer preferences, risks, and the quality of dairy products. Such databases will also support negotiations at WTO and bilaterally with importing countries so as to ensure that SPS/TBT measures are based on sound scientific principles.

Recommendation: National and international databases of production, production conditions, prices, quality requirements, consumer preferences, and demand for dairy products to identify niche markets, and information on strategic marketing of dairy products and bilateral and multilateral negotiations in the dairy sector should be developed.

Agencies like EIC and CITA, and the Ministries of Agriculture and Commerce can play a vital role in this direction.

2. *Infrastructure Requirements*

In view of the impending urgency to conform to the international quality parameters and non-tariff barriers, the potential exporters of dairy products need to be aware of the stipulated requirements of the SPS and TBT agreements and the development of necessary mechanisms/infrastructure to cover possible risks.

Recommendation: Enactment of "Central Prevention of Infectious and Contagious Diseases in Animals" Bill, enforcement of stringent quality considerations, conduct of research on improving quality and shelf life of traditional Indian dairy products, and development of infrastructure for clean milk production. Implementation of HACCP would be made mandatory to all the export units under the supervision of a coordination agency.

Coordinated efforts of EIC, CITA and the Ministry of Agriculture are necessary for this purpose.

3. *Human Resource Development Efforts*

Promotion of dairy exports in the present context requires planning of competent human resource with expertise on understanding of WTO regime and various opportunities provided by it for the Indian dairy industry.

Awareness Creation

All persons associated with production, processing and export of dairy products need to be sensitised to the newly emerging issues concerning WTO and their implication on the Indian dairy industry.

Recommendation:

i) Awareness programme for the producers and exporters on the implication of trade agreement under the WTO regime on the Indian dairy sector should be effectively organised.

ii) At the institutions of higher learning, such as the IIMs, training academies, etc., long- and short-term courses may be introduced for training of the bureaucrats and technocrats on policy planning.

Agencies associated with trade like CII and the Dairy Industry, and enforcement agencies like EIC and CITA should take active part in this awareness creation campaign.

Infrastructure Development for International Marketing: Economic and Policy Considerations

In order to compete in the global market, the exportable dairy products should be made cost effective and of high quality meeting the international standards. Production of quality products in a cost-competitive manner requires certain basic infrastructure facilities for production, processing, packaging, handling, and marketing. Major constraints that hamper export of dairy products of internationally acceptable quality include infrastructure related problems like inadequate transport (road, rail or ports), power, cold storage, packaging facilities, etc. While India now has a few of the state-of-the-art dairy plants, comparable to the very best in the world, a majority of them are outdated and worn-out. They need to be upgraded to meet the emerging needs. Besides restructuring of the existing infrastructure, it also becomes necessary to create new infrastructure for meeting certain specific requirements. Some of the major concerns associated with either strengthening of the existing or creation of new up-to-date infrastructure facilities, that are necessary for ensuring the quality of international standard, are presented below.

1. Policy Support

Suitable economic and policy considerations relating to infrastructure development become necessary for promoting export of dairy products by the industry.

Recommendation:

i) Export responsive infrastructure development policies, and increase in investment by public and private sector as partners are required. For this purpose, Export Promotion Council needs to be established.

ii) Stable and long-term infrastructure investment policies by way of providing 10 years tax holiday are necessary to promote export of dairy products.

iii) In order to facilitate export of milk products by the dairy industry, single window clearance system for export to be developed on a priority basis.

In order to implement these recommendations in a more meaningful way, agencies like EIC, CITA and IDA should initiate appropriate action.

2. *Infrastructure Requirements*

There is an acute need to set up demonstration units at the rural level to educate the farmers regarding the recommended good practices for clean milk production consistent with the international sanitary and phytosanitary requirements of WTO.

2.1 Model Villages

Since the animal breeds and feeding practices differ from region to region, there is a need to establish model villages focused on dairy development/diversification of agriculture in each of the 16 agro-climatic zones of the country. These model villages could, in turn, serve as the nuclei for orchestrating future 'integrated rural developmental process' and address wider issues of education, rural hygiene, drinking water, immunisation, social issues, gender issues, e-communication, environmental issues, sustainable development, etc. These model villages would feature community training centres for model breeding, housing, feeding, health care, machine milking, milk chilling, as well as for roads, water, power, and hygiene in the villages to produce and handle quality milk.

Recommendation:

i) Develop regionally differentiated, socially acceptable, economically competitive, ecologically sound milk production systems for the 'crop-live stock production system' prevalent in India.

ii) Model villages be established in the 16 agroclimatic zones of the country focused on dairy development. A community centre would feature model housing, machine milking, milk chilling, veterinary services, training facilities for ensuring clean milk production, and promote dairy development programmes.

2.2 Transportation from Farm to Dairy Plant

While transporting milk from the production site to the factory, care should be taken to avoid spoilage by providing appropriate cooling facilities as followed in case of horticultural produce.

Recommendation: The dairy plants should ensure that within three hours of milking, milk is pooled and chilled to 4°C and transported under refrigerated condition to the processing plants.

3. *Human Resource Development Efforts*

Through appropriate training programmes, the manpower engaged in quality control and marketing should be developed to meet the emerging challenges.

3.1 Milk Collection

Group efforts for the collection of milk in a more hygienic way should be promoted.

Recommendation: Networking of women groups for regional milk collection centres by spreading the Amul model, could be adopted.

3.2 Quality Certification

Steps need to be taken to obviate the multiplicity of agencies to issue quality certification.

Recommendation: Workshop for the exporters and certifying agencies to focus on single window system for quality certification should be put in place.

3.3 International Marketing

Specialised skills are required for becoming an exporter of dairy products in the face of highly competitive global market.

Recommendation: Concerted efforts should be made to train the marketing personnel on attractive packaging and labelling to compete globally. Most importantly, the exporters should be trained in marketing management, particularly on export management including various export laws and regulations.

Institutions such as EIC, CITA, NDRI and NDDB, should be encouraged to take up these activities in a concerted manner.

R&D Support for Product Diversification

Since the establishment of GATT, way has been paved for expanding the international trade of agriculture/dairy commodities. Most of the global dairy products exporting countries have been planning crucial strategies for milk production aimed at reducing subsidies for dairy farmers in a phased manner. Over the years, considerable expertise has been developed in dairy products on strengthening internal economies *vis-à-vis* international marketing. For instance, the East European countries have consciously reduced milk production by 30 per cent compared to 1970 levels. On the other hand, France has maintained steady growth, while reducing the number of dairy cattle. In view of the glut in butter stocks, Australia and Israel have been breeding dairy cattle for milk with low fat content.

1. Policy Support

For the dairy sector to become globally competitive by: i) bringing down the cost of production; ii) maintaining the quality to meet international standards; iii) creating a niche for dairy products of Indian ethnic origin having greater demand in the Western countries; and iv) developing dairy products from genetically modified processes, it becomes imperative for providing the necessary R&D support.

1.1 Globally Competitive Dairy Sector

At the present levels of quality of dairy products, the dairy sector is cost-competitive only if all the subsidies given by the developed countries are eliminated. Improvements in quality to match international standards will raise the cost of production, unless efficiencies of processes are also improved. A primary requirement of research is, therefore, to reduce the cost of production, raise process efficiency and simultaneously, improve the product quality.

Recommendation: The competitive ability of the dairy sector should be increased through research that leads to increase in productivity of raw milk, improvement in quality and efficiency, and reduction in costs, at all levels of the milk chain.

1.2 Product Diversification

Demand for dairy products is influenced by a number of dietary, social, economic and industrial factors, which tend to be unique for any importing country. Indian dairy industry needs to develop market intelligence for the type of dairy products that have potential demand abroad. Product diversification plans must match the anticipated demands of the market abroad. Concerted R&D efforts are required to provide necessary support to the industry. Considerable scope exists for the application of biotechnology to innovate a wide range of dairy foods having unique nutritional and therapeutic attributes. Likewise, a wide range of food ingredients having unmatched functional properties may be derived from milk for application in inimitable food formulations.

Recommendation:

i) Drive towards self-reliance in R&D for developing cutting-edge technologies targeted at export promotion should be promoted.

ii) R&D for diversification of dairy products with emphasis on nutraceutical, functional and therapeutic value through biotechnological applications represents newer opportunities for creating niche markets.

iii) Dairy foods with biofortification could be developed to overcome the problem of 'hidden hunger' or inadequate levels of micronutrients such as iron, zinc, iodine, and vitamin A, as health foods and tools for public health among the poor.

1.3 Monitoring Quality

Under the SPS/TBT agreements, it is necessary to constantly monitor the quality of raw milk, production conditions and the quality of products for pesticides, pathogens and veterinary drugs. This requires quick and accurate tests that can be carried out by monitoring agencies. The present testing procedures are cumbersome and time consuming. Research must, therefore, focus on developing quality-testing kits for application under field conditions to meet urgent needs of the industry.

Recommendation: Intensive R&D efforts need to be directed for the development of sensitive field-testing kits for on-the-spot detection of objectionable contaminants of chemical or microbial origin.

1.4 Enhancement of Shelf life

For international trade/marketing/distribution of dairy products and shelf life considerations are of vital importance, since they need to stay fresh during shipping and storage. This requires research on improving the shelf life and product packaging.

Recommendation: Innovative technologies need to be developed, utilising, preferably, innate anti-microbial components of milk coupled with natural metabolites of microbial/biotechnological origin which represent newer opportunities for the enhancement of shelf life of dairy products intended for export markets.

1.5 Traditional Indian Dairy Products

A niche global market has strongly emerged for ethnic Indian dairy products. Besides the NRIs, estimated to be 15 million, certain developed countries (mainly Canada, Australia, New Zealand and UK) have already taken major initiatives to cater to this market. Indications are, that the market is fast growing with considerable future potential. Large populations of Indian origin settled in Oceania, Middle East, Western Europe, and North America represent a lucrative export market of over US $ 1.5 billion. There is an opportunity to take advantage of this niche market by developing dairy products of Indian ethnic origin meeting the quality and standards required for the global market.

Recommendation:

i) A niche market for Indian ethnic dairy products should be developed through research on their quality parameters, manufacturing processes, preservation, and packaging.

ii) There is a need to develop an illustrated catalogue enlisting the physical, rheological, chemical and microbial profile of ethnic Indian dairy products for developing product standards for export and IPR related issues.

iii) Initiatives need to be taken through prompt R&D support coupled with government bodies on food legislation (PFA, BIS) to develop appropriate Codex standards for certain ethnic dairy products that have immediate export market (*viz.*, *rasogulla, gulabjamun, burfi, peda, shrikhand*, etc.).

iv) Pilot plant facilities may be established for the development of equipments, processes, packaging systems, and quality assurance systems for the dissemination of R&D on ethnic Indian dairy products having stake of industry, equipment manufacturers, packaging material manufacturers, and R&D institutions.

v) Special attention needs to be paid to the R&D for improvement of the shelf life, consistent with the requirement of the marketing and distribution network at international level in conjunction with the packaging conforming to the international standards.

1.6 Genetically Modified Dairy Products

Dairy products/food ingredients from genetically modified microbial processes are fast entering and are being increasingly utilised by the process industry. Biotechnology and international trade are at the forefront of global debate, as United States and European Union continue to hold dramatically different positions over this contentious issue. Differing trans-Atlantic perceptions on food safety keep developing countries from achieving food security and benefiting from free trade. There is a need to resolve this issue in respect of their safety from the perspectives of both public health and the environment.

Recommendation:

i) Suitable regulations for dealing with genetically modified dairy products should be developed through research on their bio-safety aspects, both from public health and environmental perspectives.

ii) There is need for a Science Summit of the representatives from all relevant sectors to resolve the issues concerning international trade of dairy commodities.

Necessary legislative R&D support to carry out all these activities should be provided by institutions like IFPRI, Codex, WTO, NDRI, NDDB, and SAUs.

1.7 Anticipatory Research

Importing countries, particularly the developed countries, are likely to increasingly resort to non-tariff barriers to restrict trade in dairy products, as evidenced by the deliberations of the Cancun meet of September, 2003. It is necessary to be prepared for and address such developments through futuristic research on market developments, consumer preferences and country policies with respect to dairy products.

Recommendation: The dairy sector should be prepared to deal with changing strategic deployment of non-tariff barriers by developed countries through futuristic research on global trends in market development and access, environmental and trade issues, and consumer preferences.

Requisite R&D support to carry out all these activities needs to be provided by institutions like NDRI, NDDS, and SAUs.

2. *Infrastructure Requirements*

In order to create potential market for newer dairy products, necessary infrastructure facilities need to be developed.

2.1 Demonstration Centres

For demonstration and dissemination of newer technologies as well as for manpower development, suitable infrastructure is required.

Recommendation: Semi-commercial Technology Demonstration and Dissemination Centres be established for the dissemination of newer products, processes and equipment, as well as for the training of dairy personnel towards export promotion of dairy products.

Agencies like EIC and CITA should develop suitable agendas to take necessary steps for fulfilling this strategic requirement.

2.2 Packaging Systems

Extensive use of plastic materials for packaging dairy/food products poses increasingly insurmountable environmental problems for disposal after use. Furthermore, options for packaging materials/equipments are exceedingly narrow for the indigenous dairy sector, besides being very costly. New packaging concepts/biodegradable materials aimed at product protection, enhanced functionality and food safety, need to be developed as essential components of product development, especially for the export market.

Recommendation:

i) Depending on product range, operating and marketing priorities, and consumer preferences, appropriate packaging systems need to be developed.

ii) There is a need to evolve environmental-friendly, biodegradable packaging materials to meet the demands of future, through intensive financial and scientific inputs.

iii) In this context, functional linkages need to be established between the dairy industry, packaging industry, packaging machine manufacturers, and R&D organisations by establishing Technology Development and Dissemination Centres.

Creation of new infrastructure facilities and the strengthening of existing facilities require support from the financial institutions, such as NABARD, IDBI, EXIM Bank, and IDA.

3. Human Resource Development Efforts

Technical personnel associated with production, processing and business of dairy products need to be exposed to newly emerging issues concerning WTO regime. Also, they should be equipped with newer knowledge and be given opportunities to develop and upgrade their skills through various kinds of awareness creation and training programmes, both short-term and refresher courses, on a continuing basis. Concerted efforts should be made to develop and implement appropriate HRD strategies to meet the needs of various groups of people.

3.1 Curriculum Modification

To keep pace with the rapidly changing global scenario in the processing industry, relevant modification in the curriculum of Dairy Education programmes should be effected.

Recommendation: The curriculum should be strengthened with focus on quality management, information technology, and agribusiness with special emphasis on international trade and marketing.

3.2 Awareness Creation

Awareness programme should be planned for imparting new knowledge in the latest global developments/trends in the dairy sector.

Recommendation: Training programmes on hygienic milk production, processing, packaging, and marketing should be organised.

3.3 Skill Development

For industrial application of modern/innovative technologies, the manufacturers should be equipped with necessary knowledge/skills through well-structured training programmes.

Recommendation: Training the manufacturers by exposing them to the concept of computer integrated manufacture for achieving TQM through statistical quality control (SQC)/statistical process control (SPC).

3.4 Development of Business Ethics

Paradoxically, the image of Indian dairy industry is very low among the trading countries. This is a very serious issue, which needs to be addressed through well-orchestrated campaigns at the national level. Personnel engaged in export should be encouraged to bestow attention to ethics and value system while doing business.

Recommendation: Training of personnel in business and enforcement agencies in business ethics and developing internal value systems that might enforce their implementation.

3.5 Interinstitutional Linkage

In order to bring about synergism in various HRD efforts, strong working partnerships need to be built among the agencies concerned with export of dairy products.

Recommendation: Synergistic/supportive linkages between the public sector R&D institutions and industrial sector should be further strengthened by offering contract research/regular training in production and processing technology/agribusiness including manufacturing and marketing. Besides, interinstitutional task force is to be established for evolving policy guidelines and maintaining effective strategic markets at global level.

All these suggestions are possible to realise with the active participation of institutions like NDRI, SAUs, EIC, CITA, and Dairy Industry.

Quality Management for Global Market

Consequent to the WTO Agreement on Agriculture, export subsidies and tariff barriers are being reduced on imports by all the member nations. However, non-tariff barriers are being increasingly imposed to restrict trade through strategic deployment of the WTO Agreements, SPS measures, TBT. The SPS Agreement recognises the rights of national governments to restrict trade to protect human, animal or plant health by requiring the signatories to provide clear, scientific evidence of food safety standards, as well as standards for animal and plant health. The TBT Agreement includes those provisions not covered by the SPS Agreement to restrict trade so long as they are justified by a legitimate objective i.e. national security, safety for consumers or environment, animal and plant health, or fairness of trade. The measures must be transparent and in compliance with international standards. Price and quality, largely influence the global market for dairy products. The latter is also a key factor in strategically deploying the SPS and TBT Agreements by the developing countries to restrict imports.

In principle, the SPS Agreement authorises only those measures that are based on an objective risk analysis and rejects those that constitute a 'disguised trade restriction' or discrimination between the members. The 'Principle of Equivalence' is central to this Agreement. By this principle, a WTO member shall accept other countries' SPS measures as equivalent, even if they differ from its own, or other members trading in that product. The exporting member will need to demonstrate to the importing member that its measures meet the latter's level of SPS protection. Both SPS and TBT do not directly involve the WTO in setting standards or their enforcement. Despite the multilateral WTO Agreements, it becomes necessary for the exporting countries to negotiate with individual nations on specific product quality requirements.

The SPS and TBT Agreements necessitate that issues relating to quality of dairy products need to be addressed across the entire milk chain—from production of raw milk to its processing and marketing, both in raw form as well as industrial dairy products. Codex standards specify maximum permissible limits for chemical contaminants including heavy metals (lead), mycotoxins, pesticide residues, and veterinary drugs in milk and milk products. It recommends the General Principles of Food Hygiene (GPH) based on the Hazard

Analysis and Critical Control Points (HACCP) system for milk production and processing. The PFA standards are mandatory, that prescribe minimum compositional standards and maximum levels of contaminants for pesticide residues, heavy metals and aflatoxins. The BIS prescribes standards and guidelines for dairy products and processes, and also, for food hygiene and safety. But these are, however, not mandatory.

Milk production in India spreads over several million diverse and dispersed small farms. Much of the dairy industry in India is in the small-scale cooperative sector. Within the dairy sector, there is a need for regulation to ensure implementation of the GHP and HACCP systems for TQM. But these are effective largely for keeping microbiological content within permissible standards. For controlling the chemical contaminants (pesticides, heavy metals, etc.) a large scale environmental clean-up and investment in rural infrastructure (clean water, power, etc.) would be required. Similarly, animal health issues would also need to be addressed at a much broader scale than that of the dairy industry by ensuring proper preventive and quarantine measures through regulation. Some of the important quality related issues requiring immediate attention of producers, exporters and policy makers are described below.

1. Policy Support

Policy guidelines to tackle issues relating to environmental protection and animal health are required to ensure increased export of dairy products.

1.1 Environment

The quality of raw milk is the primary concern in the export of dairy products to overcome non-tariff (SPS/TST) barriers. This will be determined by the feed quality, sanitation, environmental pollution, and the availability of power and clean water. The quality standards of PFA need to be harmonised with those of the Codex standards that govern the international trade of dairy products. It is necessary to sensitise the Environmental Protection Agencies on the implications of the WTO Agreements related to dairy sector.

Recommendation: The Environmental Protection Agencies should be sensitised to the urgency of the needs of the dairy sector to develop and implement environmental regulations that may effectively lead to lowering the quantities of pesticides and chemical contaminants in raw milk to internationally acceptable standards.

1.2 Animal Health

The present regulations and programmes for animal health aim more at curing than prevention of diseases. Preventive care is important for the export of dairy products, as the SPS Agreement includes clauses that provide for ensuring that areas from where the dairy products are exported are pest- and disease-free. This would not be possible to achieve on a countrywide basis, as the dairy sector is dispersed over millions of small farms across the entire country. It is, therefore, necessary to create Export Promotion Zones in the country that can conform to the international standards of quality measures. Effective monitoring and enforcement of environmental standards would also be possible in such zones.

Recommendation: Efforts should be made to identify and develop Export Promotion Zones for dairy products, which can be monitored and regulated to be relatively free from diseases and pests.

Agencies like NDRI, NDDB, IDA, EIC, and CITA should attend to these issues.

2. *Infrastructure Requirements*

In order to produce and market dairy products of international standard, certain infrastructural facilities need to be created, besides strengthening the existing facilities.

2.1 Monitoring Animal Health

The existing infrastructure for monitoring animal health is far too inadequate and does not meet the international norms. It needs to be augmented, modernised and networked to facilitate effective monitoring of contaminants in dairy products.

Recommendation: International standards for monitoring animal health, for at least the identified Export Promotion Zones, to begin with, in terms of number of veterinary supervisors, veterinary surgeons, veterinary clinics, and veterinary hospitals for the specified number of milch animals, needs to be imposed.

2.2 Monitoring Chemical Residues/Microbial Toxins

There is an increasing trend towards public awareness/concern towards deleterious effects of undesirable agrochemicals/veterinary drugs/microbial toxins in dairy/food products on human health. The Indian dairy industry needs to galvanise into action the strategies to dispel the adverse image of Indian dairy products, among countries abroad. This requires a meaningful review and new infrastructure through networking of public health laboratories, dairy plants and food-testing agencies in private sector. Besides, the existing laboratories need to be upgraded by creating necessary facilities for carrying out all the required tests.

Recommendations:

i) There is need to establish a National Surveillance/Monitoring Agency to carry out region-wise testing of milk and milk products for the presence of agrochemical residues/veterinary drugs/microbial toxins.

ii) Establishment of a few state-of-the-art labs, with appropriate accreditation to estimate contaminants, in different parts of the country would help the industry to ensure quality of dairy products of international standard.

2.3 Export Inspection Agency

Presently, the legislation enforcing authorities with responsibilities to regulate and maintain the quality of international standard (Codex) do not have adequate testing facilities. Apparently, there is an urgent need for establishing a network of quality control labs at the regional and national level.

Recommendations: At least, every district should have a lab for PFA enforcement, and every port should have a suitable lab for testing import/export consignments.

2.4 Rural Infrastructure

Clean water, adequate power supply and sanitation are important to implement the above recommendations.

Recommendation: Adequate rural infrastructure in terms of power, roads, clean water supplies, and sanitation, should be ensured at least in the Export Promotion Zones.

2.5 Quality Management

Globally, quality management systems like GMP, GHP, HACCP and

ISO 9000, are being adopted by the dairy sector to assure the consumers of safe and high-quality dairy products. The Ministry of Health in India has also recommended a programme on 'Food safety management through GHP, GMP and HACCP'.

Recommendation: The Dairy industry should adopt TQM processes based on the Codex standards and integrate the GHP, HACCP and ISO systems into milk collection, processing, product packaging, storage, transport, and shipping of dairy products.

Necessary action may be initiated by agencies such as DAH, IDA, CITA, NDRI, and NDDB, to create the required infrastructure facilities.

3. *Human Resource Development Efforts*

For effective management of quality related infrastructure facilities as well as the quality management systems, it becomes necessary to develop the required manpower with relevant skills through appropriate training programmes.

Recommendation:

i) Training for skill development of people to effectively man various quality-testing laboratories attached to PFA, BIS, Agmark, CSIR, EICI, APEDA, ICMR, and NDRI, is required.

ii) Training of suitable manpower on quality management systems (QMS) such as good hygienic practices (GHP) and good manufacturing practices (GMP), and application of the concept and principles of sanitary and phytosanitary (SPS), and hazard analysis and critical control point (HACCP) systems throughout the food chain from the farmer to consumer, becomes important.

The responsibility to develop the required manpower may be entrusted to institutions such as NDRI, EIC and CITA.

ANNEXURE

Details of the Policy Papers

Paper No.	Title and Year of Publication	Convenor
1.	Agricultural Scientists' Perceptions on National Water Policy, 1995	P.B.S. Sarma
2.	Fertiliser Policy Issues (2000-2025), 1997	J.S. Kanwar
3.	Harnessing and Management of Water Resources for Enhancing Agricultural Production in the Eastern Region, 1998	K. Pradhan
4.	Conservation, Management and Use of Agrobiodiversity, 1998	R.S. Paroda
5.	Sustainable Agricultural Export, 1999	K. Pradhan
6.	Reorienting Land Grant System of Agricultural Education in India, 1999	H.K. Jain
7.	Diversification of Agriculture for Human Nutrition, 2001	Mahtab S. Bamji
8.	Sustainable Fisheries and Aquaculture for Nutritional Security, 2001	T.J. Pandian
9.	Strategies for Agricultural Research in the North-East, 2001	G.L. Kaul
10.	Globalisation of Agriculture: R&D in India, 2001	K.V. Peter
11.	Empowerment of Women in Agriculture, 2001	V.P. Gupta
12.	Sanitary and Phytosanitary Agreement of the World Trade Organization: Advantage India, 2001	Anupam Varma
13.	Hi-Tech Horticulture in India, 2001	K.L. Chadha
14.	Conservation and Management of Genetic Resources of Livestock, 2001	P.N. Bhat
15.	Prioritisation of Agricultural Research, 2001	I.P. Abrol
16.	Agriculture-Industry Interface: Value Added Farm Products, 2002	Akshey K. Gupta
17.	Scientists' Views on Good Governance of an Agricultural Research Organisation, 2002	J.C. Katyal
18.	Agricultural Policy: Redesigning R&D to Achieve it's Objectives, 2002	I.P. Abrol
19.	Intellectual Property Rights in Agriculture, 2003	Mangala Rai
20.	Dichotomy between Grain Surplus and Widespread Endemic Hunger, 2003	M.S. Swaminathan

...contd. ...

Paper No.	Title and Year of Publication	Convenor
21.	Priorities of Research and Human Resource Development in Fisheries Biotechnology, 2003	T.J. Pandian
22.	Seaweed Cultivation and Utilisation, 2003	S. Ayyappan
23.	Export Potential of Dairy Products, 2003	B.N. Mathur
24.	Biosafety of Transgenic Rice, 2003	V.L. Chopra
25.	Stakeholders' Perceptions on Employment Oriented Agricultural Education, 2004	J.C. Katyal
26.	Peri-urban Vegetable Cultivation in the NCR Delhi, 2004	S. Nagarajan
27.	Disaster Management in Agriculture, 2004	Panjab Singh
28.	Impact of Inter River Basin Linkages on Fisheries, 2004	S. Ayyappan
29.	Transgenic Crops and Biosafety Issues Related to their Commercialisation in India, 2004	R.P. Sharma
30.	Organic Farming: Approaches and Possibilities in the Context of Indian Agriculture, 2005	P.K. Chhonkar
31.	Redefining Agricultural Education and Extension System in Changed Scenario, 2005	Panjab Singh
32.	Emerging Issues in Water Management: The Question of Ownership, 2005	S.S. Acharya
33.	Policy Options for Efficient Nitrogen Use, 2005	Y.P. Abrol
34.	Guidelines for Improving the Quality of Indian Journals & Professional Societies in Agriculture and Allied Sciences, 2006	R.K. Singh
35.	Low and Declining Crop Response to Fertilisers, 2006	Rajendra Prasad
36.	Belowground Biodiversity in Relation to Cropping Systems, 2006	B.N. Johri
37.	Employment Opportunities in Farm and Non-farm Sectors through Technological Interventions with Emphasis on Primary Value Addition, 2006	V. Prakash
38.	WTO and Indian Agriculture: Implications for Policy and R&D, 2006	Mruthyunjaya
39.	Highlights & Recommendations on Ten Years after Beijing Conference: Gender, Science and Agriculture, 2006	Gita Kulshrestha

Index

B

C

D

O

P

T

U

V

W

Y